D0147590

Princeton Guide
to Advanced Physics

Princeton Guide
to Advanced Physics

Alan C. Tribble

Princeton University Press ■ Princeton, New Jersey

Copyright © 1996 by Princeton University Press
Published by Princeton University Press, 41 William Street,
Princeton, New Jersey 08540
In the United Kingdom: Princeton University Press,
Chichester, West Sussex

All Rights Reserved

Library of Congress Cataloging-in-Publication Data

Tribble, Alan C., 1961–
Princeton guide to advanced physics / Alan C. Tribble.
p. cm.
Includes bibliographical references and index.
ISBN 0-691-02670-X (cloth : alk. paper).
ISBN 0-691-02662-9 (pbk. : alk. paper)
1. Physics. I. Title.
QC21.2.T75 1996
530—dc20 96-5613

This book has been composed in Times Roman

The publisher would like to acknowledge the author of this volume for
providing the camera-ready copy from which this book was printed

Text design provided by C. Alvarez

Princeton University Press books are printed on acid-free paper and meet the
guidelines for permanence and durability of the Committee on Production
Guidelines for Book Longevity of the Council on Library Resources

Printed in the United States of America

10 9 8 7 6 5 4 3 2 1

10 9 8 7 6 5 4 3 2 1
(Pbk.)

To Beth,

For supporting this project when nobody else did

■ Contents

Chapter 10
Nuclear Physics

Chapter 11
Statistical Physics

Chapter 12
Solid State Physics

■ Preface and Acknowledgments

As the title implies, this book is designed to serve as a guide to the various subjects that are encountered in a review of advanced physics. One of the problems that students of the physical sciences encounter is that there is no single manageable reference that shows the evolution of the relevant equations and illustrates the relationship, and consistency, between the various subfields. This manuscript is designed to fill that void by summarizing the more important topics encountered in a survey of graduate level courses and is intended to provide a feel for the breadth and depth of understanding that is expected of upper level students. As such, it is ideally suited as a reference manual for students studying for comprehensive examinations, either as an independent study guide or as a text in a pre-exam review course. It is also intended to provide practicing physicists with a refresher of the more fundamental topics that were mastered in the past, but the details of which may have waned with time.

This manuscript had its origins in a series of review notes that I compiled while studying for my own comprehensive exams some years back. Since that time, my handwritten stack of notes has evolved through the helpful suggestions of various technical reviewers who are too numerous to thank individually here. As is often the case with projects of this nature, defining the technical content is only half of the battle. My wife, Beth, provided support in the typing of the manuscript, but more importantly she has consistently provided the encouragement to pursue publication of the final product. My editor, Trevor Lipscombe, is to be credited with having the foresight to recognize this as a worthwhile endeavor and the perseverance to keep the product moving through the slow early stages. Alice Calaprice and Carmina Alvarez of Princeton University Press are also to be credited with designing a very readable final product.

To be certain, complete coverage of all the important points in all of physics would require orders of magnitude more length than is available here. We therefore, humbly beseech the reader to bear in mind the words of the wise man: "In this work, when it shall be found that much is omitted, let it not be forgotten that much likewise is performed." (Samuel Johnson, preface to *Dictionary of the English Language*, 1755).

Long Beach, California
April 1996

Princeton Guide
to Advanced Physics

1 Mathematical Methods

1.1 Special Functions

1.1.1 Taylor Series Expansion

A function is said to be analytic if it possesses a derivative at z and at all points of some neighborhood of z. Every function $f(z)$ analytic at $z = a$ can be expanded in a Taylor series of the form

$$f(z) = \sum_{n=0}^{\infty} c_n (z-a)^n, \tag{1.1}$$

valid in some neighborhood of point a. The coefficients c_n are given by

$$c_n = \frac{1}{n!} \frac{d^n f(z)}{dz^n} \bigg|_{z=a}. \tag{1.2}$$

1.1.2 Fourier Series Expansion

A periodic function $f(x)$ can be represented by the series

$$f(x) = \frac{a_o}{2} + \sum_{n=1}^{\infty} (a_n \cos nx + b_n \sin nx) = \sum_{n=-\infty}^{\infty} c_n e^{inx}. \tag{1.3}$$

The series converges uniformly in the interval $-\pi \leq x \leq \pi$, and the coefficients can be determined from

$$a_o = \frac{1}{\pi} \int_{-\pi}^{+\pi} f(x)\,dx, \tag{1.4}$$

$$a_n = \frac{1}{\pi} \int_{-\pi}^{+\pi} f(x) \cos nx \, dx, \tag{1.5}$$

and

$$b_n = \frac{1}{\pi} \int_{-\pi}^{+\pi} f(x) \sin nx \, dx. \tag{1.6}$$

Note that $c_o = a_o/2$; $c_{+n} = (a_n - ib_n)/2$; and $c_{-n} = (a_n + ib_n)/2$ for $n > 0$. The Fourier series expansion may also be used on discontinuous functions.

1.1.3 Fourier Transforms

One of the more useful integral transforms is the Fourier transform given by

$$F(\omega) = \frac{1}{\sqrt{2\pi}} \int_{-\infty}^{\infty} f(t) e^{i\omega t} \, dt, \tag{1.7}$$

which has as its inverse transform

$$f(x) = \frac{1}{\sqrt{2\pi}} \int_{-\infty}^{\infty} F(\omega) e^{-ik\omega} \, d\omega. \tag{1.8}$$

A slight modification of equation 1.7 provides the Fourier cosine transform,

$$F_c(\omega) = \frac{1}{\sqrt{2\pi}} \int_{0}^{\infty} f(t) \cos \omega t \, dt, \tag{1.9}$$

and the Fourier sine transform,

$$F_s(\omega) = \frac{1}{\sqrt{2\pi}} \int_{0}^{\infty} f(t) \sin \omega t \, dt. \tag{1.10}$$

These functions are related to the original by the relation

$$f(x) = \frac{1}{\sqrt{2\pi}} \left[\int_0^\infty F_c(\omega)\cos\omega t d\omega + \int_0^\infty F_s(\omega)\sin\omega t d\omega \right]. \quad (1.11)$$

Fourier transforms are used extensively in optics and quantum mechanics. For example, Fourier transforms are often used to construct a wave packet from a superposition of momentum states. Since there is a continuum of possible values of momentum, the superposition takes the form of an integral and the wave packet is

$$\psi(x) = \frac{1}{\sqrt{2\pi\hbar}} \int_{-\infty}^{+\infty} \phi(p)e^{ipx/\hbar} dp. \quad (1.12)$$

In three dimensions, the Fourier transform and its inverse are

$$\psi(\vec{r}) = \frac{1}{(2\pi\hbar)^{3/2}} \int_{-\infty}^{+\infty} \phi(\vec{p}) \exp\left(\frac{i\vec{p}\cdot\vec{r}}{\hbar}\right) d^3\vec{p} \quad , \quad (1.13)$$

and

$$\phi(\vec{p}) = \frac{1}{(2\pi\hbar)^{3/2}} \int_{-\infty}^{+\infty} \psi(\vec{r}) \exp\left(\frac{-i\vec{p}\cdot\vec{r}}{\hbar}\right) d^3\vec{r}, \quad (1.14)$$

where $\psi(\vec{r})$ corresponds to a wave function in coordinate space, and $\phi(\vec{p})$ is a wave function in momentum space.

1.1.4 Dirac Delta Function

Consider an arbitrary function $f(x)$ and its Fourier transform $F(\omega)$. From the definition of the Fourier transform, we have

$$f(x) = \frac{1}{2\pi} \int_{-\infty}^{+\infty} dk e^{ikx} \int_{-\infty}^{+\infty} f(x')e^{-ikx'} dx'. \quad (1.15)$$

Interchanging the order of integration, we have

$$f(x) = \frac{1}{2\pi} \int\limits_{-\infty}^{+\infty} f(x') \left\{ \int\limits_{-\infty}^{+\infty} e^{ik(x-x')} dk \right\} dx' . \tag{1.16}$$

We must have $f(x) = f(x)$, so we introduce the Dirac delta function

$$\delta(x - x') = \int\limits_{-\infty}^{+\infty} e^{ik(x-x')} dk , \tag{1.17}$$

and define it to have the value zero everywhere except at $x = x'$, where it has the value one. Consequently, we see that

$$f(x) = \int\limits_{-\infty}^{+\infty} dx' f(x')\delta(x - x') . \tag{1.18}$$

Some properties of the Dirac delta function are

$$\delta(x) = \delta(-x) \tag{1.19a}$$

$$x\delta(x) = 0 \tag{1.19b}$$

$$\delta(ax) = a^{-1}\delta(x),\ a > 0 \tag{1.19c}$$

$$\delta(x^2 - a^2) = \frac{1}{2a}\{\delta(x - a) = \delta(x + a)\},\ a > 0 \tag{1.19d}$$

$$\int \delta(a - x)\,\delta(x - b)dx = \delta(a - b) \tag{1.19e}$$

$$f(x)\delta(x - a) = f(a)\delta(x - a) \tag{1.19f}$$

1.2 Vector Analysis

1.2.1 Gradient Theorem

The line integral along a curve C between two points a and b is given by

$$\int_a^b \left(\vec{\nabla}\varphi\right)\cdot d\vec{l} = \varphi(b) - \varphi(a). \tag{1.20}$$

Proof:

$$\begin{aligned}
\int_a^b \left(\vec{\nabla}\varphi\right)\cdot d\vec{l} &= \int_a^b \left(\frac{\partial\varphi}{\partial x}dx + \frac{\partial\varphi}{\partial y}dy + \frac{\partial\varphi}{\partial y}dy\right) \\
&= \int_a^b \left(\frac{\partial\varphi}{\partial x}\frac{dx}{dt} + \frac{\partial\varphi}{\partial y}\frac{dy}{dt} + \frac{\partial\varphi}{\partial y}\frac{dz}{dt}\right)dt \\
&= \int_a^b \left(\frac{d\varphi}{dt}\right)dt \\
&= \varphi(b) - \varphi(a).
\end{aligned} \tag{1.21}$$

1.2.2 Stokes's Theorem

$$\oint_{Line} \vec{v}\cdot d\vec{x} = \int_{Surface} \left(\vec{\nabla}\times\vec{v}\right)\cdot d\vec{A}. \tag{1.22}$$

PROOF:

With the help of figure 1.1, we see that

$$\oint_{Line} \vec{v}\cdot d\vec{x} = \int_I + \int_{II} + \int_{III} + \int_{IV}, \tag{1.23}$$

where

$$\int_I + \int_{III} = \int_{-a}^{+a} v_x dx \Big|_{y=-b} + \int_{+a}^{-a} v_x dx \Big|_{y=+b}, \tag{1.24}$$

and similarly for paths *II* and *IV*.

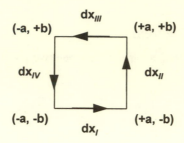

Figure 1.1 Line integral around a surface element.

Expanding in a Taylor series gives

$$v_x(y=0) = v_x(y=\pm b) \mp b \frac{dv_x}{dy}\bigg|_{y=0} + \dots \qquad (1.25)$$

The expressions for paths *I* and *III* become

$$\int_{-a}^{+a} \left\{ v_x(y=0) - b \frac{dv_x}{dy}\bigg|_{y=0} \right\} dx + \int_{+a}^{-a} \left\{ v_x(y=0) + b \frac{dv_x}{dy}\bigg|_{y=0} \right\} dx. \qquad (1.26)$$

Consequently,

$$\int_{I} + \int_{III} = \int_{-a}^{+a} (-2b) \frac{dv_x}{dy}\bigg|_{y=0} dx \rightarrow \int \frac{\partial v_x}{\partial y} \Delta y \Delta x, \qquad (1.27)$$

and similarly for paths *II* and *IV*. The total expression reduces to

$$\oint_{Line} \vec{v} \cdot d\vec{x} = \int_{Surface} (\vec{\nabla} \times \vec{v}) \cdot d\vec{A}. \qquad (1.28)$$

1.2.3 Gauss's Theorem

$$\oint_{Surface} (\vec{v} \cdot d\vec{A}) = \int_{Volume} (\vec{\nabla} \cdot \vec{v}) dV \qquad (1.29)$$

Proof:

with the help of figure 1.2, we see that

$$\oint_{Surface}(\vec{v}\cdot d\vec{A}) = \int_{I}(\vec{v}\cdot d\vec{A})+\ldots+\int_{VI}(\vec{v}\cdot d\vec{A}), \qquad (1.30)$$

which is equivalent to

$$\oint_{Surface}(\vec{v}\cdot d\vec{A}) = \int_{I}(v_x dydz)\Big|_{x=-a} + \int_{IV}(v_x dydz)\Big|_{x=+a} + \ldots \qquad (1.31)$$

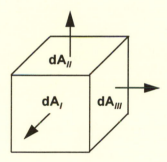

Figure 1.2 Surface geometry.

Using a Taylor series expansion for v_x, we obtain

$$v_x(x = \pm a) = v_x(x = 0) \mp a\frac{dv_x}{dx}\Big|_{x=0} + \ldots \qquad (1.32)$$

Using this in the previous expression, we see that $\oint_{Surface}(\vec{v}\cdot d\vec{A})$ reduces to

$$\int_{I}\left[v_x(x=0)+a\frac{dv_x}{dx}\Big|_{x=0}\right]dy\,dz - \int_{IV}\left[v_x(x=0)-a\frac{dv_x}{dx}\Big|_{x=0}\right] + \ldots, (1.33)$$

where we have made us of the fact that dx for region I is opposite in direction to the dx for Region IV. This further reduces to

$$\int 2a\frac{dv_x}{dx}\,dy\,dz \rightarrow \int \frac{\partial v_x}{\partial x}\,dx\,dy\,dz. \qquad (1.34)$$

When the other terms are added, we obtain

$$\oint (\vec{v}\cdot d\vec{A}) = \int (\vec{\nabla}\cdot\vec{v})dV. \qquad (1.35)$$

Surface   Volume

1.2.4 Green's Theorem

Green's identity states that for two scalar functions f_1 and f_2,

$$\int_{Volume} (f_1\nabla^2 f_2 - f_2\nabla^2 f_1)d^3x = \oint_{Surface} [(f_1(\nabla f_2) - f_2(\nabla f_1))]\cdot d\vec{S}. \quad (1.36)$$

PROOF:

It can be shown that

$$\nabla\cdot(f_1\nabla f_2) = \nabla f_1\cdot\nabla f_2 + f_1\nabla^2 f_2, \qquad (1.37)$$

and

$$\nabla\cdot(f_2\nabla f_1) = \nabla f_2\cdot\nabla f_1 + f_2\nabla^2 f_1, \qquad (1.38)$$

so that

$$\int_{Volume} (f_1\nabla^2 f_2 - f_2\nabla^2 f_1)d^3x = \int_{Volume} \nabla\cdot[(f_1\nabla f_2) - (f_2\nabla f_1)]d^3x. \quad (1.39)$$

From Gauss's theorem, we can easily see that this reduces to equation 1.34.

1.2.5 Fundamental Theorem of Vector Fields

If $\vec{\nabla}\times\vec{v} = \vec{C}(\vec{x})$ and $\vec{\nabla}\cdot\vec{v} = S(\vec{x})$, then $\vec{v} = -\nabla\varphi + \vec{\nabla}\times\vec{A}(\vec{x})$, where

$$\varphi(\vec{x}) = \frac{1}{4\pi}\int \frac{S(\vec{x}')}{|\vec{x}-\vec{x}'|}d^3\vec{x}' \text{ and } \vec{A}(\vec{x}) = \frac{1}{4\pi}\int \frac{\vec{C}(\vec{x}')}{|\vec{x}-\vec{x}'|}d^3\vec{x}'.$$

PROOF:

If $\vec{v} = -\nabla\varphi + \vec{\nabla} \times \vec{A}(\vec{x})$, then

$$\vec{\nabla}\cdot\vec{v} = -\vec{\nabla}\cdot\vec{\nabla}\varphi = -\frac{\partial^2\varphi}{\partial x_i^2}, \tag{1.40}$$

because $\vec{\nabla}\cdot\left(\vec{\nabla}\times\vec{A}\right) = 0$. Does the expression for $\varphi(\vec{x})$ satisfy $\vec{\nabla}\cdot\vec{v} = -\nabla^2\varphi$ $= S(\vec{x})$? Because ∇^2 operates on x it commutes with the integration over x'. We have

$$-\nabla^2\varphi(\vec{x}) = -\frac{1}{4\pi}\int S(\vec{x}')\left\{\nabla^2\frac{1}{|\vec{x}-\vec{x}'|}\right\}d^3\vec{x}'. \tag{1.41}$$

Consider only the expression in brackets. Integrating only this term gives

$$\int\nabla^2\frac{1}{|\vec{x}-\vec{x}'|}\,d^3\vec{x}' = \int\vec{\nabla}\cdot\vec{\nabla}\frac{1}{|\vec{x}-\vec{x}'|}\,dV' = \oint\vec{\nabla}\frac{1}{|\vec{x}-\vec{x}'|}\cdot d\vec{A}'. \tag{1.42}$$

By definition,

$$\vec{\nabla}\frac{1}{|\vec{x}-\vec{x}'|} = -\frac{(\vec{x}-\vec{x}')}{|\vec{x}-\vec{x}'|^3}, \tag{1.43}$$

so

$$\int\nabla^2\frac{1}{|\vec{x}-\vec{x}'|}\,d^3\vec{x} = -\oint\frac{(\vec{x}-\vec{x}')\cdot\hat{n}}{|\vec{x}-\vec{x}'|^3}\,dA \tag{1.44}$$

Because $\hat{n} = \dfrac{\vec{x}-\vec{x}'}{|\vec{x}-\vec{x}'|}$, the above expression reduces to

$$\int\nabla^2\frac{1}{|\vec{x}-\vec{x}'|}\,d^3x = -\oint\frac{(\vec{x}-\vec{x}')^2}{|\vec{x}-\vec{x}'|^4}\,dA = -\oint\frac{(\vec{x}-\vec{x}')^4}{|\vec{x}-\vec{x}'|^4}\,d\Omega = -4\pi. \tag{1.45}$$

Consequently, we see that $\nabla^2\dfrac{1}{|\vec{x}-\vec{x}'|} = -4\pi\delta(\vec{x}-\vec{x}')$ and

$$\vec{\nabla} \cdot v(\vec{x}) = -\frac{1}{4\pi} \int S(\vec{x}') \left\{ \nabla^2 \frac{1}{|\vec{x} - \vec{x}'|} \right\} d^3\vec{x}' = S(\vec{x}). \qquad (1.46)$$

Similarly, because $\vec{\nabla} \times \varphi(\vec{x}) = 0$, $\vec{\nabla} \times \vec{v}(\vec{x}) = \vec{\nabla} \times (\vec{\nabla} \times \vec{A}) = \vec{\nabla}(\vec{\nabla} \cdot \vec{A}) - \nabla^2 \vec{A}$, where

$$\vec{\nabla}(\vec{\nabla} \cdot \vec{A}) = \frac{1}{4\pi} \vec{\nabla} \left[\vec{\nabla} \cdot \int \frac{\vec{C}(\vec{x}')}{|\vec{x} - \vec{x}'|} d^3\vec{x}' \right], \qquad (1.47)$$

and

$$\nabla^2 \vec{A} = \frac{1}{4\pi} \nabla^2 \int \frac{\vec{C}(\vec{x}')}{|\vec{x} - \vec{x}'|} d^3\vec{x}'. \qquad (1.48)$$

From the definition $\vec{\nabla} \times \vec{v} = \vec{C}(\vec{x})$, we see that $\vec{\nabla} \cdot \vec{C} = 0$ and equation 1.47 reduces to zero. Similarly, equation 1.48 reduces to

$$\nabla^2 \vec{A} = \frac{1}{4\pi} \int \vec{C}(\vec{x}') \left[\nabla^2 \frac{1}{|\vec{x} - \vec{x}'|} \right] d^3\vec{x}' = \vec{C}(\vec{x}). \qquad (1.49)$$

Consequently, the theorem is proved.

1.2.6 Additional Properties of Vector Fields (Potential Theory)

The following three statements are equivalent

1. For every vector $\vec{v}(\vec{x})$, there is a scalar u, such that $\vec{v} = -\nabla u$.

2. $\oint \vec{v} \cdot d\vec{x} = 0$.

3. $\vec{\nabla} \times \vec{v} = 0$, or $\dfrac{\partial v_i}{\partial x_j} = \dfrac{\partial v_j}{\partial x_i}$.

PROOF: (THAT 1 IMPLIES 2)

$$\oint \vec{v} \cdot d\vec{x} = -\oint \vec{\nabla} u(\vec{x}) \cdot d\vec{S} = -\oint \left(\frac{\partial u}{\partial x} dx + \frac{\partial u}{\partial y} dy \right) = -\oint du = 0. \quad (1.50)$$

PROOF: (THAT 1 IMPLIES 3)

$$\frac{\partial v_i}{\partial x_j} = -\frac{\partial}{\partial x_j}\frac{\partial u}{\partial x_i} = -\frac{\partial}{\partial x_i}\frac{\partial u}{\partial x_j} = \frac{\partial v_j}{\partial x_i}. \tag{1.51}$$

1.2.7 Properties of Vector Spaces

A set V of elements $\phi, x, \psi, \ldots$ is called a *linear vector space* of the sum $\phi + \psi$ of any two elements $\phi, \psi \in V$, and the product $a\phi$ of any element $\phi \in V$ with any complex numbers a is defined and has the following properties:

1. If $\phi, \psi \in V$, then $\phi + \psi \in V$.

2. $\phi + \psi = \psi + \phi$.

3. $(\phi + \psi) + x = \phi + (\psi + x)$.

4. There exists a zero element such that $\phi + 0 = \phi$, for all $\phi \in V$.

5. If $\phi \in V$, then $a\phi \in V$.

6. $a(b\phi) = (ab)\phi$.

7. $1 \cdot \phi = \phi$.

8. $0 \cdot \phi = 0$.

9. $a(\phi + \psi) = a\phi + a\psi$.

10. $(a+b)\phi = a\phi + b\phi$.

A norm on a vector space V is a relation that assigns to any element ϕ of V a real number denoted by $\|\phi\|$, such that

1. $\|\phi + \psi\| \le \|\phi\| + \|\psi\|$ for all $\phi, \psi \in V$.

2. $\|\phi\| = 0$ if and only if $\phi = 0$.

3. $\|a\phi\| = |a|\|\phi\|$.

The inner product in a vector space V is a relation that assigns to any elements $\phi, \psi \varepsilon V$ a value denoted (ϕ, ψ) with the properties

1. $(\phi, \phi) \geq 0$ and $(\phi, \phi) = 0$ if and only if $\phi = 0$.

2. $(\phi, a\psi) = a(\phi, \psi)$.

3. $(\phi + x, \psi) = (\phi, \psi) + (x, \psi)$.

4. $(\psi, \phi) = (\phi, \psi)^*$.

In the inner product space, the norm is defined by $\|\phi\| = \sqrt{(\phi, \phi)}$. A complete normed vector space with an inner product is called a *Hilbert space*.

1.3 Solving Differential Equations

1.3.1 The Substitution Method

Any differential equation of the form

$$a(x)\frac{dy}{dx} + b(x)y = c(x) \tag{1.52}$$

can be rewritten as

$$y' + A(x)y = B(x). \tag{1.53}$$

To find the solution to $y(x)$, let $y = \tilde{y}g$ so that

$$y' = \tilde{y}'g + \tilde{y}g'. \tag{1.54}$$

Substituting this in the previous expression gives

$$\tilde{y}'g + \tilde{y}g' + A\tilde{y}g = B, \tag{1.55}$$

or

$$\tilde{y}'g + \tilde{y}(g' + Ag) = B. \tag{1.56}$$

If we choose g so that $g' + Ag = 0$, then

$$\frac{dg}{g} + A\,dx = 0, \tag{1.57}$$

or

$$\ln g + \int A\,dx = C, \tag{1.58}$$

which has solution

$$g = C\exp\left(\int A\,dx\right). \tag{1.59}$$

As a result, equation 1.54 reduces to

$$\tilde{y} = \int \frac{B}{g}\,dx + C'. \tag{1.60}$$

A solution for $y = \tilde{y}g$ can now be obtained from equations 1.59 and 1.60.

1.3.2 The Laplace Transform Method

The Laplace transform, $F(s)$, is defined by

$$F(s) = L\{f(t)\} = \int_0^\infty e^{-st} f(t)\,dt. \tag{1.61}$$

The Laplace transform method is often used to solve a differential equation when initial conditions are specified. Selected properties of the Laplace transform are shown below and in table 1.1.

1. $L\{e^{-at} f(t)\} = F(s+a),$ \hfill (1.62)

2. $L\{f(t-a)\} = e^{-as} L\{f(t)\},$ \hfill (1.63)

3. Derivatives:

$$L\{f'(t)\} = \int_0^\infty e^{-st} f'(t)dt$$

$$= \int_0^\infty \left\{ \frac{d}{dt}\left(e^{-st}f'(t)\right) - \frac{d\left(e^{-st}\right)}{dt}f'(t) \right\} dt$$

$$= e^{-st}f'(t)\Big|_0^\infty + s\int_0^\infty e^{-st}f'(t)dt$$

$$= sL\{f(t)\} - f(0). \tag{1.64}$$

Similarly,

$$L\{f''(t)\} = s^2 L\{f(t)\} - sf(0) - f'(0). \tag{1.65}$$

4. Convolution theorem:

$$L^{-1}\{F(s)G(s)\} = f*g, \tag{1.66}$$

where

$$f*g = \int_0^t f(\tau)g(t-\tau)d\tau. \tag{1.67}$$

5. Periodicity:

When $V(t)$ is periodic,

$$L\{V(t)\} = \int_0^\infty e^{-st}V(t)dt$$

$$= \sum_{n=0}^\infty \int_{nT}^{(n+1)T} e^{-st}V(t)dt$$

$$= \sum_{n=0}^{\infty} \int_0^T e^{-s(nT+\tau)} V(nT+\tau) d\tau$$

$$= \sum_{n=0}^{\infty} e^{-snT} \int_0^T e^{-s\tau} V(\tau) d\tau$$

$$= \frac{1}{1 - e^{-sT}} \int_0^T e^{-s\tau} V(\tau) d\tau \qquad (1.68)$$

6. Primitive function theorem:

$$L\left\{ \int_o^t f(\tau) d\tau \right\} = \frac{1}{s} F(s). \qquad (1.69)$$

Table 1.1
Laplace Transforms of Simple Functions

Function	Transform	Constraint
$f(t) = 1$	$F(s) = \int_0^{\infty} e^{-st} dt = \frac{1}{s}$	$s > 0$
$f(t) = t^n$	$F(s) = \int_0^{\infty} e^{-st} t^n dt = \frac{n!}{s^{n+1}}$	$s > 0, n > -1$
$f(t) = e^{at}$	$F(s) = \int_0^{\infty} e^{-(s-a)t} dt = \frac{1}{s-a}$	$s > a$
$f(t) = \sin kt$	$F(s) = \frac{k}{s^2 + k^2}$	$s > 0$
$f(t) = \cos kt$	$F(s) = \frac{s}{s^2 + k^2}$	$s > 0$

Example 1.1

Find the general solution to the differential equation $\ddot{x} + \dot{x} - 2x = f(t)$, *with initial conditions* $x(0) = 0$ *and* $\dot{x}(0) = \gamma$ = *constant. Solve for:* *(1)* $f(t) = 0$, *and (2)* $f(t) = 1$, *for* $0 \leq t \leq 1$, *and* $f(t) = 0$, *for* $t > 1$.

(1) Solution for $f(t) = 0$.

Applying the Laplace transform to the differential equation gives

$$s^2 F(s) - sx(0) - \dot{x}(0) + sF(s) - x(0) - 2F(s) = 0. \qquad (1.70)$$

With the initial conditions given, this reduces to

$$F(s) = \frac{\gamma}{\left(s^2 + s - 2\right)}. \qquad (1.71)$$

Using partial fractions, it can be shown that

$$\frac{1}{\left(s^2 + s - 2\right)} = \frac{1}{(s-1)(s+2)} = \frac{1}{3}\frac{1}{(s-1)} - \frac{1}{3}\frac{1}{(s+2)}, \qquad (1.72)$$

so that

$$x(t) = L^{-1}\left\{\frac{\gamma}{3}\left[\frac{1}{(s-1)} - \frac{1}{(s+2)}\right]\right\},$$

$$= \frac{\gamma}{3}\left[L^{-1}\left\{\frac{1}{(s-1)}\right\} - L^{-1}\left\{\frac{1}{(s+2)}\right\}\right],$$

$$= \frac{\gamma}{3}\left[e^t - e^{-2t}\right]. \qquad (1.73)$$

It is easily verified that this solution satisfies the original equation and the initial conditions.

(2) Solution for $f(t) = 1$, for $0 \leq t \leq 1$, and $f(t) = 0$, for $t > 1$.

Applying the Laplace transform now gives

$$s^2 F(s) - sx(0) - \dot{x}(0) + sF(s) - x(0) - 2F(s) = L\{f(t)\}, \qquad (1.74)$$

which has solution

$$F(s) = \frac{\gamma + L\{f(t)\}}{\left(s^2 + s - 2\right)}.$$ (1.75)

Using partial fractions as before, the expression for $x(t)$ is now

$$x(t) = \frac{\gamma}{3}\left[e^t - e^{-2t}\right] + \frac{1}{3}L^{-1}\left\{\frac{L\{f(t)\}}{(s-1)} - \frac{L\{f(t)\}}{(s+2)}\right\}.$$ (1.76)

The last term is of the form $L^{-1}\{F(s)G(s)\}$, so we can solve equation 1.76 using the convolution theorem. We have

$$L^{-1}\left\{\frac{L\{f(t)\}}{(s-1)}\right\} = \int_0^t f(\tau)\,e^{(t-\tau)}d\tau = e^t\int_0^t f(\tau)\,e^{-\tau}d\tau,$$ (1.77)

and

$$L^{-1}\left\{\frac{L\{f(t)\}}{(s+2)}\right\} = \int_0^t f(\tau)\,e^{-2(t-\tau)}d\tau = e^{-2t}\int_0^t f(\tau)\,e^{2\tau}d\tau.$$ (1.78)

For $t < 1$, these expressions reduce to

$$L^{-1}\left\{\frac{L\{f(t)\}}{(s-1)}\right\} = e^t\int_0^t e^{-\tau}d\tau = -e^t\left(e^{-t} - 1\right) = \left(e^t - 1\right),$$ (1.79)

and

$$L^{-1}\left\{\frac{L\{f(t)\}}{(s+2)}\right\} = e^{-2t}\int_0^t e^{2\tau}d\tau = \frac{1}{2}e^{-2t}\left(e^{2t} - 1\right) = \frac{1}{2}\left(1 - e^{-2t}\right),$$ (1.80)

respectively. The solution for $f(t)$, in the range $0 \le t \le 1$, is therefore

$$x(t) = \frac{\gamma}{3}\left[e^t - e^{-2t}\right] - \frac{1}{6}\left[2e^t - 3 + e^{-2t}\right].$$ (1.81)

For $t > 1$, equation 1.77 becomes

$$L^{-1}\left\{\frac{L\{f(t)\}}{(s-1)}\right\} = e^t \int_0^t e^{-\tau} d\tau = -e^t(e^{-1} - 1), \qquad (1.82)$$

and equation 1.78 becomes

$$L^{-1}\left\{\frac{L\{f(t)\}}{(s+2)}\right\} = e^{-2t} \int_0^t e^{2\tau} d\tau = \frac{1}{2} e^{-2t}(e^2 - 1). \qquad (1.83)$$

Consequently, in the range , the solution for $x(t)$ is

$$x(t) = \frac{\gamma}{3}\left[e^t - e^{-2t}\right] + \frac{1}{3}\left[e^t\left(1 - e^{-1}\right) - \frac{e^{-2t}}{2}\left(e^2 - 1\right)\right]. \qquad (1.84)$$

Plugging these solutions back into the original equation will verify that they are correct.

1.3.3 The Green's Function Method

Suppose we have a differential equation of the form $O\bar{u} = \bar{f}$, where O is an operator. If we can find G such that $OG = \delta$, then we can solve for $\bar{u}$ from $\bar{u} = \int G\bar{f} d\bar{x}'$. G is known as the *influence function*, or simply as the *Green's function*. The Green's function method is usually used when boundary conditions, rather than initial conditions, are specified.

If $OG = \delta(\bar{x} - \bar{x}')$ in one dimension, then $OG(\bar{x} < \bar{x}') = OG_< = 0$ and $OG(\bar{x} > \bar{x}') = OG_> = 0$. We may solve for $G_<$, $G_>$ independently, and then join the two solutions at $\bar{x} = \bar{x}'$ subject to

1. $G(\bar{x}, \bar{x}')$ must be continuous at $\bar{x}'$,

2. $\displaystyle\int_{x'-\varepsilon}^{x'+\varepsilon} OG(\bar{x}, \bar{x}') \, d\bar{x} = \int_{x'-\varepsilon}^{x'+\varepsilon} \delta(\bar{x} - \bar{x}') \, d\bar{x} = 1.$

Example 1.2

Find the solution to the differential equation $\dfrac{d^2u}{dx^2} - k^2u = f(x)$ on the interval $0 \le x \le L$, with $u(0) = u(L) = 0$, for a general function f(x).

We first solve the simpler problem $\dfrac{d^2G}{dt^2} - k^2G = \delta(x - x')$. For any arbitrary point $x < x'$,

$$\frac{d^2G_<}{dt^2} - k^2G_< = 0, \tag{1.85}$$

and similarly for $G_>$. By inspection,

$$G_< = Ae^{kx} + Be^{-kx}, \tag{1.86}$$

with $A + B = 0$ due to the boundary condition $u(0) = 0$. This reduces to

$$G_< = A\left(e^{kx} - e^{-kx}\right). \tag{1.87}$$

Similarly, for any arbitrary point $x > x'$

$$G_> = Ce^{kx} + De^{-kx}, \tag{1.88}$$

with $Ce^{kL} + De^{-kL} = 0$ due to the boundary condition $u(L) = 0$. We define $C = C'e^{-kL}$, $D = D'e^{kL}$, and the expression for $G_>$ reduces to

$$G_> = C'\left(e^{k(x-L)} - e^{-k(x-L)}\right). \tag{1.89}$$

Continuity of G at the point $x = x'$ gives

$$A\left(e^{kx'} - e^{-kx'}\right) = C'\left(e^{k(x'-L)} + e^{-k(x'-L)}\right). \tag{1.90}$$

A second constraint is obtained by integrating

$$\int_{x'-\varepsilon}^{x'+\varepsilon} \left[\frac{d^2G}{dx^2} - k^2 G \right] dx = \int_{x'-\varepsilon}^{x'+\varepsilon} \delta(x-x') dx = 1. \tag{1.91}$$

Continuity of G requires that

$$- \int_{x'-\varepsilon}^{x'+\varepsilon} k^2 G dx = -k^2 (G_> - G_<) = 0. \tag{1.92}$$

As a result, equation 1.91 reduces to

$$\int_{x'-\varepsilon}^{x'+\varepsilon} \frac{d^2G}{dx^2} dx = \frac{dG_>}{dx} - \frac{dG_<}{dx} = 1. \tag{1.93}$$

Knowing the form of $G_<$, $G_>$ we see that

$$\left. \frac{dG_>}{dx} \right|_{x=x'} = C'k \left(e^{k(x'-L)} + e^{-k(x'-L)} \right), \tag{1.94}$$

and

$$\left. \frac{dG_<}{dx} \right|_{x=x'} = Ak \left(e^{kx'} + e^{-kx'} \right), \tag{1.95}$$

so that the equation 1.83 is

$$C'k \left(e^{k(x'-L)} + e^{-k(x'-L)} \right) - Ak \left(e^{kx'} + e^{-kx'} \right) = 1. \tag{1.96}$$

Equations 1.90 and 1.96 can be used to solve for A and C'. After algebraic reduction, it can be shown that

$$A = \frac{1}{2k} \frac{\sinh k(x'-L)}{\sinh kL}, \tag{1.97}$$

and

$$C' = \frac{1}{2k} \frac{\sinh kx'}{\sinh kL}, \tag{1.98}$$

so that

$$G(x,x') = \frac{1}{k} \frac{\sinh kx \sinh k(x'-L)}{\sinh kL}. \tag{1.99}$$

The Green's function can now be combined with $f(x)$ to obtain $u(x)$.

1.3.3.1 The Sturm-Liouville Problem

An example of an equation that is often solved with Green's functions is the nonhomogeneous differential equation

$$\frac{d}{dx}\left[p(x)\frac{dy}{dx}\right] - S(x)y = f(x), \tag{1.100}$$

which is known as the *Sturm-Liouville equation*. Note that any differential equation of the form

$$A(x)\frac{d^2y}{dx^2} + B(x)\frac{dy}{dx} + C(x)y = D(x) \tag{1.101}$$

can be brought to the above form if it is multiplied by

$$H(x) = \frac{1}{A(x)} \exp\left[\int_0^x \frac{B(\xi)}{A(\xi)} d\xi\right]. \tag{1.102}$$

The Green's function satisfies

$$\frac{d}{dx}\left[p(x)\frac{dG}{dx}\right] - S(x)G = \delta(x-x'). \tag{1.103}$$

If we integrate this expression, we obtain

$$p(x)\frac{dG}{dx}\bigg|_{x'-\varepsilon}^{x'+\varepsilon} - \int_{x'-\varepsilon}^{x'+\varepsilon} S(x)G(x,x')dx = 1. \tag{1.104}$$

Since $G(x,x')$ and $S(x)$ must be continuous over the interval, we find that

$$p(x')\left[\frac{dG_>}{dx} - \frac{dG_<}{dx}\right] = 1. \tag{1.105}$$

This constraint, together with the solutions to the homogeneous equations

$$\frac{d}{dx}\left[p(x)\frac{dG_{<,>}}{dx}\right] - S(x)G_{<,>} = 0 \tag{1.106}$$

and the boundary conditions, are usually sufficient to solve for $G(x,x')$.

1.3.3.2 Free Particle Propagator

Making use of the properties of the Fourier transform, (eq. 1.10), if we plug $\phi(p)$ into the expression for $\psi(x)$ at a later time t we obtain

$$\psi(x,t) = \frac{1}{(2\pi\hbar)^{1/2}} \int\limits_{-\infty}^{+\infty} \phi(p)\exp^{\left(-\frac{ipx}{\hbar} - i\omega(p)t\right)} dp,$$

$$= \frac{1}{2\pi\hbar} \int\limits_{-\infty}^{+\infty}\int\limits_{-\infty}^{+\infty} \psi_o(x')\exp^{\left(-\frac{ip(x-x')}{\hbar} - \frac{ip^2t}{2m\hbar}\right)} dx,$$

$$= \int\limits_{-\infty}^{+\infty} \psi_o(x')\,G(x',x;t)dx', \tag{1.107}$$

where

$$G(x',x;t) = \frac{1}{2\pi\hbar} \int\limits_{-\infty}^{+\infty} \exp^{\left(-\frac{ip(x-x')}{\hbar} - \frac{ip^2t}{2m\hbar}\right)} dp. \tag{1.108}$$

Thus, the Green's function is also the *free particle propagator*. It is the probability amplitude that a particle originally at $x' = 0$ at $t = 0$ will propagate to the point x in the time interval t. In general,

$$G(x',t';x,t) = \frac{1}{2\pi\hbar} \int\limits_{-\infty}^{+\infty} \exp^{\left(+\frac{ip(x-x')}{\hbar} - \frac{ip^2(t'-t)}{2m\hbar}\right)} dp. \tag{1.109}$$

This integral may be evaluated as follows. The argument of the exponent is seen to be

$$-\frac{i(t'-t)}{2m\hbar}\left[p^2 - \frac{2m\hbar}{i(t'-t)}\frac{i(x-x')}{\hbar}p\right]$$

$$= -\frac{i(t'-t)}{2m\hbar}\left[p^2 - \frac{2m(x-x')}{(t'-t)}p + \frac{m^2(x'-x)^2}{(t'-t)^2}\right] + \frac{i(t'-t)}{2m\hbar}\left[\frac{m^2(x'-x)^2}{(t'-t)^2}\right]$$

$$= -\frac{i(t'-t)}{2m\hbar}\left[p + \frac{m(x'-x)}{(t'-t)}\right] + \frac{i(t'-t)}{2m\hbar}\left[\frac{m^2(x'-x)^2}{(t'-t)^2}\right]. \tag{1.110}$$

If we let $y = p + \dfrac{m(x'-x)}{(t'-t)}$, $dy = dp$, the expression becomes

$$G(x',t';x,t) = \frac{1}{2\pi\hbar}\exp\left(\frac{im\,(x'-x)^2}{2\hbar\,(t'-t)}\right)\int_{-\infty}^{+\infty}\exp\left(-\frac{i(t'-t)}{2m\hbar}y^2\right)dy. \tag{1.111}$$

The integral is

$$\int_{-\infty}^{+\infty}\exp\left(-a^2y^2\right)dy = \frac{\sqrt{\pi}}{a}, \tag{1.112}$$

so the expression for G reduces to

$$G(x',t';x,t) = \frac{1}{2\pi\hbar}\sqrt{\frac{2\pi\hbar m}{i(t'-t)}}\exp\left(\frac{im\,(x'-x)^2}{2\hbar\,(t'-t)}\right)$$

$$= \sqrt{\frac{m}{i2\pi\hbar(t'-t)}}\exp\left(\frac{im\,(x'-x)^2}{2\hbar\,(t'-t)}\right). \tag{1.113}$$

Note that if $t' = t$, this reduces to $\delta(x - x')$ as expected.

1.4 Complex Variables

1.4.1 Analytic Functions

If we have an analytic function $f(z) = u(x,y) + iv(x,y)$ with $dz = dx + idy$, then by definition

$$f'(z) = \frac{\left[u(x+\Delta x, y+\Delta y) - u(x,y)\right] + i\left[v(x+\Delta x, y+\Delta y) - v(x,y)\right]}{\Delta x + i\Delta y}, \quad (1.114)$$

in the limit $\Delta x \to 0$ and $\Delta y \to 0$. If $\Delta x \to 0$ first, then

$$f'(z) = -i\frac{\partial u}{\partial y} + \frac{\partial v}{\partial y}. \quad (1.115)$$

If $\Delta y \to 0$ first, then

$$f'(z) = \frac{\partial u}{\partial x} + i\frac{\partial v}{\partial x}. \quad (1.116)$$

These expressions must be equal. Consequently, the real and imaginary terms place two constraints on $f(z)$ known as the *Cauchy-Riemann conditions*. Specifically,

$$\frac{\partial u}{\partial x} = \frac{\partial v}{\partial y}, \quad (1.117)$$

and

$$\frac{\partial v}{\partial x} = -\frac{\partial u}{\partial y}. \quad (1.118)$$

Analytic functions possess some very important properties, as discussed in the next section.

1.4.1.1 Cauchy Theorem

If $f(z)$ is analytic in a simply connected domain D, and C is a simply closed curve in D, then $\oint_c f(z)dz = 0$.

PROOF:

$$\oint_c f(z)dz = \oint_c (u\,dx - v\,dy) + i\oint_c (v\,dx + u\,dy). \qquad (1.119)$$

By examination, this is equivalent to

$$\oint_c \left(\frac{\partial u}{\partial y} - \frac{\partial v}{\partial x}\right) dy\,dx + i\oint_c \left(\frac{\partial v}{\partial y} + \frac{\partial u}{\partial x}\right) dx\,dy, \qquad (1.120)$$

which is zero by the Cauchy-Riemann conditions.

1.4.2 Cauchy Integral Formula

Consider the integral

$$I = \oint_c \frac{1}{z-a}\,dz. \qquad (1.121)$$

The value of the integral must be independent of the path C. If C does not enclose the point $z = a$, the function is analytic and the integral is zero. If the curve C does enclose the point $z = a$, consider the case where C is a circle of radius R centered at $z = a$. Make the substitutions $z = a + Re^{i\theta}$, $dz = i\,Re^{i\theta}\,d\theta$, and equation 1.121 is seen to reduce to

$$I = \oint_c \frac{1}{Re^{i\theta}} i\,Re^{i\theta}\,d\theta = 2\pi i. \qquad (1.122)$$

It can be shown that if $f(z)$ is analytic inside and on C and if the point $z = a$ is in the interior of C, then

$$\oint_c \frac{f(z)}{z-a}\,dz = 2\pi i f(a). \qquad (1.123)$$

This is an important property because it allows us to use contour integration to evaluate integrals between two points. For example, the integral from $-x$ to $+x$ is equal to the closed path integral over the curve C minus the line integral above (or below) the real axis as shown in figure 1.3.

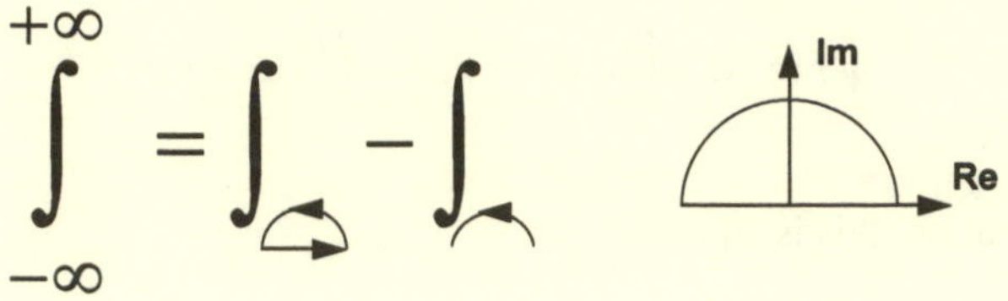

$$\int\limits_{-\infty}^{+\infty} = \int\limits_{\rightrightarrows} - \int\limits_{\curvearrowright} \qquad \text{(Im, Re diagram)}$$

Figure 1.3. Complex integration along the real and imaginary axis.

1.4.3 Residue Theorem

Let $f(z)$ be analytic in some neighborhood of $z = a$, and let C be a simple closed path lying in this neighborhood and surrounding $z = a$. The integral

$$\oint\limits_c f(z)\, dz = 2\pi i \operatorname{Res} f(a) \qquad (1.124)$$

is independent of the choice of C and is called the *residue* of $f(z)$ at the point $z = a$. Evaluating many line integrals can therefore simply be a matter of finding the residue.

1.4.3.1 Isolated Singularity

If a function $f(z)$ is analytic in the neighborhood of some point $z = a$ with the exception of the point $z = a$ itself, then it is said to have an isolated singularity at $z = a$. If this is the case, the residue may be found from

$$\operatorname{Res} f(a) = \frac{\lim}{z \to a}(z-a)f(z). \qquad (1.125)$$

If there is a finite number of isolated singularities within C, then the integral is simply the sum of the residues,

$$\oint\limits_c f(z)\, dz = 2\pi i \sum_k \operatorname{Res} f(a_k). \qquad (1.126)$$

1.4.3.2 Pole of Order m

The point $z = a$ is said to be a *pole of order m* if $A_{-m} \neq 0$ and $A_{-n} = 0$ when $n > m$. It is called a *simple pole* if $m = 1$. If this is the case, the residue may be found from

$$\operatorname{Res} f(a) = \frac{1}{(m-1)!} \lim_{z \to a} \left\{ \frac{d^{m-1}}{dz^{m-1}} \left[(z-a)^m f(z) \right] \right\}. \quad (1.127)$$

1.4.3.3 Laurent Series

If $f(z)$ is analytic over a region $r_1 < |z - a| < r_2$, it may be represented there by a generalization of Taylor's series called a *Laurent series*, where

$$f(z) = \sum_{n=-\infty}^{\infty} A_n (z-a)^n, \quad (1.128)$$

and

$$A_n = \frac{1}{2\pi i} \int_C \frac{f(z)}{(z-a)^{n+1}} dz. \quad (1.129)$$

Note that finding $\operatorname{Res} f(a)$ is the equivalent of finding the coefficient A_{-1}.

Example 1.3

Evaluate the integral $I = \int\limits_{-\infty}^{+\infty} \frac{\cos ax}{x^4 + 1} dx$, *where a is a real constant.*

By inspection, we see that

$$x^4 + 1 = \left(x + e^{i\pi/4} \right)\left(x + e^{i3\pi/4} \right)\left(x + e^{i5\pi/4} \right)\left(x + e^{i7\pi/4} \right), \quad (1.130)$$

so that, depending on whether we choose to evaluate the residue in the upper or lower half-plane we will have simple poles at $e^{i\pi/4}$ and $e^{i3\pi/4}$ or $e^{i5\pi/4}$ and $e^{i7\pi/4}$. The integral may be rewritten as

$$I = \frac{1}{2} \int_{-\infty}^{+\infty} \frac{\left(e^{iax} + e^{-iax}\right)}{x^4 + 1} \, dx. \tag{1.131}$$

Defining $x = re^{i\theta}$, we see that

$$\frac{e^{iax}}{x^4 + 1} = \frac{e^{ia\left(re^{i\theta}\right)}}{r^4 e^{i4\theta} + 1} = \frac{e^{iar\cos\theta} e^{-ar\sin\theta}}{r^4 e^{i4\theta} + 1}. \tag{1.132}$$

Because $\sin\theta$ is positive from 0 to π, the integral around the curve will go to zero if evaluated above the real axis. Similarly, the integral around the curve of the e^{-iax} term will go to zero if evaluated in the lower half-plane. Consequently, the integral I is equivalent to

$$\frac{1}{2} \int_{-\infty}^{+\infty} \frac{e^{iax}}{x^4 + 1} \, dx = \frac{1}{2} \oint_c f(z) \, dz = \pi i \left\{ \text{Res} \left(e^{i\pi/4}\right) + \text{Res} \left(e^{i3\pi/4}\right) \right\}, \tag{1.133}$$

closed in the upper half-plane, plus

$$\frac{1}{2} \int_{-\infty}^{+\infty} \frac{e^{-iax}}{x^4 + 1} \, dx = \pi i \left\{ \text{Res} \left(e^{i5\pi/4}\right) + \text{Res} \left(e^{i7\pi/4}\right) \right\}, \tag{1.134}$$

closed in the bottom half-plane. Note also that in the upper half-plane we may use the simplification

$$x^4 + 1 = \left(x + e^{i\pi/4}\right)\left(x + e^{i3\pi/4}\right)\left(x - e^{i3\pi/4}\right)\left(x - e^{i\pi/4}\right), \tag{1.135}$$

while in the lower half-plane we will use

$$x^4 + 1 = \left(x - e^{i7\pi/4}\right)\left(x - e^{i5\pi/4}\right)\left(x + e^{i5\pi/4}\right)\left(x + e^{i7\pi/4}\right). \tag{1.136}$$

From equations 1.125 and 1.126 it is easily seen that

$$\text{Res} \left(e^{i\pi/4}\right) = \lim_{z \to e^{i\pi/4}} \frac{(z - e^{i\pi/4}) e^{iaz}}{(z + e^{i\pi/4})(z + e^{i3\pi/4})(z - e^{i3\pi/4})(z - e^{i\pi/4})}$$

$$= \frac{\lim}{z \to e^{i\pi/4}} \frac{e^{iaz}}{(z+e^{i\pi/4})(z+e^{i3\pi/4})(z-e^{i3\pi/4})}$$

$$= \frac{e^{iae^{i\pi/4}}}{(2e^{i\pi/4})(e^{i\pi/4}+e^{i3\pi/4})(e^{i\pi/4}-e^{i3\pi/4})}$$

$$= \frac{e^{iae^{i\pi/4}}}{(2e^{i\pi/4})(e^{i\pi/2}+e^{i\pi}-e^{i\pi}-e^{i6\pi/4})}$$

$$= \frac{e^{iae^{i\pi/4}}}{(2e^{i\pi/4})(i-(-i))}$$

$$= \frac{1}{4i} \frac{e^{iae^{i\pi/4}}}{e^{i\pi/4}}. \tag{1.137}$$

Similarly,

$$\mathrm{Res}\left(e^{i3\pi/4}\right) = -\frac{1}{4i} \frac{e^{ia\left(e^{i3\pi/4}\right)}}{e^{i3\pi/4}}, \tag{1.138}$$

$$\mathrm{Res}\left(e^{i5\pi/4}\right) = \frac{1}{4i} \frac{e^{ia\left(e^{i5\pi/4}\right)}}{e^{i5\pi/4}} = -\frac{1}{4i} \frac{e^{ia\left(-e^{i\pi/4}\right)}}{e^{i\pi/4}}, \tag{1.139}$$

$$\mathrm{Res}\left(e^{i7\pi/4}\right) = -\frac{1}{4i} \frac{e^{ia\left(e^{i7\pi/4}\right)}}{e^{i7\pi/4}} = \frac{1}{4i} \frac{e^{ia\left(-e^{i3\pi/4}\right)}}{e^{i3\pi/4}}. \tag{1.140}$$

After simplification, the result is

$$I = \frac{\pi}{4}\left\{ \frac{e^{iae^{i\pi/4}}}{e^{i\pi/4}} - \frac{e^{ia\left(e^{i3\pi/4}\right)}}{e^{i3\pi/4}} - \frac{e^{ia\left(-e^{i\pi/4}\right)}}{e^{i\pi/4}} + \frac{e^{ia\left(-e^{i3\pi/4}\right)}}{e^{i3\pi/4}} \right\}, \tag{1.141}$$

which will further reduce to

$$I = \frac{\pi}{\sqrt{2}} \exp\left(-\frac{a}{\sqrt{2}}\right)\left[\sin\left(\frac{a}{\sqrt{2}}\right) + \cos\left(\frac{a}{\sqrt{2}}\right)\right]. \qquad (1.142)$$

1.5 Tensor Analysis

1.5.1 General Properties

The desire to express the fundamental equations of physics in a form invariant method provides much of the motivation behind tensor analysis. A tensor of rank r is simply defined as an object that transforms as a tensor. That is,

$$T'_{ij\ldots n} = T_{pq\ldots t}\, g_{ip} g_{jq} \cdots g_{nt}, \qquad (1.143)$$

where g_{ip}, g_{jq}, . . . are the cosines of the angles between the new (primed) and the old (unprimed) coordinate axes. The symbol g_{ij} is called the *metric matrix*. Note that virtually all tensor analysis relies on the Einstein summation convention. If an index is repeated, there is an implicit sum over that index. A scalar is a tensor of rank zero, a vector is a tensor of rank one, a tensor of rank two may be represented by a square array, and so on. In Cartesian coordinates, even skew ones,

$$g_{ij} = \frac{\partial x'_i}{\partial x_j} = \frac{\partial x_j}{\partial x'_i}. \qquad (1.144)$$

In general this is not true for curvilinear coordinates, and the order of the partial derivative must be observed. A tensor transforming according to the relation

$$A'^i = \frac{\partial x'_i}{\partial x_j} A^j = g_{ij} A^j \qquad (1.145)$$

is defined as a *contravariant tensor*. Contravariant indices are indicated with superscripts. A tensor transforming according to the relation

$$A'_i = \frac{\partial x_j}{\partial x'_i} A_j = g_{ij} A_j, \qquad (1.146)$$

is defined as a *covariant tensor*. Covariant indices are indicated with subscripts. A tensor that possesses both contravariant and covariant indices is called a *mixed tensor*. A tensor may be differentiated with respect to a parameter that is independent of its coordinates according to

$$\frac{dA_i'}{dt} = \frac{dA_k}{dt} g_{ik}. \tag{1.147}$$

Similarly, a tensor may be differentiated with respect to a parameter that is related to coordinates according to

$$\frac{\partial A_i'}{\partial s} = \frac{\partial A_i'}{\partial x_m} \frac{\partial x_m}{\partial s} = \frac{\partial A_k}{\partial x_m} \frac{\partial x_m}{\partial s} g_{ik}. \tag{1.148}$$

However, differentiating a tensor with respect to coordinates is a more complicated process. From equation 1.146 the derivative of a covariant tensor is given by

$$\frac{\partial A_i'}{\partial x_m'} = \frac{\partial A_k}{\partial x_m'} g_{ik} + A_k \frac{\partial g_{ik}}{\partial x_m'} = \frac{\partial A_k}{\partial x_m} \frac{\partial x_k}{\partial x_i'} + A_k \frac{\partial}{\partial x_m'} \left(\frac{\partial x_k}{\partial x_i'} \right), \tag{1.149}$$

which further reduces to

$$\frac{\partial A_i'}{\partial x_m'} = \frac{\partial A_k}{\partial x_n} \frac{\partial x_n}{\partial x_m'} \frac{\partial x_k}{\partial x_i'} + A_l' \frac{\partial^2 x_k}{\partial x_i' \partial x_m'} \frac{\partial x_i'}{\partial x_k}. \tag{1.150}$$

The first term in equation 1.149 is called the *tensor term* and the second is called the *affine term*. Note that the affine term is present only when the transformation coefficients are not constant, as is the case in curvilinear space. The affine term is zero in Cartesian coordinates, both orthogonal and skew. Consequently, in orthogonal coordinates only the tensor term remains and the derivative of a tensor with respect to coordinates is a tensor of rank r + 1. The affine term is usually written with the symbol

$$\Gamma_{im}^l = \frac{\partial^2 x_k}{\partial x_i' \partial x_m'} \frac{\partial x_i'}{\partial x_k} = \frac{1}{2} g^{l\sigma} \left[\frac{\partial g_{\sigma i}}{\partial x^m} + \frac{\partial g_{\sigma m}}{\partial x^i} - \frac{\partial g_{im}}{\partial x^\sigma} \right], \tag{1.151}$$

which is called the *Christoffel* symbol. From equation 1.150 we see that the total differential dA_i of any given component of a vector will consist of two parts. One part will be due to the change of coordinate axes, denoted by δA_i, and the other will be due to the intrinsic variation of the vector field. This second term is called the *covariant derivative* of a vector A_i and is often denoted with a semicolon, as in

$$A_{i;k} = dA_i - \delta A_i = \left(\frac{\partial A_i}{\partial u_k} - \Gamma^m_{ik} A_m \right) du^k. \qquad (1.152)$$

Similarly, the usual partial derivative is often abbreviated with a comma:

$$A_{i,j} = \frac{\partial A_i}{\partial x_j}. \qquad (1.153)$$

1.6 Bibliography

Abramowitz, M., and Stegun, I. A. *Handbook of Mathematical Functions.* New York: Dover Publications, 1972.

Arfken, G. *Mathematical Methods for Physicists.* 2nd ed. New York: Academic Press, 1978.

Butkov, E. *Mathematical Physics.* Reading, Mass.: Addison-Wesley, 1968.

Churchill, R. V. *Operational Mathematics.* 3rd ed. New York: McGraw-Hill, 1972.

Gradshteyn, I. S., and Ryzhik, I. M. *Table of Integrals, Series, and Products.* Orlando, Fla.: Academic Press, 1980.

Rainville, E. D., and Bedient, P. E. *Elementary Differential Equations.* 6th ed., New York: MacMillan, 1981.

2 Classical Mechanics

2.1 Fundamental Techniques

2.1.1 The Virial Theorem

The equation of motion of a system can be written in the form

$$\vec{F}_i - \dot{\vec{p}}_i = 0.$$

(2.1)

We are interested in the quantity

$$G = \sum_i \vec{p}_i \cdot \vec{r}_i.$$

(2.2)

Differentiating this expression gives

$$\frac{dG}{dt} = \sum_i \dot{\vec{r}}_i \cdot \vec{p}_i + \sum_i \dot{\vec{p}}_i \cdot \vec{r}_i = \sum_i m_i \dot{\vec{r}}_i \cdot \dot{\vec{r}}_i + \sum_i \dot{\vec{p}}_i \cdot \vec{r}_i.$$

(2.3)

From equation 2.1 this reduces to

$$\frac{dG}{dt} = 2T + \sum_i \vec{F}_i \cdot \vec{r}_i.$$

(2.4)

Averaging over a period of time τ, we obtain

$$\frac{1}{\tau}\int_0^\tau \frac{dG}{dt} dt = \frac{1}{\tau}[G(\tau) - G(0)] = \langle 2T \rangle + \left\langle \sum_i \vec{F}_i \cdot \vec{r}_i \right\rangle.$$

(2.5)

If the motion is periodic, with τ = period, then the left-hand side of equation 2.5 is zero, and we see that

$$\langle T \rangle = -\frac{1}{2} \left\langle \sum_i \vec{F}_i \cdot \vec{r}_i \right\rangle. \tag{2.6}$$

2.1.2 D'Alembert's Principle

From equation 2.1 it follows that the virtual work done by this system is also zero:

$$\sum_i \left(\vec{F}_i - \dot{\vec{p}}_i \right) \cdot \delta \vec{r}_i = 0. \tag{2.7}$$

The total force will be a combination of externally applied forces and internal constraints,

$$\vec{F}_i = \vec{F}_i'^{\,a} + \vec{f}_i, \tag{2.8}$$

so that equation 2.8 reduces to

$$\sum_i \left(\vec{F}_i'^{\,a} - \dot{\vec{p}}_i \right) \cdot \delta \vec{r}_i - \sum_i \vec{f}_i \cdot \delta \vec{r}_i = 0. \tag{2.9}$$

If we restrict our attention to rigid bodies and other systems for which the forces of constraint do no work, then we conclude that the condition for equilibrium of a system is given by D'Alembert's principle, which states

$$\sum_i \left(\vec{F}_i^a - \dot{\vec{p}}_i \right) \cdot \delta \vec{r}_i = 0. \tag{2.10}$$

2.1.3 Lagrange's Equation

If $\vec{r}_i$ is a function of independent variables q_i, then

$$\delta \vec{r}_i = \sum_j \frac{\partial \vec{r}_i}{\partial q_j} \delta q_j. \tag{2.11}$$

Dropping the superscript "a" for convenience, the first term from equation 2.10 is

$$\sum_i \vec{F}_i \cdot \delta \vec{r}_i = \sum_{i,j} \vec{F}_i \cdot \frac{\partial \vec{r}_i}{\partial q_j} \delta q_j = \sum_j Q_j \delta q_j, \qquad (2.12)$$

where Q_j is the generalized force. The second term in equation 2.10 is

$$\sum_i \dot{\vec{p}}_i \cdot \delta \vec{r}_i = \sum_i m_i \ddot{\vec{r}}_i \cdot \delta \vec{r}_i = \sum_{i,j} m_i \ddot{\vec{r}}_i \cdot \frac{\partial \vec{r}_i}{\partial q_j} \delta q_j. \qquad (2.13)$$

By definition,

$$\sum_i m_i \ddot{\vec{r}}_i \cdot \frac{\partial \vec{r}_i}{\partial q_j} = \sum_i \left(\frac{d}{dt} \left(m_i \dot{\vec{r}}_i \cdot \frac{\partial \vec{r}_i}{\partial q_j} - m_i \dot{\vec{r}}_i \cdot \frac{d}{dt} \left(\frac{\partial \vec{r}_i}{\partial q_j} \right) \right) \right), \qquad (2.14)$$

and

$$\frac{d}{dt} \left(\frac{\partial \vec{r}_i}{\partial q_j} \right) = \frac{\partial \dot{\vec{r}}_i}{\partial q_j} = \frac{\partial \vec{v}_i}{\partial q_j}. \qquad (2.15)$$

Similarly,

$$\vec{v}_i = \sum_k \frac{\partial r_i}{\partial q_k} \dot{q}_k + \frac{\partial r_i}{\partial t}, \qquad (2.16)$$

so it follows that

$$\frac{\partial \vec{v}_i}{\partial \dot{q}_j} = \frac{\partial \vec{r}_i}{\partial q_j}. \qquad (2.17)$$

Using equations 2.15 and 2.17 in equation 2.14, we find that

$$\sum_i m_i \ddot{\vec{r}}_i \cdot \frac{\partial \vec{r}_i}{\partial q_j} = \sum_i \left(\frac{d}{dt} \left(m_i \vec{v}_i \cdot \frac{\partial \vec{v}_i}{\partial \dot{q}_j} \right) - m_i \vec{v}_i \cdot \frac{\partial \vec{v}_i}{\partial q_j} \right). \qquad (2.18)$$

Equations 2.12 and 2.18 are combined to give

$$\sum_j \left(\frac{d}{dt} \left(\frac{\partial}{\partial \dot{q}_j} \left(\sum_i \frac{1}{2} m_i v_i^2 \right) \right) - \frac{\partial}{\partial q_j} \left(\sum_i \frac{1}{2} m_i v_i^2 \right) - Q_j \right) \delta q_j = 0 \quad (2.19)$$

or

$$\sum_j \left(\left(\frac{d}{dt} \left(\frac{\partial T}{\partial \dot{q}_j} \right) - \frac{\partial T}{\partial q_j} \right) - Q_j \right) \delta q_j = 0. \quad (2.20)$$

Consequently,

$$\left(\frac{d}{dt} \left(\frac{\partial T}{\partial \dot{q}_j} \right) - \frac{\partial T}{\partial q_j} \right) = Q_j. \quad (2.21)$$

If $F_i = -\nabla V_i$, then

$$Q_j = \sum_i \vec{F}_i \cdot \frac{\partial \vec{r}_i}{\partial q_j} = -\sum_i \nabla_i V \cdot \frac{\partial \vec{r}_i}{\partial q_j} = -\frac{\partial V}{\partial q_j}. \quad (2.22)$$

Equation 2.21 is equivalent to

$$\left(\frac{d}{dt} \left(\frac{\partial T}{\partial \dot{q}_j} \right) - \frac{\partial}{\partial q_j} (T - V) \right) = 0. \quad (2.23)$$

We define $L = T - V$ to be the Lagrangian. If V is not a function of time, then

$$\left(\frac{d}{dt} \left(\frac{\partial L}{\partial \dot{q}_j} \right) - \frac{\partial L}{\partial q_j} \right) = 0, \quad (2.24)$$

which is known as *Lagrange's equation*, can be used to determine the equations of motion. We also define

$$p_j = \frac{\partial L}{\partial \dot{q}_j} \quad (2.25)$$

to be the *canonical*, or *conjugate*, *momentum*. If the Lagrangian does not contain a given coordinate q_j then the coordinate is said to be cyclic or ignorable.

2.1.3.1 The Two-Body Central Force Problem

As an example of the application of Lagrange's equation, consider a system of two mass points m_1 and m_2 subject to an interaction potential V, where V is any function of the vector between the particles. The kinetic energy is

$$T = \frac{1}{2}m_1\dot{r}_1^2 + \frac{1}{2}m_2\dot{r}_2^2. \qquad (2.26)$$

The kinetic energy can also be written as the kinetic energy of the center of mass plus the kinetic energy about the center of mass. We define $\vec{R}$ = position of the center of mass, and $\vec{r} = \vec{r}_1 - \vec{r}_2$ = vector between m_1 and m_2. The kinetic energy of the center of mass is given by

$$T_{cm} = \frac{1}{2}(m_1 + m_2)\dot{R}^2. \qquad (2.27)$$

Relative to the center of mass, the positions of m_1 and m_2 are given by

$$\vec{r}_1' = \frac{-m_2}{m_1 + m_2}\vec{r}, \qquad (2.28)$$

and

$$\vec{r}_2' = \frac{m_1}{m_1 + m_2}\vec{r}. \qquad (2.29)$$

Therefore, the kinetic energy about the center of mass is given by

$$T' = \frac{1}{2}m_1\dot{r}_1'^2 + \frac{1}{2}m_2\dot{r}_2'^2 = \frac{1}{2}\left(\frac{m_1 m_2}{m_1 + m_2}\right)\dot{r}^2. \qquad (2.30)$$

Consequently, the Lagrangian is

$$L = \frac{1}{2}(m_1 + m_2)\dot{R}^2 + \frac{1}{2}\left(\frac{m_1 m_2}{m_1 + m_2}\right)\dot{r}^2 - V(r, \dot{r}, \ldots). \qquad (2.31)$$

We see immediately that because the potential is only a function of the vector between the particles, the conjugate momentum of R (the momentum of the center of mass) is constant. That is, the motion of the center of mass has no effect on the motion about the center of mass. This also implies that there will be no out-of-plane motion.

2.1.3.2 The Inverse Square Law of Forces

When V is a function of r only, as is the case for gravitational or electrostatic forces, equation 2.31 may be expressed in polar coordinates as

$$L = \frac{1}{2}\mu\left(\dot{r}^2 + r^2\dot{\theta}^2\right) - V(r). \qquad (2.32)$$

Note that we have chosen to ignore the term describing the motion of the center of mass since it has no effect on other parameters, and we have introduced the definition

$$\mu = \frac{m_1 m_2}{m_1 + m_2}, \qquad (2.33)$$

where μ is termed the *reduced mass*. The equations of motion are found from Lagrange's equation (eq. 2.24). For the variable $q = \theta$, we have

$$\frac{d}{dt}\left(\frac{\partial L}{\partial \dot{\theta}_i}\right) - \frac{\partial L}{\partial \theta_i} = \frac{d}{dt}(\mu r^2 \dot{\theta}) = 0, \qquad (2.34)$$

while for $q = r$ we have

$$\frac{d}{dt}\left(\frac{\partial L}{\partial \dot{r}}\right) - \frac{\partial L}{\partial r} = \mu\ddot{r} - \mu r\dot{\theta}^2 + \frac{\partial V}{\partial r} = 0. \qquad (2.35)$$

For many problems of interest, such as orbital mechanics, $m_1 \gg m_2$ and $\mu \to m_2$. The physical consequence is that the smaller particle, m_2, is subjected to the largest perturbation in its motion. From this point forward we will follow the usual convention and replace μ with the symbol m with the understanding that it refers to the motion of the smaller of m_1 and m_2 about the center of mass.

Equation 2.34 is the statement of conservation of angular momentum. That is,

$$l = mr^2\dot{\theta} \qquad (2.36)$$

is a constant. Equation 2.36 can be rewritten in the form

$$l\,dt = mr^2 d\theta, \qquad (2.37)$$

which implies

$$\frac{d}{dt} = \frac{l}{mr^2}\frac{d}{d\theta}, \qquad (2.38)$$

and

$$\frac{d^2}{dt^2} = \frac{l}{mr^2}\frac{d}{d\theta}\left(\frac{l}{mr^2}\frac{d}{d\theta}\right). \qquad (2.39)$$

The area swept out by a moving body is given by

$$A = \frac{1}{2}r(r\theta). \qquad (2.40)$$

It follows that

$$\frac{dA}{dt} = \frac{l}{mr^2}\frac{d}{d\theta}\left[\frac{1}{2}r(r\theta)\right] = \frac{l}{2}. \qquad (2.41)$$

Because angular momentum is conserved, dA/dt is also constant. This is Kepler's second law, which states that the planets sweep out equal areas in equal times.

From the definition of l, equation 2.35 becomes

$$m\ddot{r} - \frac{l^2}{mr^3} = -\frac{\partial V}{\partial r}, \qquad (2.42)$$

or, because V is only a function of r,

$$m\ddot{r} = -\frac{d}{dr}\left(V + \frac{1}{2}\frac{l^2}{mr^2}\right). \tag{2.43}$$

The particle moves in an effective potential given by

$$V_{eff} = \left(V + \frac{l^2}{2mr^2}\right). \tag{2.44}$$

Equation 2.43 reduces to

$$m\ddot{r} = -\frac{dV_{eff}}{dr}, \tag{2.45}$$

thus

$$m\dot{r}\ddot{r} = \frac{d}{dt}\left(\frac{1}{2}m\dot{r}^2\right) = -\frac{dr}{dt}\frac{dV_{eff}}{dr} = -\frac{dV_{eff}}{dt}. \tag{2.46}$$

Consequently,

$$\frac{d}{dt}\left(\frac{1}{2}m\dot{r}^2 + V_{eff}\right) = 0, \tag{2.47}$$

which is the statement that energy is conserved. From this equation, we also find that

$$\dot{r} = \pm\sqrt{\left(\frac{2}{m}\left(E - V_{eff}\right)\right)} = \pm\sqrt{\left(\frac{2}{m}\left(E - V - \frac{l^2}{2mr^2}\right)\right)}, \tag{2.48}$$

or simply

$$dt = \frac{dr}{\sqrt{\frac{2}{m}\left(E - V - \frac{l^2}{2mr^2}\right)}}. \tag{2.49}$$

Substituting the relation between dt and $d\theta$, equation 2.37, and we find that

$$d\theta = \frac{1}{\sqrt{\dfrac{2mE}{l^2} - \dfrac{2mV}{l^2} - \dfrac{1}{r^2}}} \frac{dr}{r^2}. \qquad (2.50)$$

Consider the case when the potential is of the form $V = -\dfrac{k}{r} = -ku$. Equation 2.50 becomes

$$d\theta = \frac{-1}{\sqrt{\dfrac{2mE}{l^2} + \dfrac{2mk}{l^2}u - u^2}} du. \qquad (2.51)$$

Integrating this expression gives

$$\theta_f - \theta_i = \arcsin \frac{-2u + \dfrac{2mk}{l^2}}{\sqrt{\dfrac{8mE}{l^2} + \left(\dfrac{2mk}{l^2}\right)^2}} = \arcsin \frac{\left(1 - \dfrac{l^2}{mk}\dfrac{1}{r}\right)}{\sqrt{\dfrac{2El^2}{mk^2} + 1}}. \qquad (2.52)$$

Inverting this expression gives

$$\frac{1}{r} = \frac{mk}{l^2}\left\{1 - \left[\sqrt{1 + \frac{2El^2}{mk^2}}\right]\sin\left(\theta_f - \theta_i\right)\right\}. \qquad (2.53)$$

This is usually written in the form

$$\frac{1}{r} = \frac{mk}{l^2}\left\{1 - e\sin\left(\theta_f - \theta_i\right)\right\}, \qquad (2.54)$$

where

$$e = \sqrt{1 + \frac{2El^2}{mk^2}}. \qquad (2.55)$$

This is the equation for a conic section having several classes of solutions:

1. If $e > 1$, and $E > 0$, the orbit is a hyperbola.

2. If $e = 1$, and $E = 0$, the orbit is a parabola.

3. If $e < 1$, and $E < 0$, the orbit is an ellipse.

4. If $e = 0$, and $E = -\dfrac{mk^2}{2l^2}$, the orbit is a circle.

2.2 Variational Techniques

2.2.1 The Calculus of Variations

Consider a function $f(y, \dot{y}, x)$ defined on a path $y = y(x)$ between x_1 and x_2 where $\dot{y} = dy / dx$. We wish to find a particular path $y(x)$ such that the integral

$$J = \int_{x_1}^{x_2} f(y, \dot{y}, x) dx \qquad (2.56)$$

has a stationary value relative to paths differing infinitesimally from the correct function $y(x)$. Since J must have a stationary value for the correct path relative to any neighboring path, the variation must be zero relative to some particular set of neighboring paths. Such a set of paths can be denoted by

$$y(x, \alpha) = y(x, 0) + \alpha \eta(x), \qquad (2.57)$$

where $y(x, 0)$ is the correct path and $\eta(x_1) - \eta(x_2) = 0$. Explicitly,

$$J(\alpha) = \int_{x_1}^{x_2} f(y(x, \alpha), \dot{y}(x, \alpha), x) dx. \qquad (2.58)$$

A necessary condition for a stationary point is

$$\left(\frac{dJ}{d\alpha} \right)\bigg|_{\alpha=0} = 0. \qquad (2.59)$$

From equation 2.58 ,

$$\frac{dJ}{d\alpha} = \int_{x_1}^{x_2} \left(\frac{\partial f}{\partial y} \frac{\partial y}{\partial \alpha} + \frac{\partial f}{\partial \dot{y}} \frac{\partial \dot{y}}{\partial \alpha} \right) dx .$$

(2.60)

It is easily seen that

$$\frac{\partial f}{\partial \dot{y}} \frac{\partial \dot{y}}{\partial \alpha} = \frac{\partial f}{\partial \dot{y}} \frac{\partial^2 y}{\partial x \partial \alpha} ,$$

(2.61)

so that

$$\int_{x_1}^{x_2} \frac{\partial f}{\partial \dot{y}} \frac{\partial \dot{y}}{\partial \alpha} dx = \int_{x_1}^{x_2} \frac{\partial f}{\partial \dot{y}} \frac{\partial^2 y}{\partial x \partial \alpha} dx = \left. \frac{\partial f}{\partial \dot{y}} \frac{\partial y}{\partial \alpha} \right|_{x_1}^{x_2} - \int_{x_1}^{x_2} \frac{d}{dx} \left(\frac{\partial f}{\partial \dot{y}} \right) \frac{\partial y}{\partial \alpha} dx .$$

(2.62)

From the boundary conditions, the first term on the right-hand side vanishes and equation 2.60 reduces to

$$\frac{dJ}{d\alpha} = \int_{x_1}^{x_2} \left(\frac{\partial f}{\partial y} - \frac{d}{dx} \frac{\partial f}{\partial \dot{y}} \right) \frac{\partial y}{\partial \alpha} dx .$$

(2.63)

The fundamental lemma of the calculus of variations states that if

$$\int_{x_1}^{x_2} M(x)\eta(x)dx = 0$$

(2.64)

for all $\eta(x)$ continuous through the second derivative, then $M(x)$ must be identically zero on the interval. Thus J is stationary only if

$$\frac{\partial f}{\partial y} - \frac{d}{dx} \left(\frac{\partial f}{\partial \dot{y}} \right) = 0.$$

(2.65)

If f is a function of many independent variables, then

$$\frac{\partial f}{\partial y_i} - \frac{d}{dx}\left(\frac{\partial f}{\partial \dot{y}_i}\right) = 0. \qquad (2.66)$$

2.2.2 Hamilton's Principle

Hamilton's principle states that the motion, in configuration space, of a system where all nonconstraining forces are derivable from a generalized scalar potential that may be a function of coordinates, velocities, and time is such that the integral

$$I = \int_{t_1}^{t_2} L\,dt \qquad (2.67)$$

has a stationary value for the correct path of the motion. That is,

$$\delta I = \delta \int_{t_1}^{t_2} L\,dt = 0. \qquad (2.68)$$

The integral I is termed the action, and Hamilton's principle states that the variation in I is zero. In other words, the action is minimized. By comparison with equation 2.65, it follows that

$$\frac{\partial L}{\partial q_i} - \frac{d}{dt}\frac{\partial L}{\partial \dot{q}_i} = 0, \qquad (2.69)$$

which is Lagrange's equation.

2.2.3 Lagrange Multipliers

D'Alembert's principle, and the resulting form of Lagrange's equation, assume no constraint forces. Consider a treatment when the equations of constraint can be put in the form

$$\sum_k a_{lk}dq_k + a_{lt} = 0, \qquad (2.70)$$

where the a_{lk} and a_{lt}'s may be functions of a, t. For virtual displacements it follows that

$$\sum_k a_{lk}\delta q_k = 0. \tag{2.71}$$

If this is true, then it must also follow that

$$\lambda_l \sum_k a_{lk}\delta q_k = 0, \tag{2.72}$$

where the λ_l are undetermined coefficients called *Lagrange multipliers*. From equations 2.63, 2.68, and 2.69 it is seen that Hamilton's principle is equivalent to

$$\int \sum_k \left(\frac{\partial L}{\partial q_k} - \frac{d}{dt}\frac{\partial L}{\partial \dot{q}_k} \right) \delta q_k = 0. \tag{2.73}$$

By the same process, equation 2.72 is equivalent to

$$\int \sum_{k,l} \lambda_l a_{lk}\delta q_k dt = 0. \tag{2.74}$$

We combine these two relations to obtain

$$\int \sum_{k=1}^n \left(\frac{\partial L}{\partial q_k} - \frac{d}{dt}\frac{\partial L}{\partial \dot{q}_k} + \sum_l \lambda_l a_{lk} \right) \delta q_k dt = 0. \tag{2.75}$$

The δq_k's are not necessarily independent, but because the values of the λ_l's are undetermined we may choose them such that

$$\frac{d}{dt}\left(\frac{\partial L}{\partial \dot{q}_k} \right) - \frac{\partial L}{\partial q_k} = \sum_l \lambda_l a_{lk} . \tag{2.76}$$

These equations, together with equation 2.70, can be used to determine the equations of motions for systems with constraining forces.

Example 2.1

Consider the case of a ladder of length L that is inclined against a frictionless wall and floor, as shown in figure 2.1. Find the equations of motion.

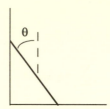

Figure 2.1 An inclined ladder.

The position of the center of mass of the ladder, and its orientation, can be described with the variables x, y, θ. The motion of the ladder is constrained by the wall and floor. We have the two constraints,

$$x = \frac{L}{2}\sin\theta \qquad (2.77)$$

and

$$y = \frac{L}{2}\cos\theta. \qquad (2.78)$$

From equation 2.72 it follows that these give the constraining relations

$$\lambda_1\left[dx - \frac{L}{2}\cos\theta\, d\theta\right] = 0, \qquad (2.79)$$

and

$$\lambda_2\left[dy + \frac{L}{2}\sin\theta\, d\theta\right] = 0, \qquad (2.80)$$

respectively. By inspection, the kinetic energy is

$$T = \frac{1}{2}m\left(\dot{x}^2 + \dot{y}^2\right) + \frac{1}{2}I\dot{\theta}^2, \qquad (2.81)$$

where $I = \dfrac{1}{12} mL^2$. Similarly, the potential energy is

$$V = mgy, \tag{2.82}$$

so that the Lagrangian is

$$L = \frac{1}{2} m\left(\dot{x}^2 + \dot{y}^2 \right) + \frac{mL^2}{24} \dot{\theta}^2 - mgy. \tag{2.83}$$

From equations 2.76, 2.79, and 2.80 the equations of motion are

$$m\ddot{x} = \lambda_1, \tag{2.84}$$

$$m\ddot{y} + mg = \lambda_2, \tag{2.85}$$

and

$$\frac{mL^2}{24} \ddot{\theta} = -\lambda_1 \left(\frac{L}{2} \cos\theta \right) + \lambda_2 \left(\frac{L}{2} \sin\theta \right), \tag{2.86}$$

respectively. From equations 2.77 and 2.78 we see that

$$\ddot{x} = -\frac{L}{2} \sin\theta \dot{\theta}^2 + \frac{L}{2} \cos\theta \ddot{\theta} \tag{2.87}$$

and

$$\ddot{y} = \frac{L}{2} \cos\theta \dot{\theta}^2 - \frac{L}{2} \sin\theta \ddot{\theta}. \tag{2.88}$$

From equations 2.84 and 2.87 we see that

$$\lambda_1 = m \left(-\frac{L}{2} \sin\theta \dot{\theta}^2 + \frac{L}{2} \cos\theta \ddot{\theta} \right), \tag{2.89}$$

while from equations 2.85 and 2.88 we see that

$$\lambda_2 = m \left(\frac{L}{2} \cos\theta \dot{\theta}^2 - \frac{L}{2} \sin\theta \ddot{\theta} + g \right). \tag{2.90}$$

When these are combined with equation 2.86 and simplified we obtain

$$\frac{1}{6}\ddot{\theta} = \left(-\sin\theta\dot{\theta}^2 + \cos\theta\ddot{\theta}\right)(\cos\theta) - \left(\cos\theta\dot{\theta}^2 - \sin\theta\ddot{\theta} + g\right)(\sin\theta), \quad (2.91)$$

which is equivalent to

$$-\frac{5}{6}\ddot{\theta} = -2\sin\theta\cos\theta\dot{\theta}^2 - g\sin\theta. \tag{2.92}$$

2.3 Rigid Body Motion

2.3.1 Rotations

A rigid body in space needs six independent generalized coordinates to specify its configuration. For example, three coordinates are needed to specify the location of the center of mass relative to some external axes, and three other coordinates are needed to specify the orientation of the body relative to a coordinate system parallel to the external axes. The orientation is specified by stating the direction cosine of the body axes relative to the external axes. That is, if the prime denotes body axes, then

$$\hat{i}' = \left(\hat{i}'\cdot\hat{i}\right)\hat{i} + \left(\hat{i}'\cdot\hat{j}\right)\hat{j} + \left(\hat{i}'\cdot\hat{k}\right)\hat{k} \tag{2.93}$$

or

$$\hat{i}' = \alpha_1\hat{i} + \alpha_2\hat{j} + \alpha_3\hat{k} \tag{2.94}$$

or

$$\hat{i}' = \cos\left(\hat{i}',\hat{i}\right)\hat{i} + \cos\left(\hat{i}',\hat{j}\right)\hat{j} + \cos\left(\hat{i}',\hat{k}\right)\hat{k}. \tag{2.95}$$

Similarly,

$$\bar{x}' = \bar{\bar{R}}\bar{x}, \tag{2.96}$$

where

$$\vec{R} = \begin{pmatrix} \alpha_1 & \alpha_2 & \alpha_3 \\ \beta_1 & \beta_2 & \beta_3 \\ \gamma_1 & \gamma_2 & \gamma_3 \end{pmatrix}. \tag{2.97}$$

Because

$$\|\vec{x}'\| = \|\vec{x}\|, \tag{2.98}$$

we have

$$\vec{R}\vec{R}^t = \vec{I}, \tag{2.99}$$

or

$$\alpha_1^2 + \beta_1^2 + \gamma_1^2 = 1, \tag{2.100}$$

for $l = 1, 2, 3$. That is,

$$x_i' = \alpha_{ij} x_j, \tag{2.101}$$

and

$$x_i' x_i' = \alpha_{ij}\alpha_{ik} x_j x_k = x_i x_i. \tag{2.102}$$

Therefore, $\alpha_{ij}\alpha_{ik} = 1$ if $j = k$, $\alpha_{ij}\alpha_{ik} = 0$ otherwise. In two dimensions,

$$\vec{R} = \begin{pmatrix} \cos\phi & \sin\phi \\ -\sin\phi & \cos\phi \end{pmatrix}. \tag{2.103}$$

2.3.1.1 General Properties of Rotations

If $\vec{G} = \vec{A}\vec{F}$ and we transform to a new coordinate system, then $\vec{B}\vec{G} = \vec{B}\vec{A}\vec{F} = \vec{B}\vec{A}\vec{B}^{-1}\vec{B}\vec{F}$ and we say that $\vec{B}\vec{A}\vec{B}^{-1}$ is the form of $\vec{A}$ in the new coordinate system. $\vec{A}' = \vec{B}\vec{A}\vec{B}^{-1}$ defines a similarity transformation. In some coordinate system,

$$\ddot{A}' = \begin{pmatrix} \cos\phi & \sin\phi & 0 \\ -\sin\phi & \cos\phi & 0 \\ 0 & 0 & 1 \end{pmatrix}, \qquad (2.104)$$

so that $Tr\ddot{A} = 1 + 2\cos\phi$. This property holds true in all coordinate systems.

Example 2.2
Find the axis of rotation and compute the angle of rotation for

$$\ddot{R} = \begin{pmatrix} \dfrac{3}{4} & \dfrac{\sqrt{6}}{4} & \dfrac{1}{4} \\ -\dfrac{\sqrt{6}}{4} & \dfrac{1}{2} & \dfrac{\sqrt{6}}{4} \\ \dfrac{1}{4} & -\dfrac{\sqrt{6}}{4} & \dfrac{3}{4} \end{pmatrix}.$$

It can easily be verified that $\ddot{R}\ddot{R}^T = \ddot{I}$, so $\ddot{R}$ satisfies the requirements of a rotation matrix. The axis of rotation can be defined in terms of the vector that remains unchanged by $\ddot{R}$. That is, if $\bar{x} = \ddot{R}\bar{x}$, then $\bar{x}$ is the axis of rotation. We must solve

$$\begin{pmatrix} \dfrac{3}{4} & \dfrac{\sqrt{6}}{4} & \dfrac{1}{4} \\ -\dfrac{\sqrt{6}}{4} & \dfrac{1}{2} & \dfrac{\sqrt{6}}{4} \\ \dfrac{1}{4} & -\dfrac{\sqrt{6}}{4} & \dfrac{3}{4} \end{pmatrix} \begin{pmatrix} x_1 \\ x_2 \\ x_3 \end{pmatrix} = (x_1 \quad x_2 \quad x_3), \qquad (2.105)$$

which gives the equations

$$3x_1 + \sqrt{6}x_2 + x_3 = 4x_1, \qquad (2.106)$$

$$-\sqrt{6}x_1 + 2x_2 + \sqrt{6}x_3 = 4x_2, \qquad (2.107)$$

$$x_1 - \sqrt{6}x_2 + 3x_3 = 4x_3. \qquad (2.108)$$

These three equations can be solved to show that $x_1 = x_3$ and $x_2 = 0$. The normalized axis of rotation is therefore $\frac{1}{\sqrt{2}}(1,0,1)$. By examination, $\mathrm{Tr}\,\ddot{R} = 1 + 2\cos\theta = 2$, so that $\theta = 60°$.

2.3.1.2 The Euler Angles

Rather than specify the nine independent elements of the rotation matrix, we may describe the orientation in terms of three Euler angles. For example, rotate the initial system of axes, $x\,y\,z$, by an angle ψ counterclockwise about z. This defines the $\xi\,\eta\,\zeta$ axes. Next, rotate about the ξ axis by an angle θ in the counterclockwise direction. This defines the $\xi'\,\eta'\,\zeta'$ axes. Finally rotate counterclockwise by an angle ϕ about the ζ' axes. This defines the $\xi'\,\psi'\,\zeta'$ axes. In matrix form,

$$\vec{x}' = \ddot{A}\vec{x}, \tag{2.109}$$

where

$$\ddot{A} = \ddot{B}\ddot{C}\ddot{D}, \tag{2.110}$$

and

$$\ddot{B} = \begin{pmatrix} \cos\psi & \sin\psi & 0 \\ -\sin\psi & \cos\psi & 0 \\ 0 & 0 & 1 \end{pmatrix}, \tag{2.111}$$

$$\ddot{C} = \begin{pmatrix} 1 & 0 & 0 \\ 0 & \cos\theta & \sin\theta \\ 0 & -\sin\theta & \cos\theta \end{pmatrix}, \tag{2.112}$$

$$\ddot{D} = \begin{pmatrix} \cos\phi & \sin\phi & 0 \\ -\sin\phi & \cos\phi & 0 \\ 0 & 0 & 1 \end{pmatrix}. \tag{2.113}$$

2.3.1.3 The Cayley-Klein Parameters

Consider a general linear transformation in two-dimensional space,

$$\vec{u}' = \alpha\vec{u} + \beta\vec{v} \tag{2.114}$$

and

$$\vec{v}' = \gamma\vec{u} + \delta\vec{v}, \tag{2.115}$$

where the transformation matrix is

$$\ddot{Q} = \begin{pmatrix} \alpha & \beta \\ \gamma & \delta \end{pmatrix}. \tag{2.116}$$

Note that α, β, γ, δ may be complex. If we require $\ddot{Q}\ddot{Q}^t = \vec{I}$ and $\left|\ddot{Q}\right| = +1$, we find that $\beta = -\gamma^*$ and $\delta = \alpha^*$, that is,

$$\ddot{Q} = \begin{pmatrix} \alpha & \beta \\ -\beta^* & \alpha^* \end{pmatrix}. \tag{2.117}$$

Consider a matrix of the form

$$\vec{P} = \begin{pmatrix} z & x - iy \\ x + iy & -z \end{pmatrix}, \tag{2.118}$$

such that

$$\vec{P}' = \ddot{Q}\,\vec{P}\,\ddot{Q}^t. \tag{2.119}$$

The hermitian property and the trace of a matrix are unaffected by similarity transformations. Consequently, $\vec{P}'$ is of the form

$$\vec{P}' = \begin{pmatrix} z' & x' - iy' \\ x' + iy' & -z' \end{pmatrix}. \tag{2.120}$$

If we let $x_+ = x + iy$ and $x_- = x - iy$, then

$$\vec{P}' = \begin{pmatrix} z' & x_-' \\ x_+' & z' \end{pmatrix} = \begin{pmatrix} \alpha & \beta \\ \gamma & \delta \end{pmatrix}\begin{pmatrix} z & x_- \\ x_+ & -z \end{pmatrix}\begin{pmatrix} \delta & -\beta \\ -\gamma & \alpha \end{pmatrix}. \tag{2.121}$$

In this way, we may define a nine-element rotation matrix in terms of four Cayley-Klein parameters.

2.3.2 The Rate of Change of a Vector

The rate of change of a vector $\vec{r}$ as seen by an observer in the body system of axes will differ from the corresponding change as seen by an observer fixed in space. If the body axes are rotating with angular velocity ω, the general solution is

$$\left(\frac{d\vec{r}}{dt}\right)_{space} = \left(\frac{d\vec{r}}{dt}\right)_{body} + \vec{\omega} \times \vec{r}. \qquad (2.122)$$

We have

$$\vec{v}_s = \vec{v}_b + \vec{\omega} \times \vec{r}, \qquad (2.123)$$

and a successive application of equation 2.122 gives

$$\vec{a}_s = \vec{a}_b + 2(\vec{\omega} \times \vec{v}_b) + \vec{\omega} \times (\vec{\omega} \times \vec{r}) + \frac{d\vec{\omega}}{dt} \times \vec{r}. \qquad (2.124)$$

If the angular velocity of the body is constant, Newton's law is

$$\vec{F} = m\vec{a}_s = m\vec{a}_b + 2m(\vec{\omega} \times \vec{v}_b) + m\vec{\omega} \times (\vec{\omega} \times \vec{r}). \qquad (2.125)$$

To an observer in the rotating system, it appears as though the particle is moving under an effective force

$$\vec{F}_{eff} = m\vec{a}_b = \vec{F} - 2m(\vec{\omega} \times \vec{v}_b) - m\vec{\omega} \times (\vec{\omega} \times \vec{r}). \qquad (2.126)$$

The second term on the right-hand side is called the *Coriolis force* and the last term on the right-hand side is called the *centrifugal force*.

Example 2.3
Show that if a particle is thrown up vertically with initial speed v_o and reached a height h, it will experiment a Coriolis deflection that is opposite in direction and four times greater in magnitude than the deflection it would experience if it were dropped at rest from the same maximum height.

The rate of rotation of the Earth, ω, is identical for each scenario. The difference is the initial velocity and position of the particle. From equation 2.126, the Coriolis force is $2(\vec{\omega} \times \vec{v}_b)$. Defining the coordinate axes as shown in figure 2.2, we have $\omega_x = \omega_y = 0$ and $\omega_z = \omega$, while $v_x = 0$, $v_y = v\cos\theta$, and $v_z = v\sin\theta$.

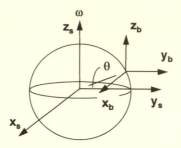

Figure 2.2 Coriolis geometry.

As a result,

$$-2(\vec{\omega} \times \vec{v}_b) = +2(\omega v\cos\theta)\hat{i} + 2(0)\hat{j} + 2(0)\hat{k} \qquad (2.127)$$

As a result, the acceleration due to the Coriolis force is simply

$$\ddot{x} = 2\omega v\cos\theta. \qquad (2.128)$$

A body falling under the influence of gravity satisfies the condition

$$v = v(0) - gt, \qquad (2.129)$$

where $v(0)$ is defined as the velocity at time $t = 0$. Substituting this into equation 2.128 and integrating gives

$$\dot{x} = 2\omega\cos\theta\left[v(0)t - \frac{1}{2}gt^2\right]. \qquad (2.130)$$

Integrating a second time gives

$$x = 2\omega\cos\theta\left[\frac{1}{2}v(0)t^2 - \frac{1}{6}gt^3\right]. \qquad (2.131)$$

To solve the problem at hand, we let t be the time it takes for the particle to complete the trip. The position of the particle at any time is given by

$$h(t) = h(0) + v(0)t - \frac{1}{2}gt^2. \tag{2.132}$$

In one case, the particle starts at rest, $v(0) = 0$, at height h. From the definition of the problem, it can be seen that the time required to complete the drop to the ground is

$$t_1 = \sqrt{\frac{2h}{g}} = \frac{v_o}{g}. \tag{2.133}$$

Consequently, the deflection when dropped from rest at height h is

$$x_1 = 2\omega \cos\theta \left[-\frac{1}{6}g\left(\frac{v_o}{g}\right)^3 \right] = -\left(\frac{1}{3}\right)\left(\frac{v_o}{g}\right)^3 g\omega \cos\theta. \tag{2.134}$$

If, instead, we let t_2 be the time it takes to fall from height h, we have $v_i = v_o$ and $t_2 = 2t_1$. The deflection is

$$x_2 = 2\omega \cos\theta \left[\frac{1}{2}v_o\left(\frac{2v_o}{g}\right)^2 - \frac{1}{6}g\left(\frac{2v_o}{g}\right)^3 \right] = +\left(\frac{4}{3}\right)\frac{v_o^3}{g^2}\omega \cos\theta. \tag{2.135}$$

Thus, the deflection when thrown from the ground up is opposite in direction and four times greater in magnitude than then deflection when dropped from rest.

2.3.3 The Rigid Body Equations of Motion

We have previously seen that the total kinetic energy of a rigid body may be expressed as the sum of the kinetic energy of the entire body as if concentrated at the center of mass, plus the kinetic energy of the motion about the center of mass. If a rigid body moves with one point stationary, the angular momentum about the center of mass is

$$\vec{L} = m_i \ (\vec{r}_i \times \vec{v}_i). \tag{2.136}$$

Since r_i is a fixed vector in the body,

$$\vec{v}_i = \vec{\omega} \times \vec{r}_i. \tag{2.137}$$

Thus

$$\vec{L} = m_i (\vec{r}_i \times (\vec{\omega} \times \vec{r}_i)) = m_i (\vec{\omega} r_i^2 - \vec{r}_i \, (\vec{r}_i \cdot \vec{\omega})). \qquad (2.138)$$

The x-component is

$$L_x = \omega_x m_i (r_i^2 - x_i^2) - \omega_y m_i x_i y_i - \omega_z m_i x_i z_i \qquad (2.139)$$

or

$$L_x = I_{xx}\omega_x + I_{xy}\omega_y + I_{xz}\omega_z. \qquad (2.140)$$

We define

$$I_{jk} = \int_v \rho(\vec{r})(r^2 \delta_{jk} - x_j x_k)dV \qquad (2.141)$$

to be the inertia tensor so that

$$\vec{L} = \vec{I} \, \vec{\omega}. \qquad (2.142)$$

The kinetic energy of motion about a point is

$$T = \frac{1}{2} m_i v_i^2 = \frac{1}{2} m_i \vec{v}_i \cdot (\vec{\omega} \times \vec{r}_i), \qquad (2.143)$$

or

$$T = \frac{\vec{\omega} \cdot \vec{L}}{2} = \frac{1}{2} \vec{\omega} \cdot \vec{I} \cdot \vec{\omega}. \qquad (2.144)$$

The moment of inertia, $\vec{I}$, about some given axis is related to the moment about a parallel axis through the center of mass, as shown in figure 2.3. Let the vector from the origin to the center of mass be $\vec{R}$, and let the radii vectors from the origin and the center of mass to the ith particle be $\vec{r}_i$ and $\vec{r}_i{}'$, respectively. That is

$$\vec{r}_i = \vec{R} + \vec{r}_i{}'. \qquad (2.145)$$

The moment of inertia about axis a is

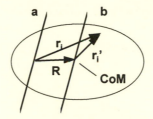

Figure 2.3. Moment of inertia about a parallel axis.

$$I_a = m_i (\vec{r}_i \times \hat{n})^2 = m_i \, [(\vec{R} + \vec{r}_i{}') \times \hat{n}]^2, \qquad (2.146)$$

or

$$I_a = M(\vec{R} \times \hat{n})^2 + m_i \, (\vec{r}_i{}' + \hat{n})^2 + 2m_i (\vec{R} \times \hat{n})(\vec{r}_i{}' \times \hat{n}). \qquad (2.147)$$

The last term is $2(\vec{R} \times \hat{n}) \cdot (\hat{n} \times m_i \vec{r}_i{}')$, but $\sum_i m_i r_i{}' = 0$ and $\vec{I}_b = m_i (\vec{r}_i{}' \times \hat{n})^2$,

so

$$\vec{I}_a = \vec{I}_b + M(\vec{R} \times \hat{n})^2. \qquad (2.148)$$

2.3.3.1 The Euler Equations and Torque-Free Motion

By definition,

$$\left(\frac{d\vec{L}}{dt} \right)_s = \left(\frac{d\vec{L}}{dt} \right)_b + \vec{\omega} \times \vec{L} = \vec{N}. \qquad (2.149)$$

In the body system, using $L_i = I_{ii} \, \omega_i$, we find

$$I_i \frac{d\omega_i}{dt} + \varepsilon_{ijk} \omega_j \omega_k I_k = N_i. \qquad (2.150)$$

The Euler equations are

$$I_1 \dot{\omega}_1 - \omega_2 \omega_3 \, (I_2 - I_3) = N_1, \qquad (2.151)$$

$$I_2 \dot{\omega}_2 - \omega_3 \omega_1 \, (I_3 - I_1) = N_2, \qquad (2.152)$$

$$I_3\dot{\omega}_3 - \omega_1\omega_2\,(I_1 - I_2) = N_3. \tag{2.153}$$

The principle axes of a system are those where the tensor $\bar{I}$ is diagonal. If this is the case,

$$T = \frac{1}{2}\frac{L_1^2}{I_1} + \frac{1}{2}\frac{L_2^2}{I_2} + \frac{1}{2}\frac{L_3^2}{I_3}. \tag{2.154}$$

Because T is constant, the relation defines an ellipsoid fixed in the body axes. Because angular momentum is conserved, $\bar{L}$ must be on a fixed sphere defined by

$$L^2 = L_1^2 + L_2^2 + L_3^2. \tag{2.155}$$

For the given initial conditions (kinetic energy and angular momentum) the path of $\bar{L}$ is constrained to be the intersection of the sphere, L^2, and the ellipsoid, T. If we have a body symmetrical about the z-axis so that $I_1 = I_2$, the Euler equations are, in the absence of torques,

$$I_1\dot{\omega}_1 = (I_1 - I_3)\,\omega_3\,\omega_1, \tag{2.156}$$

$$I_2\dot{\omega}_2 = -(I_1 - I_3)\,\omega_3\,\omega_1, \tag{2.157}$$

$$I_3\dot{\omega}_3 = 0. \tag{2.158}$$

We have $\omega_3 =$ constant, and

$$\dot{\omega}_1 = -\Omega\omega_2, \quad \dot{\omega}_2 = \Omega\omega_1, \tag{2.159}$$

where

$$\Omega = \frac{I_3 - I_1}{I_1}\omega_3. \tag{2.160}$$

Thus,

$$\ddot{\omega}_1 = -\Omega\omega_1, \tag{2.161}$$

which has solutions $\omega_1 = A\cos\Omega t$ and $\omega_2 = A\sin\Omega t$. Hence, the total angular velocity is constant in magnitude but precesses with frequency Ω about the ζ-axis. We may solve for A and ω_3 by noting that

$$T = \frac{1}{2}I_1 A^2 + \frac{1}{2}I_3\omega_3^2 \tag{2.162}$$

and

$$L^2 = I_1^2 A^2 + I_3^2\omega_3^2. \tag{2.163}$$

2.3.3.2 The Heavy Symmetrical Top

Consider a heavy top with a symmetry axis taken to be the z-axis of the coordinate system fixed in the body. The three Euler angles are θ = inclination of the z-axis from the vertical; ϕ = azimuth of the top about the vertical; and ψ = rotation angle of the top about its own z-axis. The kinetic energy is

$$T = \frac{1}{2}I_1(\omega_1^2 + \omega_2^2) + \frac{1}{2}I_3\omega_3^2, \tag{2.164}$$

or

$$T = \frac{1}{2}I_1(\dot\theta^2 + \dot\phi^2 \sin^2\theta) + \frac{1}{2}I_3(\dot\psi + \dot\phi\cos\theta)^2. \tag{2.165}$$

The potential energy is

$$V = mgl\cos\theta, \tag{2.166}$$

so that the Lagrangian is given by

$$L = \frac{1}{2}I_1(\dot\theta^2 + \dot\phi^2 \sin^2\theta) + \frac{1}{2}I_3(\dot\psi + \dot\phi\cos\theta)^2 - mgl\cos\theta. \tag{2.167}$$

We have

$$p_\psi = \frac{\partial L}{\partial\dot\psi} = I_3(\dot\psi + \dot\phi\cos\theta) = I_3\omega_3 = I_1 a, \tag{2.168}$$

$$p_\phi = \frac{\partial L}{\partial \dot{\phi}} = (I_1 \sin^2 \theta + I_3 \cos^2 \theta)\dot{\phi} + I_3 \dot{\psi} \cos\theta = I_1 b, \quad (2.169)$$

and

$$E = T + V = \frac{1}{2} I_1 (\dot{\theta}^2 + \dot{\phi}^2 \sin^2 \theta) + \frac{1}{2} I_3^2 \omega_3^2 + Mgl \cos\theta. \quad (2.170)$$

Combining these first two expressions gives

$$\dot{\phi} = \frac{b - a\cos\theta}{\sin^2 \theta} \quad (2.171)$$

and

$$\dot{\psi} = \frac{I_1 a}{I_3} - \cos\theta \frac{b - a\cos\theta}{\sin^2 \theta}. \quad (2.172)$$

2.4 Oscillatory Motion

2.4.1 Oscillations

Consider a system that is subjected to a potential that is only a function of coordinates. If we expand the coordinates q_i about their equilibrium position according to

$$q_i = q_{io} + \eta_i, \quad (2.173)$$

and then perform a Taylor series expansion on the potential, we obtain

$$V(q_i) = V(q_{io}) + \left(\frac{\partial V}{\partial q_i}\right)\Bigg|_o \eta_i + \frac{1}{2}\left(\frac{\partial^2 V}{\partial q_i \partial q_j}\right)\Bigg|_o \eta_i \eta_j + \dots . \quad (2.174)$$

The first term on the right-hand side may be redefined to be zero by shifting the zero potential to be the equilibrium value. Similarly, if the generalized forces are zero, the second term is also zero. As a result, the potential can be approximated by the matrix relation

$$V(q_i) \approx \frac{1}{2} \left(\frac{\partial^2 V}{\partial q_i \partial q_j} \right)\bigg|_o \eta_i \eta_j. \tag{2.175}$$

Similarly, we can define

$$T = \frac{1}{2} m_{ij} \dot{q}_i \dot{q}_j = \frac{1}{2} m_{ij} \dot{\eta}_i \dot{\eta}_j = \frac{1}{2} T_{ij} \dot{\eta}_i \dot{\eta}_j, \tag{2.176}$$

so that the Lagrangian is

$$L = \frac{1}{2} \left(T_{ij} \dot{\eta}_i \dot{\eta}_j - V_{ij} \eta_i \eta_j \right). \tag{2.177}$$

The equations of motion are given by

$$T_{ij} \ddot{\eta}_j + V_{ij} \eta_j = 0. \tag{2.178}$$

If we try a solution of the form $\eta_i = Ca_i \exp^{-i\omega t}$ we find that

$$V_{ij} a_j - \omega^2 T_{ij} a_j = 0, \tag{2.179}$$

which is equivalent to

$$\left| V - \omega^2 T \right| = 0. \tag{2.180}$$

Example 2.4
Two particles move in one dimension at the junction of three springs as shown in figure 2.4. The springs all have unstretched length a and force constants as shown. Find the frequencies of the normal modes of oscillation.

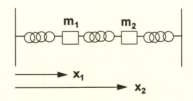

Figure 2.4 Geometry of the two mass - three spring problem.

We define the coordinates x_1 and x_2 to describe the displacement of the two blocks, relative to the left attachment point. In these coordinates,

$$T = \frac{1}{2}m\dot{x}_1^2 + \frac{1}{2}m\dot{x}_2^2. \tag{2.181}$$

Expanding about equilibrium, we define $x_1 = a + \eta_1$, $x_2 = 2a + \eta_2$. Equation 2.181 is equivalent to

$$T = \frac{1}{2}m\dot{\eta}_1^2 + \frac{1}{2}m\dot{\eta}_2^2 = \frac{1}{2}T_{ij}\dot{\eta}_i\dot{\eta}_j, \tag{2.182}$$

where

$$T_{ij} = \begin{pmatrix} m & 0 \\ 0 & m \end{pmatrix}. \tag{2.183}$$

Similarly,

$$V = \frac{1}{2}k(x_1 - a)^2 + \frac{1}{2}3k[x_2 - x_1 - a]^2 + \frac{1}{2}k(2a - x_2)^2, \tag{2.184}$$

or

$$V = \frac{1}{2}V_{ij}\eta_i\eta_j, \tag{2.185}$$

with

$$V_{ij} = \left(\frac{\partial^2 V}{\partial x_i \partial x_j}\right)\bigg|_a. \tag{2.186}$$

By examination,

$$\frac{\partial V}{\partial x_1} = k(x_1 - a) - 3k(x_2 - x_1 - a), \tag{2.187}$$

$$\frac{\partial^2 V}{\partial x_1^2} = k + 3k = 4k, \tag{2.188}$$

and

$$\frac{\partial V}{\partial x_2} = 3k(x_2 - x_1 - a) - k(2a - x_2),$$
(2.189)

$$\frac{\partial^2 V}{\partial x_2^2} = 3k + k = 4k.$$
(2.190)

Likewise, the cross terms are

$$\frac{\partial^2 V}{\partial x_1 \partial x_2} = \frac{\partial^2 V}{\partial x_2 \partial x_1} = -3k,$$
(2.191)

so that

$$V_{ij} = \begin{pmatrix} 4k & -3k \\ -3k & 4k \end{pmatrix}.$$
(2.192)

The secular equation is

$$\left| V - \omega^2 T \right| = \begin{vmatrix} 4k - \omega^2 m & -3k \\ -3k & 4k - \omega^2 m \end{vmatrix} = 0,$$
(2.193)

which reduces to

$$\left(4k - \omega^2 m \right)^2 - 9k = m^2 \omega^4 - 8km\omega^2 + 7k^2 = 0.$$
(2.194)

From the quadratic equation, this has solutions

$$\omega_1 = \sqrt{\frac{k}{m}},$$
(2.195)

and

$$\omega_2 = \sqrt{\frac{7k}{m}}.$$
(2.196)

2.5 Hamilton's Equations

2.5.1 Legendre Transformations and Hamilton's Equations of Motion

Consider a function f of two variables such that

$$df = u\,dx + v\,dy, \tag{2.197}$$

where $u = \dfrac{\partial f}{\partial x}$ and $v = \dfrac{\partial f}{\partial y}$. We wish to change from the variables x, y to the variables u, y. We define a function

$$g = f - ux. \tag{2.198}$$

We have

$$dg = df - x\,du - u\,dx = v\,dy - x\,du, \tag{2.199}$$

so that $x = -\dfrac{\partial g}{\partial u}$ and $v = \dfrac{\partial g}{\partial y}$. This is an example of a Legendre transformation. It is frequently used in thermodynamics.

Recall that Lagrange's equations are

$$\frac{d}{dt}\left(\frac{\partial L}{\partial \dot{q}_i}\right) - \frac{\partial L}{\partial q_i} = 0, \tag{2.200}$$

where $L = L(q_i, \dot{q}_i, t)$ and the conjugate momenta are

$$p_i = \frac{\partial L(q_i, \dot{q}_i, t)}{\partial \dot{q}_i}. \tag{2.201}$$

We can transform from the variables $(q_i, \dot{q}_i, t)$ to the variables (q_i, p_i, t) by the use of

$$H(q_i, p_i, t) = \dot{q}_i p_i - L(q_i, \dot{q}_i, t), \tag{2.202}$$

where H is called the *Hamiltonian*. We have

$$dH = \frac{\partial H}{\partial q_i} dq_i + \frac{\partial H}{\partial p_i} dp_i + \frac{\partial H}{\partial t} dt \,. \tag{2.203}$$

However, from the definition,

$$dH = p_i d\dot{q}_i + \dot{q}_i dp_i - \frac{\partial L}{\partial q_i} dq_i - \frac{\partial L}{\partial \dot{q}_i} d\dot{q}_i - \frac{\partial L}{\partial t} dt \,, \tag{2.204}$$

or simply,

$$dH = \dot{q}_i dp_i - \frac{\partial L}{\partial q_i} dq_i - \frac{\partial L}{\partial t} dt \,. \tag{2.205}$$

Comparing the two expressions for dH, we obtain Hamilton's equations,

$$\frac{\partial H}{\partial q_i} = -\frac{\partial L}{\partial q_i} = -\dot{p}_i, \tag{2.206}$$

$$\frac{\partial H}{\partial p_i} = \dot{q}_i, \tag{2.207}$$

$$\frac{\partial H}{\partial t} = -\frac{\partial L}{\partial t}\,. \tag{2.208}$$

If the equations defining the generalized coordinates don't depend on time explicitly, and if the forces are derivable from a conservative potential V, then

$$H = T + V = E \,. \tag{2.209}$$

If we define a column matrix $\vec{\eta}$ with $2n$ elements such that $\eta_i = q_i$, $\eta_{i+n} = p_i$, for $i \leq n$, then

$$\left(\frac{\partial H}{\partial \eta_i} \right) = \frac{\partial H}{\partial q_i}, \tag{2.210}$$

and

$$\left(\frac{\partial H}{\partial \vec{\eta}_{i+n}} \right) = \frac{\partial H}{\partial p_i}\,. \tag{2.211}$$

If we define $\bar{J} = \begin{pmatrix} 0 & 1 \\ 1 & 0 \end{pmatrix}$, then Hamilton's equations may be written in the form

$$\dot{\bar{\eta}} = \bar{J} \frac{\partial H}{\partial \bar{\eta}}. \tag{2.212}$$

This is called *symplectic notation*.

Example 2.5
The Lagrangian for a simple spring is given by

$$L = \frac{1}{2} m \dot{x}^2 - \frac{1}{2} k x^2.$$

Find the Hamiltonian and the equations of motion using the Hamiltonian formulation. Identify any conserved quantities.

From the definition of p_i, we have

$$p_x = \frac{\partial L}{\partial \dot{x}} = m \dot{x}. \tag{2.213}$$

From the definition of the Hamiltonian (eq. 2.202), we see that

$$H = \dot{x}(m\dot{x}) - \frac{1}{2} m \dot{x}^2 + \frac{1}{2} k x^2, \tag{2.214}$$

or simply,

$$H = \frac{1}{2m} p_x^2 + \frac{1}{2} k x^2. \tag{2.215}$$

From Hamilton's equations, we have

$$\frac{\partial H}{\partial q_i} = kx = -\dot{p}_x, \tag{2.216}$$

$$\frac{\partial H}{\partial p_i} = \frac{p_x}{m} = \dot{x}, \tag{2.217}$$

$$\frac{\partial H}{\partial t} = 0. \tag{2.218}$$

The first equation is the equation of motion in one dimension,

$$m\ddot{x} + kx = 0, \tag{2.219}$$

the second equation is the definition of momentum, and the last equation is the statement of conservation of energy.

2.5.2 Canonical Transformations

If a generalized coordinate q_i has constant conjugate momenta it is said to be cyclic. If this is the case, then $p_i = 0$, which tells us that the Hamiltonian is independent of that p_i. If all coordinates q_i are cyclic, the conjugate momenta can be defined by $p_i = \alpha_i$. Consequently,

$$\dot{q}_i = \frac{\partial H}{\partial \alpha_i} = \omega_i, \tag{2.220}$$

or

$$q_i = \omega_i t + \beta_i. \tag{2.221}$$

A problem is often easier to solve if we can find a system where the number of cyclic coordinates is maximum. How do we transform to this set of coordinates? We need a new set of coordinates Q_i, P_i, where $Q_i = Q_i(q_i, p_i, t)$, $P_i = P_i(q_i, p_i, t)$. We require Q_i, P_i to be canonical coordinates. Therefore, some function $K = K(Q_i, P_i, t)$ exists such that

$$\dot{Q}_i = \frac{\partial K}{\partial P_i} \tag{2.222}$$

and

$$\dot{P}_i = \frac{\partial K}{\partial Q_i}. \tag{2.223}$$

If Q_i, P_i are canonical coordinates, they must satisfy a modified Hamilton's principle that can be put in the form

$$\delta \int_{t_1}^{t_2} \left(P_i \dot{Q}_i - K(Q_i, P_i, t) \right) dt = 0, \qquad (2.224)$$

because the old coordinates satisfy

$$\delta \int_{t_1}^{t_2} \left(p_i \dot{q}_i - H(q_i, p_i, t) \right) dt = 0. \qquad (2.225)$$

Both requirements can be satisfied if we require a relation

$$\lambda \left(p_i \dot{q}_i - H \right) = P_i \dot{Q}_i - K + \frac{dF}{dt}, \qquad (2.226)$$

where λ = constant and F is any function of the phase space coordinates continuous through the second derivative. λ is related to a scale transformation. If $\lambda = 1$, the relation defines a canonical transformation. The function F is termed the *generating function*. It may be a function of q_i, p_i, Q_i, P_i, t and defines the transformation.

2.5.3 Symplectic Transformations and Poisson Brackets

Recall that Hamilton's equations can be written in the from

$$\dot{\vec{\eta}} = \vec{J} \frac{\partial H}{\partial \vec{\eta}}. \qquad (2.227)$$

If we have a canonical transformation from $\eta \rightarrow \xi = \xi(\eta)$, then

$$\dot{\xi}_i = \frac{\partial \xi_i}{\partial \eta_j} \dot{\eta}_j. \qquad (2.228)$$

In matrix form,

$$\dot{\vec{\xi}} = \vec{M} \dot{\vec{\eta}}, \qquad (2.229)$$

where $M_{ij} = \dfrac{\partial \xi_i}{\partial \eta_j}$. We have

$$\dot{\vec{\xi}} = \vec{M}\vec{J}\frac{\partial H}{\partial \vec{\eta}},$$ (2.230)

also

$$\frac{\partial H}{\partial \eta_i} = \frac{\partial H}{\partial \xi_j}\frac{\partial \xi_j}{\partial \eta_i},$$ (2.231)

and

$$\frac{\partial H}{\partial \vec{\eta}} = \tilde{\vec{M}}\frac{\partial H}{\partial \vec{\xi}}.$$ (2.232)

Consequently,

$$\dot{\vec{\xi}} = \vec{M}\,\vec{J}\,\tilde{\vec{M}}\frac{\partial H}{\partial \vec{\xi}} = \vec{J}\frac{\partial H}{\partial \vec{\xi}}.$$ (2.233)

Therefore, a transformation is canonical if

$$\vec{M}\,\vec{J}\,\tilde{\vec{M}} = \vec{J}.$$ (2.234)

The Poisson bracket of a function is defined by

$$[u,v]_{PB} = \frac{\partial u}{\partial q_i}\frac{\partial v}{\partial p_i} - \frac{\partial u}{\partial p_i}\frac{\partial v}{\partial q_i}.$$ (2.235)

We have

$$[\vec{\eta},\vec{\eta}]_{PB} = \vec{J},$$ (2.236)

and

$$[\vec{\xi},\vec{\xi}]_{PB} = \frac{\partial \vec{\xi}}{\partial \vec{\eta}}\,\vec{J}\,\frac{\partial \vec{\xi}}{\partial \vec{\eta}} = \tilde{\vec{M}}\,\vec{J}\,\vec{M} = \vec{J}.$$ (2.237)

In other words, the fundamental Poisson brackets are invariant under canonical transformations. We also have

$$\frac{du}{dt} = \frac{\partial u}{\partial q_i} \dot{q}_i + \frac{\partial u}{\partial p_i} \dot{p}_i + \frac{\partial u}{\partial t},$$
(2.238)

or

$$\frac{du}{dt} = [u, H]_{PB} + \frac{\partial u}{\partial t}.$$
(2.239)

Similarly,

$$\dot{q}_i = [q_i, H]_{PB}, \quad \dot{p}_i = [p_i, H]_{PB}.$$
(2.240)

2.6 Continuous Systems

2.6.1 The Transition from a Discrete to a Continuous System

Consider an infinitely long elastic rod that can undergo small longitudinal vibrations. We approximate this by an infinite chain of equal mass points a distance a apart and connected by uniform massless springs having force constraints k. The kinetic energy is

$$T = \frac{1}{2} \sum_i m \dot{\eta}_i^2,$$
(2.241)

where m is the mass of each particle and η_i is the location of the ith particle. The potential energy is

$$V = \frac{1}{2} \sum_i k(\eta_{i+1} - \eta_i)^2.$$
(2.242)

We have

$$L = \frac{1}{2} \sum_i \left[m \dot{\eta}_i^2 - k(\eta_{i+1} - \eta_i)^2 \right],$$
(2.243)

or

$$L = \frac{1}{2} \sum_i a \left[\frac{m}{a} \dot{\eta}_i^2 - ka \left(\frac{\eta_{i+1} - \eta_i}{a} \right)^2 \right].$$ (2.244)

The equations of motion are

$$\frac{m}{a} \ddot{\eta}_i - ka \left(\frac{\eta_{i+1} - \eta_i}{a^2} \right) + ka \left(\frac{\eta_i - \eta_{i-1}}{a^2} \right) = 0.$$ (2.245)

We have $m/a = \mu =$ mass/unit length. Hooke's law states that the extension of a rod/unit length is proportional to the force, i.e.,

$$F = Y\xi,$$ (2.246)

where $\xi = \left(\frac{\eta_{i+1} - \eta_i}{a} \right)$ is the extension/unit length. The force necessary to stretch the string by an amount x is

$$F = k(\eta_{i+1} - \eta_i) = ka \left(\frac{\eta_{i+1} - \eta_i}{a} \right).$$ (2.247)

Consequently, $ka = Y =$ Young's modulus. Note that

$$\frac{\eta_{i+1} - \eta_i}{a} = \frac{\eta(x + a) - \eta(x)}{a} \rightarrow \frac{d\eta}{dx},$$ (2.248)

as $a \rightarrow 0$. Also as $a \rightarrow 0$, the summation over the particles becomes an integral, and equation 2.244 becomes

$$L = \frac{1}{2} \int \left(\mu \dot{\eta}^2 - Y \left(\frac{d\eta}{dx} \right)^2 \right) dx.$$ (2.249)

The equation of motion is

$$\mu \frac{d^2\eta}{dt^2} - Y \frac{d^2\eta}{dx^2} = 0.$$ (2.250)

This is a wave equation with wave propagation velocity $v = \sqrt{Y/\mu}$. Equation 2.249 is said to define a Lagrangian density,

$$\tilde{L} = \mu \left(\frac{d\eta}{dt} \right)^2 - Y \left(\frac{d\eta}{dx} \right)^2. \tag{2.251}$$

2.6.2 The Lagrangian Formulation

Consider a Lagrangian density

$$\tilde{L} = \tilde{L} \left(\eta, \frac{d\eta}{dx}, \frac{d\eta}{dt}, x, t \right). \tag{2.252}$$

Hamilton's principle is

$$\delta I = \delta \int\limits_1^2 \int \tilde{L} \, dx \, dt = 0. \tag{2.253}$$

We choose value of η such that

$$\eta(x,t;\alpha) = \eta(x,t;0) + \alpha \xi(x,t), \tag{2.254}$$

where $\eta(x,t;0)$ is the function that satisfies Hamilton's principle. We have

$$\delta I = \left(\frac{dI}{d\alpha} \right)_{\alpha=0} = 0, \tag{2.255}$$

where

$$\frac{dI}{d\alpha} = \int\limits_{t_1}^{t_2} \int\limits_{x_1}^{x_2} dx dt \left\{ \frac{\partial \tilde{L}}{\partial \eta} \frac{\partial \eta}{\partial \alpha} + \frac{\partial \tilde{L}}{\partial \left(\frac{d\eta}{dx} \right)} \frac{\partial}{\partial \alpha} \left(\frac{d\eta}{dx} \right) + \frac{\partial \tilde{L}}{\partial \left(\frac{d\eta}{dt} \right)} \frac{\partial}{\partial \alpha} \left(\frac{d\eta}{dt} \right) \right\}. \tag{2.256}$$

This expression may be simplified as follows. First,

$$\int\limits_{t_1}^{t_2} dt \left\{ \frac{\partial \tilde{L}}{\partial \left(\frac{d\eta}{dt} \right)} \frac{\partial}{\partial \alpha} \left(\frac{d\eta}{dt} \right) \right\} = - \int\limits_{t_1}^{t_2} \frac{d}{dt} \left[\frac{\partial \tilde{L}}{\partial \left(\frac{d\eta}{dt} \right)} \right] \frac{\partial \eta}{\partial \alpha} \, dt, \tag{2.257}$$

plus a boundary term that goes to zero. Also,

$$\int_{x_1}^{x_2} dx \left\{ \frac{\partial \tilde{L}}{\partial \left(\frac{d\eta}{dx}\right)} \frac{\partial}{\partial \alpha} \left(\frac{d\eta}{dx}\right) \right\} = -\int_{x_1}^{x_2} \frac{d}{dx} \left[\frac{\partial \tilde{L}}{\partial \left(\frac{d\eta}{dx}\right)} \right] \frac{\partial \eta}{\partial \alpha} dx, \quad (2.258)$$

plus another boundary term that goes to zero. Therefore,

$$\delta I = \int_{t_1}^{t_2} \int_{x_1}^{x_2} dx dt \left\{ \frac{\partial \tilde{L}}{\partial \eta} - \frac{d}{dt} \left(\frac{\partial \tilde{L}}{\partial \left(\frac{d\eta}{dx}\right)} \right) - \frac{d}{dx} \left(\frac{\partial \tilde{L}}{\partial \left(\frac{d\eta}{dt}\right)} \right) \right\} \left(\frac{\partial \eta}{\partial \alpha} \right)_{\alpha=0} = 0, \quad (2.259)$$

and

$$\frac{d}{dt} \left(\frac{\partial \tilde{L}}{\partial \left(\frac{d\eta}{dx}\right)} \right) + \frac{d}{dx} \left(\frac{\partial \tilde{L}}{\partial \left(\frac{d\eta}{dt}\right)} \right) - \frac{\partial \tilde{L}}{\partial \eta} = 0. \quad (2.260)$$

Similarly for other Lagrangians.

2.6.3 Noether's Theorem

A formal description of the connection between invariance or symmetry properties and conserved quantities is contained in Noether's theorem. We consider transformations where

$$x_\mu \to x'_\mu = x_\mu + \delta x_\mu, \quad (2.261)$$

$$\eta_\rho(x_\mu) \to \eta'_\rho(x'_\mu) = \eta_\rho(x_\mu) + \eta_\rho(\delta x_\mu), \quad (2.262)$$

$$\tilde{L}\left(\eta_\rho(x_\mu), \eta_{\rho,\nu}(x_\mu), x_\mu\right) \to \tilde{L}'\left(\eta'_\rho(x'_\mu), \eta'_{\rho,\nu}(x'_\mu), x'_\mu\right). \quad (2.263)$$

We make three assumptions:

1. Four-space is Euclidean.

2. The Lagrangian density has the same functional form after transformation.

3. The magnitude of the action integral is invariant under the transformation.

From assumptions 2, 3, we have

$$\int_{\Omega'} \tilde{L}\left(\eta'_\rho, \eta'_{\rho,v}, x'_\mu\right) dx'_\mu - \int_{\Omega} \tilde{L}\left(\eta_\rho, \eta_{\rho,v}, x_\mu\right) dx_\mu = 0. \quad (2.264)$$

x'_μ is a dummy variable, so let $x'_\mu \to x_\mu$ to obtain

$$\int_{\Omega'} \tilde{L}\left(\eta'_\rho(x_\mu), \eta'_{\rho,v}(x_\mu), x_\mu\right) dx_\mu - \int_{\Omega} \tilde{L}\left(\eta_\rho(x_\mu), \eta_{\rho,v}(x_\mu), x_\mu\right) dx_\mu = 0. \quad (2.265)$$

Under the transformations, the first-order difference between the integrals consists of two parts, one is an integral over Ω and the other is an integral over $\Omega' - \Omega$. For example, in one dimension,

$$\int_{a+\delta a}^{b+\delta b} [f(x) + \delta f(x)] dx - \int_a^b f(x) dx$$

$$= \int_a^b \delta f(x) dx + \int_b^{b+\delta b} [f(x) + \delta f(x)] dx - \int_a^{a+\delta a} [f(x) + \delta f(x)] dx. \quad (2.266)$$

To first order, the last two terms are

$$\int_b^{b+\delta b} f(x) dx - \int_a^{a+\delta a} f(x) dx = \delta b f(b) - \delta a f(a). \quad (2.267)$$

The difference between integrals is

$$\int_b^b \delta f(x) dx + f(x)\delta x \Big|_a^b = \int_a^b \left[\delta f(x) + \frac{d}{dx}(\delta x f(x)) \right] dx. \quad (2.268)$$

Consequently, for the Lagrangians we have

$$\int_{\Omega'} \tilde{L}(\eta', x_\mu) dx_\mu - \int_{\Omega} \tilde{L}(\eta, x_\mu) dx_\mu$$

$$= \int_{\Omega'} \left[\tilde{L}(\eta', x_\mu) - \tilde{L}(\eta, x_\mu) \right] d x_\mu + \int_S \tilde{L}(\eta) \, \delta x_\mu dS_\mu = 0, \quad (2.269)$$

or

$$\int_{\Omega'} d x_\mu \left\{ \left[\tilde{L}(\eta', x_\mu) - \tilde{L}(\eta, x_\mu) \right] + \frac{d}{dx_\mu} \left(\tilde{L}(\eta, x_\mu) \delta x_\mu \right) \right\} = 0. \quad (2.270)$$

To first order,

$$\tilde{L}(\eta', x_\mu) - \tilde{L}(\eta, x_\mu) = \frac{\partial \tilde{L}}{\partial \eta_\rho} \bar{\delta} \eta_\rho + \frac{\partial \tilde{L}}{\partial \eta_{\rho,v}} \bar{\delta} \eta_{\rho,v}, \quad (2.271)$$

or

$$\tilde{L}(\eta', x_\mu) - \tilde{L}(\eta, x_\mu) = \frac{d}{dx_v} \left(\frac{\partial \tilde{L}}{\partial \eta_{\rho,v}} \bar{\delta} \eta_\rho \right). \quad (2.272)$$

The invariance condition is

$$\int dx_\mu \frac{d}{dx_v} \left\{ \frac{\partial \tilde{L}}{\partial \eta_{\rho,v}} \bar{\delta} \eta_\rho + \tilde{L} \delta x_v \right\} = 0. \quad (2.273)$$

We define

$$\delta x_v = \varepsilon_r X_{rv}, \quad (2.274)$$

and

$$\delta \eta_\rho = \varepsilon_r \Psi_{r\rho}. \quad (2.275)$$

Because

$$\eta'_\rho(x'_\mu) = \eta_\rho(x_\mu) + \bar{\delta} \eta_\rho(x_\mu), \quad (2.276)$$

and

$$\delta \eta_\rho = \bar{\delta} \eta_\rho + \frac{\partial \eta_\rho}{\partial x_\sigma} \delta x_\sigma, \qquad (2.277)$$

we have

$$\bar{\delta} \eta_\rho = \varepsilon_r \left(\Psi_{r\rho} - \eta_{\rho,\sigma} X_{r\sigma} \right). \qquad (2.278)$$

Equation 2.273 reduces to

$$\int \varepsilon_r \frac{d}{dx_\nu} \left\{ \left(\frac{\partial \tilde{L}}{\partial \eta_{\rho,\nu}} \eta_{\rho,\sigma} - \tilde{L} \delta_{\nu\sigma} \right) X_{r\sigma} - \frac{\partial \tilde{L}}{\partial \eta_{\rho,\nu}} \Psi_{r\rho} \right\} (dx_\mu) = 0. \quad (2.279)$$

The result is Noether's theorem, which states that

$$\frac{d}{dx_\nu} \left\{ \left(\frac{\partial \tilde{L}}{\partial \eta_{\rho,\nu}} \eta_{\rho,\sigma} - \tilde{L} \delta_{\nu\sigma} \right) X_{r\sigma} - \frac{\partial \tilde{L}}{\partial \eta_{\rho,\nu}} \Psi_{r\rho} \right\} = 0. \qquad (2.280)$$

2.7 Bibliography

Lanczos, C. *The Variational Principles of Mechanics.* 4th ed. Toronto: University of Toronto Press, 1970.

Goldstein, H. *Classical Mechanics.* 2d ed. Reading, Mass.: Addison-Wesley, 1980.

Symon, K. R. *Mechanics*, 3d ed. Reading, Mass.: Addison-Wesley, 1971.

3 Electrodynamics

3.1 The Electrostatic Field

3.1.1 The Maxwell Equations

Maxwell, and independently Heaviside, looked at the work of Poisson, Gauss, Ampere, and others and showed that four coupled equations, plus the Lorentz force law, unified all of electromagnetics. The Maxwell equations are summarized in table 3.1.

Table 3.1

Maxwell's Equations

Equation	mks units	cgs units
Faraday's law	$\vec{\nabla} \times \vec{E} = -\dfrac{\partial \vec{B}}{\partial t}$	$\vec{\nabla} \times \vec{E} = -\dfrac{1}{c}\dfrac{\partial \vec{B}}{\partial t}$
Poisson's equation	$\vec{\nabla} \cdot \vec{E} = \dfrac{\rho}{\varepsilon_o}$	$\vec{\nabla} \cdot \vec{E} = 4\pi\rho$
Maxwell-Ampere law	$\vec{\nabla} \times \vec{B} = \mu_o \vec{J} + \mu_o \varepsilon_o \dfrac{\partial \vec{E}}{\partial t}$	$\vec{\nabla} \times \vec{B} = \dfrac{4\pi}{c}\vec{J} + \dfrac{1}{c}\dfrac{\partial \vec{E}}{\partial t}$
(no name)	$\vec{\nabla} \cdot \vec{B} = 0$	$\vec{\nabla} \cdot \vec{B} = 0$

Most references that are dedicated to the subject of electrodynamics use the cgs system of units rather than mks. Unless otherwise indicated, this chapter will utilize cgs units exclusively. A comparison of cgs and mks units is provided in section 3.6.

Implicit in the Maxwell equations is the continuity equation,

$$\vec{\nabla} \cdot \vec{J} + \frac{\partial \rho}{\partial t} = 0, \tag{3.1}$$

and the fact that Maxwell's equations produce wave solutions in free space. That is,

$$\nabla^2 \vec{B} = \frac{1}{c^2} \frac{\partial^2 \vec{B}}{\partial t^2},$$ (3.2)

and

$$\nabla^2 \vec{E} = \frac{1}{c^2} \frac{\partial^2 E}{\partial t^2}.$$ (3.3)

3.1.2 Lorentz Force Law

The law describing electromagnetic forces is the Lorentz force law,

$$\vec{F} = q \left(\vec{E} + \frac{1}{c} \vec{v} \times \vec{B} \right).$$ (3.4)

This equation, plus the Maxwell equations, unify all of electrodynamics.

3.1.3 Superposition of Coulomb Forces

The E-field produced by a point charge is found from Poisson's equation. With the help of equation 1.29, we see that

$$\int \vec{\nabla} \cdot \vec{E} dV = \oint \vec{E} \cdot d\vec{S} = 4\pi r^2 E,$$ (3.5)

and

$$\int 4\pi \rho dV = 4\pi q,$$ (3.6)

so that

$$\vec{E} = \frac{q}{r^2} \hat{r}.$$ (3.7)

The E-field produced by a number of discrete point charges is simply the superposition of the individual fields. This principle of superposition can be used to solve more complicated distributions.

3.1.3.1 Line Charge

Consider a uniform line of length L and charge q, aligned with the z-axis such that $-\frac{1}{2}L < z < \frac{1}{2}L$, with λ units of charge per unit length. By definition,

$$dq = \rho dV = \lambda dz. \tag{3.8}$$

At the plane $z = 0$, the field will be perpendicular to the line as shown in figure 3.1.

Figure 3.1 A line charge.

In general,

$$dE_\perp = \frac{dq}{r^2}\sin\theta = \frac{dq}{\left(z^2 + s^2\right)}\frac{s}{\left(z^2 + s^2\right)^{1/2}}, \tag{3.9}$$

where dq = charge at z. Thus,

$$dE_\perp = \frac{\lambda s}{\left(z^2 + s^2\right)^{3/2}}dz, \tag{3.10}$$

and

$$E_\perp = \lambda s \int_{-L/2}^{+L/2}\left(z^2 + s^2\right)^{-3/2}dz = \lambda s\frac{1}{s^2}\frac{L}{\left(\frac{L^2}{4} + s^2\right)^{1/2}} \tag{3.11}$$

or

$$E_\perp = \frac{2\lambda L}{s}\left(L^2 + 4s^2\right)^{-1/2}. \tag{3.12}$$

3.1.3.2 Ring Charge

A ring of charge (fig. 3.2) with radius a and λ units of charge per length has $dq = a\lambda d\varphi$.

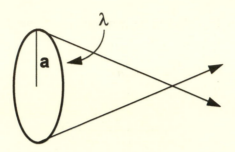

Figure 3.2 A ring of charge.

The E-field on the center axis is the sum of $E_\perp$ around the ring. That is,

$$dE_\perp = \frac{dq}{r^2}\cos\theta = \frac{dq}{\left(a^2 + z^2\right)}\frac{z}{\left(a^2 + z^2\right)^{1/2}} = \frac{z}{\left(a^2 + z^2\right)^{3/2}}a\lambda\,d\varphi, \tag{3.13}$$

and

$$E_\perp = \frac{z}{\left(a^2 + z^2\right)^{3/2}}2\pi a\lambda = \frac{qz}{\left(a^2 + z^2\right)^{1/2}}. \tag{3.14}$$

3.1.3.3 Disk of Charge

A uniform disk of charge σ per unit area can be treated as a superposition of rings, that is,

$$dq = 2\pi s\sigma ds\Big|_0^a, \tag{3.15}$$

which gives

$$dE_\perp = \frac{z}{\left(s^2 + z^2\right)^{3/2}} 2\pi s \sigma ds. \qquad (3.16)$$

Integrating gives

$$E_\perp = 2\pi\sigma z \left[\frac{-1}{\left(z^2 + s^2\right)^{1/2}} \right]_0^a = 2\pi\sigma\left\{ 1 - \frac{|z|}{\left(z^2 + a^2\right)^{1/2}} \right\}. \qquad (3.17)$$

3.1.3.4 Infinite Plane of Charge

If we let $r \to \infty$ for a disk, we obtain, from equation 3.17,

$$E = 2\pi\sigma. \qquad (3.18)$$

3.1.3.5 Sphere of Charge

Consider a homogeneous sphere of charge density $\rho = \dfrac{3q}{4\pi a^3}$. By inspection,

$$E(r \le a) = \frac{qr}{a^3}, \qquad (3.19a)$$

$$E(r \ge a) = \frac{q}{r^2}. \qquad (3.19b)$$

3.1.4 The Electrostatic Potential

The general solution to $\nabla \times \vec{E} = 0$ is $\vec{E} = -\nabla\phi(\vec{r})$. The solution to Poisson's equation is then

$$\nabla^2\phi = -4\pi\rho(\vec{r}). \qquad (3.20)$$

From the fundamental theorem of vector fields (section 1.2.5) ϕ is given by

$$\phi = \oint dV \frac{\rho}{R}. \qquad (3.21)$$

3.1.4.1 Line Charge

As in the previous section,

$$\phi = \oint dV \frac{\rho}{R} = \oint \frac{\rho}{R} \, dx \, dy \, dz = \int \frac{\lambda \, dz}{R}. \tag{3.22}$$

Making use of the relations $r^2 = z^2 + s^2$, $z = r\cos\varphi$, and $s^2 = r^2 \sin^2 \varphi$, the expression reduces to

$$\phi = \lambda \int_{r\cos\varphi - L/2}^{r\cos\varphi + L/2} \frac{dz}{\left(z^2 + r^2 \sin^2 \varphi\right)^{1/2}},$$

$$= \lambda \log\left(z + \sqrt{z^2 + r^2 \sin^2 \varphi} \right) \Big|_{r\cos\varphi - L/2}^{r\cos\varphi + L/2},$$

$$= \lambda \log\left\{ \frac{\left[r^2 \cos^2 \varphi + Lr \cos\varphi + L^2/4 + r^2 \sin^2 \varphi \right]^{1/2} + r\cos\varphi + L/2}{\left[r^2 \cos^2 \varphi - Lr \cos\varphi + L^2/4 + r^2 \sin^2 \varphi \right]^{1/2} + r\cos\varphi - L/2} \right\},$$

$$= \lambda \log\left\{ \frac{\left(r^2 + Lr \cos\varphi + \frac{1}{4}L^2 \right)^{1/2} + r\cos\varphi + L/2}{\left(r^2 - Lr \cos\varphi + \frac{1}{4}L^2 \right)^{1/2} + r\cos\varphi - L/2} \right\}. \tag{3.23}$$

A similar process may be used for other charge configurations. In general,

$$E_r = \frac{\partial \phi}{\partial r} \hat{r},$$

$$E_\theta = \frac{1}{r} \frac{\partial \phi}{\partial \theta} \hat{\theta},$$

$$E_\phi = \frac{1}{r \sin\theta} \frac{\partial \phi}{\partial \varphi} \hat{\varphi}. \tag{3.24}$$

3.1.5 Multipoles

The field arising from a finite source of spatial extent which may be arbitrarily distributed can be investigated as follows. For the geometry shown in figure 3.3, we have

$$\tilde{r}^2 = R^2 + r^2 - 2\vec{r} \cdot \vec{R},$$ (3.25)

with

$$\tilde{r} = R\sqrt{1+\varepsilon},$$ (3.26)

and

$$\varepsilon = \frac{r}{R}\left(\frac{r}{R} - 2\hat{r} \cdot \hat{R}\right).$$ (3.27)

Figure 3.3 Definition of variables.

Consequently,

$$\frac{1}{\tilde{r}} = \frac{1}{R}\frac{1}{(1+\varepsilon)^{1/2}} = \frac{1}{R}\left\{1 - \frac{1}{2}\varepsilon + \frac{3}{8}\varepsilon^2 - \frac{5}{16}\varepsilon^3 + \ldots\right\},$$ (3.28)

or

$$\frac{1}{\tilde{r}} = \frac{1}{R}\left\{1 - \frac{1}{2}\frac{r}{R}\left(\frac{r}{R} - 2\hat{r} \cdot \hat{R}\right) + \frac{3}{8}\frac{r^2}{R^2}\left(\frac{r^2}{R^2} - \frac{4r}{R}\hat{r} \cdot \hat{R} + 4(\hat{r} \cdot \hat{R})^2\right) - \ldots\right.$$ (3.29)

In terms of Legendre polynomials,

$$\frac{1}{\vec{r}} = \frac{1}{R}\sum_{n=0}^{\infty}\left(\frac{r}{R}\right)^n P_n(\cos\theta).$$ (3.30)

This gives us the multipole expansion for ϕ. From equation 3.21, we have

$$\phi = \sum_{n=0}^{\infty}\frac{1}{R}\int\frac{r^n}{R^n}P_n(\cos\theta)\rho dV,$$ (3.31)

or

$$\phi = \frac{q}{r} + \frac{\hat{r}\cdot\vec{D}}{r^2} + \frac{\hat{r}\cdot\ddot{Q}\cdot\hat{r}}{2r^3} + \cdots,$$ (3.32)

where

$$q = \text{monopole moment,}$$ (3.33a)

$$\vec{D} = \text{dipole moment} = \int\vec{r}\rho(\vec{r})dV = \oint\vec{r}dq,$$ (3.33b)

$$\ddot{Q} = \text{quadrupole moment} = dq\left(3\vec{r}\vec{r} - \vec{I}r^2\right),$$ (3.33c)

and so on. An electric dipole may be formed by having two equal and opposite point charges at some distance l apart. If they have positions $\vec{r} = \pm\frac{1}{2}\vec{l}$, then

$$\phi(\vec{r}) = \frac{q}{\left|\vec{r} - \frac{1}{2}\vec{l}\right|} - \frac{q}{\left|\vec{r} + \frac{1}{2}\vec{l}\right|}.$$ (3.34)

We have

$$\frac{1}{\left|\vec{r} - \frac{1}{2}\vec{l}\right|} = \left(r^2 - \vec{r}\cdot\vec{l} + \frac{1}{4}l^2\right) = \frac{1}{r}\left[1 - \frac{\vec{r}\cdot\vec{l}}{r^2} + \frac{l^2}{4r^2}\right]^{-1/2},$$

$$= \frac{1}{r}\left[1 + \frac{1}{2}\frac{\vec{r}\cdot\vec{l}}{r^2} - \frac{1}{8}\frac{l^2}{4r^2} + \frac{3}{8}\frac{\left(\vec{r}\cdot\vec{l}\right)^2}{r^4} - \frac{3}{8}\frac{l^4}{16r^4} + \cdots\right]. \quad (3.35)$$

Therefore,

$$\phi(\vec{r}) = \frac{q}{r}\left[1 + \frac{1}{2}\frac{\hat{r}\cdot\vec{l}}{r} - \frac{1}{8}\frac{l^2}{4r^2} + \frac{3}{8}\frac{\left(\hat{r}\cdot\vec{l}\right)^2}{r^2} - \frac{3}{8}\frac{l^4}{16r^4} + \cdots\right]$$

$$\frac{-q}{r}\left[1 - \frac{1}{2}\frac{\hat{r}\cdot\vec{l}}{r} - \frac{1}{8}\frac{l}{4r^2} + \frac{3}{8}\frac{\left(\hat{r}\cdot\vec{l}\right)^2}{r^2} - \frac{3}{8}\frac{l^2}{16r^4} + \cdots\right], \quad (3.36)$$

or

$$\phi(\vec{r}) = \frac{q}{r}\left[\frac{\hat{r}\cdot\vec{l}}{r} + \left(\vec{l}\cdot\hat{r}\right)\frac{5\left(\vec{l}\cdot\hat{r}\right)^2 - 3l^2}{8r^3} + \cdots\right]. \quad (3.37)$$

By inspection, the dipole term is $\vec{D} = q\vec{l}$ and it is the dominant term as $r \gg l$. We have

$$\phi_D = \frac{ql}{r^2}\cos\varphi = \frac{D}{r^2}\cos\varphi, \quad (3.38)$$

and

$$E_r = \frac{2D\cos\varphi}{r^3}, \quad (3.39)$$

$$E_\varphi = \frac{D\sin\varphi}{r^3}. \quad (3.40)$$

Example 3.1
Show that a linearly oscillating point charge will emit both dipole and quadrupole radiation and that the frequency of quadrupole radiation is twice the frequency of dipole radiation.

By inspection, a linearly oscillating charge has charge density

$$\rho(\vec{x},t) = q\left[\hat{i}\delta\left(\vec{x} - \frac{L}{2}\cos\omega t\right) + \hat{j}\delta(\vec{y}-0) + \hat{k}\delta(\vec{z}-0)\right]. \qquad (3.41)$$

The dipole moment is, from equation 3.33,

$$\vec{D} = \int \vec{r}\rho(\vec{r})dV = q\frac{L}{2}\cos\omega t. \qquad (3.42)$$

The quadrupole moment is, from equation 3.33,

$$\overset{\leftrightarrow}{Q} = q\left(3\vec{r}\vec{r} - \overset{\leftrightarrow}{I}r^2\right) = q\begin{pmatrix} 3\cos^2\omega t - \cos^2\omega t & 0 & 0 \\ 0 & -\cos^2\omega t & 0 \\ 0 & 0 & -\cos^2\omega t \end{pmatrix}. \qquad (3.43)$$

By definition, $\cos^2\alpha = \frac{1}{2}(1+\cos2\alpha)$, so

$$\overset{\leftrightarrow}{Q} = \frac{q}{2}\begin{pmatrix} 2(1+\cos2\omega t) & 0 & 0 \\ 0 & -(1+\cos2\omega t) & 0 \\ 0 & 0 & -(1+\cos2\omega t) \end{pmatrix}, \qquad (3.44)$$

which has twice the frequency of radiation of the dipole term.

3.1.6 The Energy of Electrostatic Systems

Setting up an electrostatic field requires work. If we have one point charge at the origin, bringing in another charge from infinity requires work. By definition,

$$W = \int \vec{F}\cdot d\vec{r} = \int_{\infty}^{r_f} \frac{q_1q_2}{r^2}\,dr = -\frac{q_1q_2}{r_f}. \qquad (3.45)$$

The work of establishing a system of point charges is

$$W = \frac{1}{2}\sum_{i=j} \frac{q_1q_2}{r_{ij}} = \frac{1}{2}\sum_{i} q_i\phi_i, \qquad (3.46)$$

with

$$\phi_i = \sum_{i=j} \frac{q_j}{r_{ij}}. \tag{3.47}$$

This is generalized to

$$W = \frac{1}{2} \oint dV(\vec{r}) \rho(\vec{r}) \, \phi(\vec{r}), \tag{3.48}$$

or

$$W = \frac{1}{2} \oint\oint dV(\vec{r}) dV(\vec{r}') \frac{\rho(\vec{r})\rho(\vec{r}')}{|\vec{r} - \vec{r}'|}. \tag{3.49}$$

If we have two systems of charges, $\rho(r) = \rho_1(r) + \rho_2(r)$ such that

$$q_1 = \oint dV \rho_1(r), \; q_2 = \oint dV \rho_2(r), \tag{3.50}$$

then

$$W = \frac{1}{2} \oint\oint dV(r) dV(r') \frac{[\rho_1(r) + \rho_2(r)][\rho_1(r') + \rho_2(r')]}{|r - r'|}, \tag{3.51}$$

or

$$W = W_1 + W_2 + U_{12}, \tag{3.52}$$

with

$$W_i = \frac{1}{2} \oint\oint dV(r) dV(r') \frac{\rho_i(r)\rho_i(r)}{|r - r'|}, \tag{3.53}$$

defined as the work required to set up system one and two, and

$$U_{12} = \oint\oint dV(r) dV(r') \frac{\rho_1(r)\rho_2(r')}{|r - r'|}, \tag{3.54}$$

defined as the interaction energy. Consider a singly charged particle of localized extent with $q = \oint dV(\vec{s})\rho(\vec{s})$ interacting with a potential ϕ. Let $\vec{r}$ be the center of mass of the localized charge q. We have

$$U(\vec{r}) = \oint dV(\vec{s})\,\rho(\vec{s})\,\phi(\vec{r}+\vec{s}). \qquad (3.55)$$

Expanding the expression for ϕ gives

$$\phi(\vec{r}+\vec{s}) = \phi(\vec{r}) + \vec{s}\cdot\nabla\phi(\vec{r}) + \frac{1}{2}\sum_{i,j} s_i s_j \frac{\partial^2\phi}{\partial x_i \partial x_j}, \qquad (3.56)$$

which reduces the expression for interaction energy to

$$U(\vec{r}) = q\phi(\vec{r}) + \nabla\phi\cdot\oint dq\vec{s} + \frac{1}{2}\sum_{i,j}\frac{\partial^2\phi}{\partial x_i \partial x_j}\oint dq s_i s_j, \qquad (3.57)$$

or

$$U(\vec{r}) = q\phi(\vec{r}) - \vec{D}\cdot\vec{E} - \frac{1}{6}\sum_{i,j} Q_{ij}\frac{\partial E_i}{\partial x_j}. \qquad (3.58)$$

3.2 The Electromagnetic Field

3.2.1 Describing Laplace Fields

3.2.1.1 Boundary Conditions — Green's Functions

The entire electric field arising from sources $\rho(\vec{r}_s)$ is determinable everywhere within a volume V if it is known on the enclosing surface only. Consider a one-dimensional example,

$$\frac{d^2\phi}{dx^2} = s(x). \qquad (3.59)$$

If x_o is an edge point of a region in which $s(x)$ has been given explicitly, then we can define

$$\phi = \phi_s + \phi_o, \tag{3.60}$$

where

$$\phi_s(x) = \int_{x_o}^{x} dx' \int_{x_o}^{x'} dx'' s(x''), \tag{3.61}$$

and

$$\phi_o(x) = a(x - x_o) + b, \tag{3.62}$$

with $b = \phi(x_o)$, $a = \left(\dfrac{d\phi}{dx}\right)_{x_o}$. Thus, giving these boundary conditions determines a unique $\phi(x)$. If $s(x)$ is given only in a closed region, $x_o \leq x \leq x$, a unique solution can be found by giving $\phi(x_o)$ and $\phi(x_1)$. Then

$$a = \frac{\phi(x) - \phi_s(x_1) - \phi(x_o)}{x_1 - x_o}, \tag{3.63}$$

and

$$b = \phi(x_o). \tag{3.64}$$

We make use of Green's theorem (sec. 1.2.4) to see that

$$\int_V dV |\nabla(\phi_1 - \phi_2)|^2 = \oint d\vec{S} \cdot \nabla(\phi_1 - \phi_2)^2, \tag{3.65}$$

or

$$\int_V dV |\vec{E}_1 - \vec{E}_2|^2 = \oint d\vec{S}(\phi_1 - \phi_2)\nabla_n(\phi_1 - \phi_2). \tag{3.66}$$

A unique solution $\vec{E}_1 = \vec{E}_2$ is determined as soon as

1. $\phi_1 = \phi_2$ is prescribed, the Dirichlet boundary conditions, or

2. $\nabla_n\phi_1 = \nabla_n\phi_2$ is prescribed, the Neumann boundary conditions.

3.2.1.2 Azimuthally Symmetric Laplace Fields

From equation 3.31 we see that the potential distribution can be given by

$$\phi(r,\theta) = q\sum_{l=0}^{\infty} R_l(r)P_l(\cos\theta), \qquad (3.67)$$

where R_l satisfies $\dfrac{d}{dr}r^2\dfrac{dR}{dr} = \lambda R$, and $\lambda = l(l+1)$. If we try a solution $R_l \sim r^\alpha$, we find $\alpha = +1$ or $\alpha = -(l+1)$, thus

$$\phi(r,\theta) = q\sum_{l=0}^{\infty}\left(A_l r^l + \frac{B_l}{r^{l+1}}\right)P_l(\cos\theta). \qquad (3.68)$$

If we have

$$\phi_q(\vec{r}_f) = \frac{q}{\vec{r}_f - \vec{r}_s} = \frac{q}{\left(r_f^2 - 2r_f r_s \cos\theta + r_s^2\right)^{1/2}}, \qquad (3.69)$$

we may expand this in a power series in the ratio $\rho = r_f/r_s$ for $r_f < r_s$ and $\rho = r_s/r_f$ for $r_s > r_f$. The coefficients are powers of $\cos\theta$ and are exactly the Legendre polynomials. We find

$$\phi_q(\vec{r}_f) = q\sum_{l=0}^{\infty}\frac{r_<^l}{r_>^{l+1}}P_l(\cos\theta). \qquad (3.70)$$

3.2.1.3 Cylindrical Harmonics

Appropriate Laplace fields $\phi(s,\varphi,z)$ can be sought in forms that are separable in the three cylindrical coordinates:

$$\phi = Z(s)\Phi(\varphi)F(z). \qquad (3.71)$$

$\nabla^2\phi = 0$ in cylindrical coordinate gives

$$s^2 \frac{\nabla^2 \phi}{\phi} = \frac{s}{Z} \frac{d}{ds}\left(s\frac{dZ}{ds}\right) + \frac{\Phi''(\varphi)}{\Phi} + s^2 \frac{F''(z)}{F(z)} = 0. \qquad (3.72)$$

If we let $\Phi'' = -m^2\Phi$, we have

$$-\frac{1}{Z}\frac{1}{s}\frac{d}{ds}\left(s\frac{dZ}{ds}\right) + \frac{m^2}{s^2} = \frac{F''(z)}{F(z)}. \qquad (3.73)$$

If we let $F'' = k^2 F$, and $\xi = ks$, we find

$$\frac{d^2 Z_m}{d\xi^2} + \frac{dZ_m}{\xi d\xi} + \left(1 - \frac{m^2}{\xi^2}\right)Z_m = 0. \qquad (3.74)$$

This is a Bessel's equation and has the solution

$$Z_m(\xi) = a_m J_m(\xi) + b_m N_m(\xi), \qquad (3.75)$$

with J_m = Bessel's function, and N_m = Neumann's function.

3.2.2 Field Energy

Whenever a charge element $\rho(\vec{r})dV(\vec{r})$ undergoes a displacement $\vec{u}(\vec{r})dt$ in the presence of an electric field $\vec{E}$, work $W = Fd = (\rho dv)\vec{E} \cdot \vec{u}dt$ is done by the electric force. Thus, power $\vec{E} \cdot \vec{J}$ is being expanded per unit of volume occupied by current in density $\vec{J}(\vec{r})$. The total power in a circuit is then

$$\oint dV \vec{J} \cdot \vec{E} = I \oint d\vec{r} \cdot \vec{E} = (emf)I. \qquad (3.76)$$

In accordance with the Maxwell equations,

$$\vec{J} = \frac{c}{4\pi}\left(\vec{\nabla} \times \vec{B} - \frac{1}{c}\frac{\partial \vec{E}}{\partial t}\right). \qquad (3.77)$$

If the element $\vec{J} \cdot \vec{E}dVdt$ is negative, this indicates that energy is being transferred to the field and

$$-\oint \vec{J} \cdot \vec{E} dV = \frac{1}{4\pi} \oint dV \vec{E} \cdot \left[\frac{\partial \vec{E}}{\partial t} - c \left(\vec{\nabla} \times \vec{B} \right) \right]. \qquad (3.78)$$

However,

$$\vec{\nabla} \cdot \left(\vec{E} \times \vec{B} \right) = \vec{B} \cdot \left(\vec{\nabla} \times \vec{E} \right) - \vec{E} \cdot \left(\vec{\nabla} \times \vec{B} \right) = \frac{1}{c} \vec{B} \cdot \frac{\partial \vec{B}}{\partial t} - \vec{E} \cdot \left(\vec{\nabla} \times \vec{B} \right), \qquad (3.79)$$

we have

$$-\oint \vec{J} \cdot \vec{E} dV = \frac{1}{4\pi} \oint dV \left[\vec{E} \cdot \frac{\partial \vec{E}}{\partial t} + \vec{B} \cdot \frac{\partial \vec{B}}{\partial t} + c \vec{\nabla} \cdot \left(\vec{E} \times \vec{B} \right) \right],$$

$$= \frac{d}{dt} \oint dV \frac{E^2 + B^2}{8\pi} + \frac{c}{4\pi} \oint d\vec{S} \cdot \left(\vec{E} \times \vec{B} \right). \qquad (3.80)$$

The last term vanishes on the edge of the field, and we have an energy density

$$w(\vec{r},t) = \frac{E^2 + B^2}{8\pi}. \qquad (3.81)$$

Example 3.2

A permanent magnet has circular pole pieces of radius a separated by an arbitrary distance. The magnetic field B is uniform between the pole pieces. Calculate the force between the pole pieces in terms of B and a.

From equation 3.81, in the absence of an electric field the energy density is

$$w = \frac{B^2}{8\pi}. \qquad (3.82)$$

From the definition of work,

$$F = \frac{W}{l} = \frac{WA}{lA} = \frac{W}{V} A = wA = \frac{B^2}{8\pi} \pi a^2 = \frac{B^2 a^2}{8}. \qquad (3.83)$$

3.2.3 The Poynting Vector

We may define the field energy in a volume V as

$$w(t) = \oint_V dV(\vec{r})w(\vec{r},t). \tag{3.84}$$

The previous expression was

$$\frac{dW}{dt} = -\int_V dV \vec{J} \cdot \vec{E} - \oint d\vec{S} \cdot \frac{c}{4\pi}\left(\vec{E} \times \vec{B}\right). \tag{3.85}$$

In general, the surface integral does not vanish and we may define the Poynting vector

$$\vec{N} = \frac{c}{4\pi} \vec{E} \times \vec{B}, \tag{3.86}$$

which is interpreted as an amount of field energy flowing at the point of its evaluation per unit time and per unit of cross-sectioned area transverse to the direction of $\vec{N}$. We have

$$\int_V dV \left(\frac{\partial W}{\partial t} + \nabla \cdot \vec{N} + \vec{J} \cdot \vec{E} \right) = 0, \tag{3.87}$$

or

$$\frac{\partial W}{\partial t} + \nabla \cdot \vec{N} = -\vec{J} \cdot \vec{E}. \tag{3.88}$$

3.2.4 Field Momentum

The rate at which mechanical momentum $\vec{p}$ is imparted to all the matter in some volume V is

$$\frac{d\vec{P}}{dt} = \vec{F} = \int_V dV \vec{f} = \int_V dV \left(\rho \vec{E} + \vec{J} \times \frac{\vec{B}}{c} \right), \tag{3.89}$$

where

$$\int_V dV \vec{f} = \frac{1}{4\pi} \int_V dV \left\{ \vec{E} \left(\vec{\nabla} \cdot \vec{E} \right) + \left[\left(\vec{\nabla} \times \vec{B} \right) - \frac{1}{c} \frac{\partial \vec{E}}{\partial t} \right] \times \vec{B} \right\}, \quad (3.90)$$

so that

$$\vec{F} = \frac{d}{dt} \oint dV \frac{\vec{E} \times \vec{B}}{4\pi c}. \quad (3.91)$$

The result shows that the force of reaction on the field imparts to it a momentum that is distributed over the field with density

$$\vec{g} = \frac{\vec{E} \times \vec{B}}{4\pi c} = \frac{\vec{N}}{c^2}. \quad (3.92)$$

The field momentum is

$$\vec{P}(t) = \int_V dV \frac{\vec{E} \times \vec{B}}{4\pi c}, \quad (3.93)$$

and we have

$$\frac{d\vec{P}}{dt} - \int_V dV \vec{f} = \frac{1}{4\pi} \int dV \left\{ \vec{E} \times \frac{\vec{B}}{c} + \vec{E} \left(\vec{\nabla} \cdot \vec{E} \right) - \vec{B} \times \left(\vec{\nabla} \times \vec{B} \right) \right\}. \quad (3.94)$$

If we let $\dot{\vec{B}} = -c \left(\vec{\nabla} \times \vec{E} \right)$ and add $\vec{B} \left(\vec{\nabla} \times \vec{B} \right)$ in the last integral, we may write

$$\frac{d\vec{P}}{dt} = -\vec{F} + \oint d\vec{S} \cdot \vec{T}, \quad (3.95)$$

where $\vec{T}$ is the Maxwell stress tensor, $\vec{T} = \dfrac{\vec{E}\vec{E} - \vec{B}\vec{B}}{4\pi} - \vec{I}w$, which gives

$$\frac{\partial \vec{g}}{\partial t} + \nabla \cdot \left(-\vec{T} \right) = -\vec{f}. \quad (3.96)$$

3.2.5 Field Angular Momentum

A space V in which field angular momentum is distributed with a density $\vec{g}(\vec{r},t)$ has a resultant field angular momentum,

$$\vec{J}(t) \equiv \int_V dV \vec{r} \times \vec{g}(t).$$

(3.97)

We have

$$\frac{d\vec{J}}{dt} = \int dV(\vec{r})\vec{r} \times \frac{\partial \vec{g}}{\partial t} = -\int dV \vec{r} \times \vec{f} + \int dV \vec{r} \times \left(\vec{\nabla} \cdot \vec{T}\right).$$

(3.98)

We define

$$\vec{r} \times \left(\vec{\nabla} \cdot \vec{T}\right) = -\nabla \cdot \vec{M}, \ \ \vec{M} = \vec{T} \times \vec{r},$$

(3.99)

and note that the integral $\int dV \vec{r} \times \vec{f}$ is a resultant torque being imparted by electromagnetic forces, and so

$$\int dV \vec{r} \times \vec{f} = \frac{d\vec{L}}{dt}.$$

(3.100)

As a result

$$\frac{d}{dt}\left(\vec{J} + \vec{L}\right) = -\oint d\vec{S} \cdot \vec{M},$$

(3.101)

where $\vec{J} + \vec{L}$ is the total angular momentum. Therefore, $\vec{M}$ must be a current density of field angular momentum. The continuity equation is

$$\frac{\partial}{\partial t}\left(\vec{r} \times \vec{g}\right) + \vec{\nabla} \cdot \vec{M} = -\vec{r} \times \vec{f}.$$

(3.102)

3.3 Fields of a Moving Point Charge

3.3.1 Retarded Potentials

Consider the equation

$$\left(\nabla^2 - \frac{1}{c^2}\frac{\partial^2}{\partial t^2}\right)\phi = -4\pi\rho(\vec{r},t). \tag{3.103}$$

If we have a point charge, then by the isotropy we expect a solution of the form

$$\frac{1}{R}\frac{\partial^2}{\partial R^2}R\phi - \frac{1}{c^2}\frac{\partial^2}{\partial t^2}\phi = -4\pi q(t)\delta(R) \tag{3.104}$$

everywhere except at $R = 0$. We have

$$\frac{\partial^2}{\partial R^2}R\phi = \frac{1}{c^2}\frac{\partial^2}{\partial t^2}R\phi, \tag{3.105}$$

and

$$R\phi(R,t) = f_r\left(t - \frac{R}{c}\right) + f_a\left(t + \frac{R}{c}\right), \tag{3.106}$$

where $f_r(t) + f_a(t) = q(t)$. We find two solutions,

$$\phi_-(r,t) = \frac{q(t - R/c)}{R}, \tag{3.107}$$

and

$$\phi_+(r,t) = \frac{q(t + R/c)}{R}, \tag{3.108}$$

which are known as the *retarded* and *advanced* solutions, respectively. Advanced solutions violate the principle of causality and are physically inadmissible.

3.3.2 Radiation by a Point Charge

The energy flux density from E_a, B_a is

$$\vec{N}_a = \frac{c}{4\pi}\vec{E}_a \times \vec{B}_a = \hat{R}\frac{c}{4\pi}\left|\vec{E}_a(\vec{r},t)\right|^2 = \hat{R}\frac{cB_a^2}{4\pi}, \qquad (3.109)$$

which has magnitude

$$N_a = \hat{R}\cdot\vec{N}_a = \frac{q^2}{4\pi cR^2}\frac{\left|\hat{R}\times\left[\left(\hat{R}-\vec{\beta}\right)\times\dot{\vec{\beta}}\right]\right|^2}{\left(1-\vec{\beta}\cdot\hat{R}\right)^6}. \qquad (3.110)$$

The energy per unit time emanating into a solid angle element $d\Omega$ is

$$\frac{dP(t_q)}{d\Omega} = R^2 N_a = \frac{q^2}{4\pi c}\frac{\left|\hat{R}\times\left[\left(\hat{R}-\vec{\beta}\right)\times\dot{\vec{\beta}}\right]\right|^2}{\left(1-\vec{\beta}\cdot\hat{R}\right)^5}. \qquad (3.111)$$

For low velocities, i.e., as $v \to 0$, we have

$$\vec{E}_a \to \frac{q}{c^2 R}\hat{R}\times\left(\hat{R}\times\dot{\vec{v}}\right), \quad \vec{B}_a \to \frac{q}{c^2 R}\dot{\vec{v}}\times\hat{R}. \qquad (3.112)$$

The radiation pattern is

$$\frac{dP}{d\Omega} = \frac{q^2}{4\pi c^3}\left|\hat{R}\times\dot{\vec{v}}\right|^2 = \frac{q^2\dot{v}^2}{4\pi c^3}\sin^2\varphi, \qquad (3.113)$$

with

$$P = \frac{2}{3}\frac{q^2\dot{v}^2}{c^3} \qquad (3.114)$$

for $v \ll c$. This is the *Larmor radiation formula*.

3.3.3 The Lienard-Wiechert Potentials

Consider a point charge q following some given trajectory $\vec{r}_q(t_s)$. The result for the scalar potential at a time t is not merely $\dfrac{q}{R(t - R/c)}$ where $R(t_q)$ is the distance from the charge to the field point at the retarded time $t_q = t - R/c$, because $\oint dV(\vec{r}_s)\rho(\vec{r}_s, t_s)$ is not generally equal to the charge present at any given moment.

Consider the volume element ΔV. For its contribution to the field at the given $\vec{r}$, $\rho(\vec{r}_s, t_s)$ must be evaluated with t_s ranging from $t_R = t - R/c$ to $t - (R - \Delta R)/c = t_R + \Delta R/c$. The charge is now

$$\rho(\vec{r} - \vec{R}, t_q)\Delta V = \Delta q(t_R). \tag{3.115}$$

This is the charge present in ΔV at $t_R = t - R/c$ when the sweep begins. During the duration of the sweep there may be a flux of charge across the surface of ΔV. Flux through the inner surface will be encountered, flux through the outer surface will not. The net charge entering ΔV through its inner surface per unit time is

$$\rho \vec{u} \cdot (-\hat{R})(R - \Delta R)^2 \, d\Omega. \tag{3.116}$$

To find the additions of charge, multiply by the duration time $\Delta R/c$.

As $\Delta R \to dR \to 0$, we have $\Delta V \to dV = R^2 dR \, d\Omega$. As a result $(R - \Delta R)^2 \, d\Omega \, (\Delta R/c) \to dV/c$. Thus, the charge encountered is

$$dq = \left(1 - \frac{\vec{v}}{c} \cdot \hat{R}\right)\rho(\vec{r} - \vec{R}, t_R)dV, \tag{3.117}$$

which gives

$$\phi(\vec{r}, t) = \oint \frac{dq}{R(1 - \vec{v} \cdot \hat{R}/c)}. \tag{3.118}$$

The Lienard-Wiechert potentials are

$$\phi(\vec{r},t) = \left(\frac{q}{R - \vec{\beta} \cdot \hat{R}} \right)_{t_q = t - R/c}, \tag{3.119}$$

and (as will be defined in sec. 3.5.1)

$$\vec{A}(\vec{r},t) = \left(\frac{\vec{v}}{c} \phi \right)_{t_q = t - R/c}. \tag{3.120}$$

Using the Maxwell equations gives

$$\vec{E} = \vec{E}_v + \vec{E}_a,$$

$$\vec{B} = \vec{B}_v + \vec{B}_a, \tag{3.121}$$

with

$$\vec{E}_v(\vec{r},t) = \frac{q}{R^2} \frac{1 - \beta^2}{\left(1 - \vec{\beta} \cdot \hat{R}\right)^3} \left(\hat{R} - \vec{\beta} \right),$$

$$\vec{B}_v = \vec{\beta} \times \vec{E}_v, \tag{3.122}$$

and

$$\vec{E}_a(\vec{r},t) = \frac{q}{cR} \frac{\hat{R}\left[\left(\hat{R} - \vec{\beta} \right) \times \dot{\vec{\beta}} \right]}{\left(1 - \vec{\beta} \cdot \hat{R}\right)^3},$$

$$\vec{B}_a = \hat{R} \times \vec{E}_a. \tag{3.123}$$

As $R \rightarrow \infty$ the flux, $\vec{N}$, of energy from E_v, B_v goes to zero, but not the flux, $\vec{N}$, from E_a, B_a. Consequently, a charge must be accelerated in order to radiate.

3.4 Conductors and Dielectrics

3.4.1 Conducting Surfaces

The materials classed as conductors behave as if charges placed within them shift about freely. When a conductor is placed in the presence of an electric field, the charges rearrange themselves until electrostatic equilibrium is attained. Consequently, inside conductors, $E = 0$ and ϕ = constant. Because field lines are always $\perp$ to surfaces of equal potential, the electric field must be $\perp$ to any conductor surfaces. Because $E = 0$ inside a conductor, there will be a discontinuity at the surface. Consequently, inside conductors

$$E_n = 4\pi\sigma ,$$

$$E_\parallel = 0, \tag{3.124}$$

while at the surface,

$$\Delta E_n = 4\pi\sigma ,$$

$$\Delta E_\parallel = 0. \tag{3.125}$$

3.4.2 Image Charges

The effect of a conductor in the presence of an electric field can be investigated using image charges.

3.4.2.1 Infinite Plane

Consider the situation shown in the left of figure 3.4. The potential at the plane $z = 0$ is constant, so this must be equivalent to the situation shown at right. Consequently, we can determine the electric field for all $z > 0$, since uniqueness allows only one admissible solution.

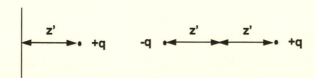

Figure 3.4 Image charges used to simulate a conducting plane.

The induced surface charge density on the conductor is found from $E = \left(-\dfrac{\partial \phi}{\partial z}\right)_{z=0}$. The result is

$$\sigma = \frac{E}{4\pi} = -\frac{ql}{2\pi\left(s^2 - l^2\right)^{3/2}} \qquad (3.126)$$

for $l = z$.

3.4.2.2 Sphere

Consider a sphere of radius a with a point charge q a distance r_o from the center, as shown in figure 3.5. In order to insure that $\phi_c = 0$, we must have an image charge inside the conductor and it must be along the line that connects the center of the sphere with q. At the point R_c we have

$$\phi = \phi_c = \frac{q}{R_c} + \frac{q'}{R_c{}'} = 0, \qquad (3.127)$$

where

$$R_c = \left[a^2 + r_o{}^2 - 2ar_o\cos\varphi\right]^{1/2},$$

$$R_c' = \left[a^2 + (r_o{}')^2 - 2ar_o{}'\cos\varphi\right]^{1/2}. \qquad (3.128)$$

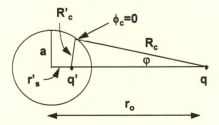

Figure 3.5 Image charges used to simulate a conducting sphere.

The inequality holds only if

$$a^2 + (r_0')^2 - 2ar_0'\cos\varphi = \left(\frac{q'}{q}\right)^2 (a^2 + r_0^2 - 2ar_0\cos\varphi). \qquad (3.129)$$

This is true independent of ϕ, so we have

$$a^2 + (r_0')^2 = \left(\frac{q'}{q}\right)^2 (a^2 + r_0^2),$$

$$r_0' = \left(\frac{q'}{q}\right)^2 r_0. \qquad (3.130)$$

We find that

$$r_0' = \frac{a^2}{r_0},$$

$$q' = -q\frac{a}{r_0}. \qquad (3.131)$$

The distribution is

$$\sigma = \left(\frac{\partial\phi}{\partial r}\right)_{r=\frac{a}{4\pi}} = \frac{E}{4\pi} = -\frac{q(r_0^2 - a^2)}{4\pi aR_c^3}. \qquad (3.132)$$

If the sphere is not grounded but is neutral, the effect may be simulated by a second image charge at the center.

Example 3.3

A particle with mass m and charge q is suspended by a string of length L. At a distance d below the point of suspension there is an infinite plane conductor. (1) Determine the frequency of the pendulum for small oscillations about its equilibrium point (neglect gravity and radiation damping); and (2) calculate the average energy loss per unit time by radiation.

(1) The equivalent of the conducting plane is an image charge $-q$ placed a distance d below the plane, as shown in figure 3.6.

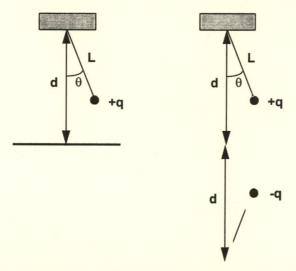

Figure 3.6 Comparison of an oscillating charge above a conductor, *left*, with an oscillating charge above an image charge, *right*.

The kinetic energy of the configuration, which is due to both charges, is

$$T = \frac{1}{2}(2m)v^2 = mL^2\dot{\theta}^2. \tag{3.133}$$

Similarly, the potential energy of the first charge is due to the electric field associated with the image charge, or

$$V = -\frac{q}{r} = -\frac{q}{2(d - L\cos\theta)}. \tag{3.134}$$

The Lagrangian is

$$L = T - V = mL^2\dot{\theta}^2 + \frac{q}{2(d - L\cos\theta)}. \tag{3.135}$$

From Lagrange's equation (eq. 2.24) the equation of motion is

$$2mL^2\ddot{\theta} - \left\{ -\frac{qL\sin\theta}{2(d - L\cos\theta)^2} \right\} = 0, \tag{3.136}$$

or

$$\ddot{\theta} + \left\{ \frac{q\sin\theta}{4mL(d - L\cos\theta)^2} \right\} = 0. \tag{3.137}$$

For small oscillations, $\sin\theta \sim \theta$, $\cos\theta \sim 1$, and

$$\ddot{\theta} + \left\{ \frac{q}{4mL(d - L)^2} \right\}\theta = 0, \tag{3.138}$$

so that the frequency of oscillation is

$$\omega = \left\{ \frac{q}{4mL(d - L)^2} \right\}^{1/2}. \tag{3.139}$$

(2) The power loss is given by the Larmor radiation formula (eq. 3.114). We have

$$\dot{v} = L\ddot{\theta} = L\omega^2\theta, \tag{3.140}$$

and

$$P = \frac{2}{3}\frac{q^2}{c^3}\frac{q}{4mL(d - L)^2}\theta. \tag{3.141}$$

3.4.3 An Ideal Dielectric

An ideal dielectric is defined as a medium that can be characterized by giving some finite continuous dipole moment density distribution $\vec{P}(\vec{r})$, called the *polarization* of the medium. The volume effect of this distribution is equivalent to a field arising from a charge distribution,

$$\sigma_p(\vec{r}_s) = -\nabla_s \cdot \vec{P}(\vec{r}_s). \tag{3.142}$$

The resultant net charge can be represented by a surface density σ_p, where $\sigma_p dS = \rho_p dV$. We find

$$\sigma_p = -\hat{n} \cdot \Delta \vec{P} = -\left(P_n^o - P_n^{\ i} \right). \tag{3.143}$$

The charges represented by ρ_p and σ_p are called *bound charges*. Charges fixed in free space or resident on conductors are *free charges* ρ', σ'. We have

$$\vec{\nabla} \cdot \vec{E} = 4\pi \left(\rho' + \rho_p \right) = 4\pi \left(\rho' - \vec{\nabla} \cdot \vec{P} \right). \tag{3.144}$$

We define

$$\nabla \cdot \left(\vec{E} + 4\pi \vec{P} \right) = 4\pi \rho', \tag{3.145}$$

and

$$\vec{D} = \vec{E} + 4\pi \vec{P}, \tag{3.146}$$

where $\vec{D}$ is the electric displacement.

3.4.4 The Dielectric Constant

We expect a connection between polarization and field of the form

$$\vec{P}(\vec{r}) = \alpha \vec{E}_o(\vec{r}), \tag{3.147}$$

where $\vec{E}_o$ is the electric force, per unit charge, on the element of dielectric at $\vec{r}$, and α is the polarizability. We have $\alpha = Na^3$, where N = the number density of molecules, and a is an effective radius of conduction. We note that $\vec{E}_o$ is not the same as the average field $\vec{E}(\vec{r})$. $\vec{E}_o$ should not include the force on the element generated by itself. If we form a spherical cavity of radius $s \to 0$, the difference of the local field $\vec{E}_o$ from the average $\vec{E}$ stems from the bound surface charges on the cavity boundary. A surface element $dS = 2\pi s^2 d(\cos\varphi)$ will have

$$\sigma_p = -P_s^{\ o} = -P\cos\varphi. \tag{3.148}$$

The field element $d\vec{E}_s$ at the center of the cavity, due to σ_p will have magnitude $\sigma_p dS / s^2$ in a radial direction. We have

$$\vec{E}_o - \vec{E} = -\hat{P}\oint\left(dS\frac{\sigma_p}{s^2}\right)\cos\varphi = 2\pi\vec{P}\int_{-1}^{+1}d(\cos\varphi)\cos^2\phi = \frac{4\pi}{3}\vec{P}. \quad (3.149)$$

or

$$\vec{P} = \frac{\alpha}{1-\dfrac{4\pi\alpha}{3}}\vec{E} = \chi_e\vec{E}, \quad (3.150)$$

where χ_e is the electric susceptibility. Using this gives

$$\vec{D} = \vec{E} + 4\pi\vec{P} = (1 + 4\pi\chi_e)\vec{E} = \varepsilon\vec{E}, \quad (3.151)$$

where ε = dielectric constant, and

$$\varepsilon = 1 + \frac{4\pi\alpha}{1 - 4\pi\alpha/3}. \quad (3.152)$$

This is the Clausius-Mossotti equation.

Example 3.4

A spherical conducting shell is supported atop a conical dielectric base as shown. The projected apex of the cone coincides with the center of the sphere. The dielectric constant of the base is ε. Find E and D everywhere when a charge Q is placed on the sphere.

By Poisson's equation, $E = D = 0$ for $r < a$. Also,

$$\int \vec{\nabla}\cdot\vec{E}dV = \oint \vec{E}\cdot d\vec{S} = 4\pi r^2 E \quad (3.153)$$

is seen to be equivalent to

$$\int \frac{4\pi\rho}{\varepsilon}dV = 4\pi\int\frac{\rho}{\varepsilon}r^2\sin\theta\,dr\,d\theta\,d\phi. \quad (3.154)$$

By inspection,

$$\rho = \frac{Q}{4\pi a^2}\delta(r-a), \quad (3.155)$$

and equation 3.154 reduces to

$$2\pi Q \left[\frac{2}{\varepsilon} \int_0^\alpha \sin\theta d\theta + \int_\alpha^\pi \sin\theta d\theta \right] = 2\pi Q \left\{ \frac{1-\cos\alpha}{\varepsilon} + (1+\cos\alpha) \right\}. \quad (3.156)$$

Combining this with equation 3.153, we see that

$$E(r) = \frac{Q}{2r^2} \left\{ \frac{1-\cos\alpha}{\varepsilon} + (1+\cos\alpha) \right\}, \quad (3.157)$$

and $D(r) = \varepsilon E(r)$ if $0 \le \theta \le \alpha$, $D(r) = E(r)$ otherwise.

3.4.5 Dielectric Effects

The capacitance of an isolated conductor is $C = \dfrac{q_c}{\Delta\phi_c}$. In the case of a conducting sphere, the potential difference between infinity and the sphere is $\phi_c = q_c / a$, which implies that $C = a$. If the sphere is filled with a dielectric medium of constant ε, we find $C_\varepsilon = \varepsilon a$.

The boundary conditions for the field at the surface of a dielectric are

$$\Delta D_n = D_n{}^o - D_n{}^i = 4\pi\sigma, \quad (3.158)$$

or

$$\Delta E_\parallel = E_\parallel^o - E_\parallel^i = 0, \quad (3.159)$$

where σ' is the area density of free charge. Values of $\sigma \ne 0$ usually occur at interfaces between conductors and dielectrics. The total charge satisfies $\sigma = E_n^o / 4\pi$ but only a part, $\sigma' = D_n^o / 4\pi$, represents free charge. On the surface separating two different dielectrics, the boundary conditions are

$$D_n^o = D_n^i, \quad (3.160)$$

and

$$E_\parallel^o = E_\parallel^i . \quad (3.161)$$

The angles of incidence and reflection are

$$\frac{\tan i}{\tan r} = \frac{E_\parallel^o / E_n^i}{E_\parallel^o / E_n^o} = \frac{D_\parallel^o / D_n^i}{D_\parallel^o / D_n^o} = \frac{\varepsilon_i}{\varepsilon_r}. \tag{3.162}$$

3.5 Magnetostatics

3.5.1 The Vector Potential

Because $\vec{\nabla} \cdot \vec{B} = 0$, we may express $\vec{B}$ in terms of a vector potential $\vec{A}$. That is,

$$\vec{B} = \vec{\nabla} \times \vec{A}. \tag{3.163}$$

From the Maxwell equations,

$$\vec{\nabla} \times \vec{E} = -\frac{1}{c}\frac{\partial \vec{B}}{\partial t} = -\frac{1}{c}\frac{\partial}{\partial t}\left(\vec{\nabla} \times \vec{A}\right) = -\vec{\nabla} \times \frac{1}{c}\frac{\partial \vec{A}}{\partial t}, \tag{3.164}$$

or

$$\vec{\nabla} \times \left(\vec{E} + \frac{1}{c}\frac{\partial \vec{A}}{\partial t}\right) = 0. \tag{3.165}$$

Consequently, we have

$$\vec{E} + \frac{1}{c}\frac{\partial \vec{A}}{\partial t} = -\nabla V, \tag{3.166}$$

or

$$\vec{E} = -\nabla V - \frac{1}{c}\frac{\partial \vec{A}}{\partial t}. \tag{3.167}$$

This being the case, the Lorentz force law (eq. 3.4) is equivalent to

$$\vec{F} = q\left(\vec{E} + \frac{1}{c}\vec{v} \times \vec{B}\right),$$

$$= q\left[\left(-\nabla V - \frac{1}{c}\frac{\partial \vec{A}}{\partial t}\right) + \frac{1}{c}\vec{v} \times \left(\vec{\nabla} \times \vec{A}\right)\right],$$

$$= -q\left[\nabla V + \frac{1}{c}\frac{\partial \vec{A}}{\partial t} - \frac{1}{c}\nabla\left(\vec{v} \cdot \vec{A}\right) + \frac{1}{c}\left(\vec{v} \cdot \vec{\nabla}\right)\vec{A}\right]. \qquad (3.168)$$

It will be shown in the chapter on fluid mechanics that

$$\frac{d\vec{A}}{dt} = \frac{\partial \vec{A}}{\partial t} + \left(\vec{v} \cdot \vec{\nabla}\right)\vec{A}; \qquad (3.169)$$

consequently,

$$\frac{d}{dt}\left[m\vec{u} + \frac{q}{c}\vec{A}\right] = -q\nabla\left[V - \frac{\vec{u} \cdot \vec{A}}{c}\right]. \qquad (3.170)$$

The term $m\vec{u} + \frac{q}{c}\vec{A}$ is defined as the conjugate momenta, $\vec{p}_c$.

3.5.2 Induced Electromotive Forces

If we have a loop carrying current I_1, the magnetic flux through the first loop, as a second current loop, with current I_2, is brought near, is

$$\Phi_{12} = \int d\vec{S}_1 \cdot \vec{B}_2(\vec{r}_1), \qquad (3.171)$$

or

$$\Phi_{12} = \int d\vec{S}_1 \cdot \left(\nabla_1 \times \vec{A}_2\right) = \oint d\vec{r}_1 \cdot \vec{A}_2 = \frac{I_2}{c}\oint\oint\frac{d\vec{r}_1 \cdot d\vec{r}_2}{R}. \qquad (3.172)$$

This defines the mutual inductance,

$$\Phi_{12} = cM_{12}I_2, \qquad (3.173)$$

with

$$M_{12} = \frac{1}{c^2} \oint \oint \frac{d\vec{r}_1 \cdot d\vec{r}_2}{R} = M_{21}. \qquad (3.174)$$

From Faraday's law, we also have

$$\oint \vec{E}_1 \cdot d\vec{r}_1 = -\frac{1}{c} \frac{d\Phi_{12}}{dt}. \qquad (3.175)$$

3.5.3 The Magnetostatic Vector Potential, Biot-Savart Law

From Ampere's law, in the absence of electric fields we have

$$\nabla^2 \vec{A} = -\frac{4\pi}{c} \vec{J}, \qquad (3.176)$$

or, from the fundamental theorem of vector fields (sec. 1.2.5)

$$\vec{A} = \oint \frac{dV(\vec{r}_s)\vec{j}(\vec{r}_s)}{c|\vec{r}_f - \vec{r}_s|}. \qquad (3.177)$$

Consider a rigid homogeneous sphere of charge that becomes a source of magnetostatic field by being set spinning around one of its diameters with some uniform angular velocity ω. We have $\vec{u} = \vec{\omega} \times \vec{r}_s$, so that by inspection

$$\vec{J}(\vec{r}_s) = \rho\vec{u} = \frac{3q}{4\pi a^3} \vec{\omega} \times \vec{r}_s, \qquad (3.178)$$

and

$$\vec{A}(\vec{r}) = \frac{3q}{4\pi a^3 c} \vec{\omega} \times \oint dV(\vec{r}_s) \frac{\vec{r}_s}{|\vec{r} - \vec{r}_s|},$$

$$= \frac{3q}{4\pi a^3 c} \vec{\omega} \times \hat{r} \oint dV(\vec{r}_s) \frac{\hat{r} \cdot \vec{r}_s}{|\vec{r} - \vec{r}_s|}. \qquad (3.179)$$

The volume integral is

$$2\pi \int_0^a dr_s r_s^2 \int_{-1}^{+1} d(\cos\varphi) \frac{r_s \cos\varphi}{\left(r^2 + r_s^2 - 2rr_s \cos\varphi\right)^{1/2}}, \qquad (3.180)$$

so that

$$\vec{A}(\vec{r}) = \frac{q}{a^3 c}(\vec{\omega} \times \hat{r}) \int_0^a dr_s r_s^3 \frac{r_<}{r_>^2}, \qquad (3.181)$$

where $r_<$ and $r_>$ are the lesser and the greater of r and r_s. The result is

$$\vec{A}(r < a) = \frac{qa^2}{5c} \vec{\omega} \times \vec{r} \frac{5a^2 - 3r^2}{2a^5},$$

$$\vec{A}(r \geq a) = \frac{qa^2}{5c} \vec{\omega} \times \frac{\hat{r}}{r^2}. \qquad (3.182)$$

Using this we are able to find $\vec{B}$.

The vector potential for idealized filamentary sources is

$$\vec{A}(\vec{r}_f) = \sum_s \frac{I_s}{c} \int_S \frac{dr_s}{|r_f - r_s|}. \qquad (3.183)$$

An expression for an element $d\vec{B}$ of magnetostatic field arising from an element of source is

$$\vec{B}(\vec{r}_f) = \nabla_f \times \vec{A}(\vec{r}_f) = \oint \frac{dV(\vec{r}_s)\vec{j}(\vec{r}_s) \times \hat{R}}{cR^2}, \qquad (3.184)$$

with $\vec{R} = \vec{r}_f - \vec{r}_s$. This is the Biot-Savart Law.

3.5.4 Magnetic Induction

Similar to the electric polarization, induced sources of magnetization can be represented by magnetic polarization, $\vec{M}(\vec{r}_s)$. This would give rise to a vector potential field

$$\vec{A}_m(\vec{r}_f) = \int_V dV(\vec{r}_s) \vec{M} \times \frac{\hat{R}}{R^2},$$

$$= \int_V dV(\vec{M} \times \nabla_s) \frac{1}{R},$$

$$= \int_V dV\left(-\nabla_s \times \frac{\vec{M}}{R} + \frac{1}{R}\nabla_s \times \vec{M}\right),$$

$$= \frac{1}{c}\int_v dV \frac{c\nabla_s \times \vec{M}}{R} + \frac{1}{c}\oint \frac{c\vec{M} \times d\vec{s}}{R}. \tag{3.185}$$

Note that the surface term will be zero interior to the surface. We define

$$\vec{J}_m(\vec{F}) = c\vec{\nabla} \times \vec{M}. \tag{3.186}$$

Ampere's law becomes

$$\vec{\nabla} \times \vec{B} = \frac{4\pi}{c}\left(\vec{J}' + c\vec{\nabla} \times \vec{M}\right), \tag{3.187}$$

or

$$\vec{\nabla} \times \left(\vec{B} - 4\pi\vec{M}\right) = \frac{4\pi}{c}\vec{J}'. \tag{3.188}$$

We define

$$\vec{H} = \vec{B} - 4\pi\vec{M}, \tag{3.189}$$

where H is the magnetic field and B is the magnetic induction. We have boundary conditions

$$\Delta B_n{}^o = B_n{}^o - B_n{}^i = 0 \tag{3.190}$$

and

$$\Delta H_\parallel = H_\parallel - H_\parallel^i = 0. \tag{3.191}$$

3.5.5 Permeability and Ferromagnetism

The functional dependence found between B and H in most materials is

$$\vec{B} = \mu \vec{H}, \tag{3.192}$$

where μ is a constant called the permeability and is characteristic of the substance in question. If $\mu > 1$ the substance is paramagnetic, if $\mu < 1$ the substance is diamagnetic. If $\mu < 1$ then,

$$\vec{M} = \frac{\vec{B} - \vec{H}}{4\pi} = \frac{(\mu - 1)\vec{H}}{4\pi}, \tag{3.193}$$

is antiparallel to the inducing fields and reduces the field. We expect a distribution of microscopic dipole moments antiparallel to the inducing field. Paramagnetic effects arise from permanent microscopic current distribution. We define the magnetic susceptibility,

$$k = \frac{\vec{M}}{\vec{H}} = \frac{\mu - 1}{4\pi}. \tag{3.194}$$

In these terms, the Maxwell equations are

$$\vec{\nabla} \times \vec{E} = -\frac{1}{c} \frac{\partial \vec{B}}{\partial t},$$

$$\vec{\nabla} \cdot \vec{D} = 4\pi \rho',$$

$$\vec{\nabla} \times \vec{H} = \frac{4\pi}{c} \vec{J}' + \frac{1}{c} \frac{\partial \vec{D}}{\partial t},$$

$$\vec{\nabla} \cdot \vec{B} = 0. \tag{3.195}$$

with $\vec{D} = \vec{E} + 4\pi\vec{P}$ and $\vec{H} = \vec{B} - 4\pi\vec{M}$.

3.6 A Comparison of mks and cgs Units

The key parameters in electrodynamics, in both mks and cgs units, are compared in table 3.2.

Table. 3.2

Comparison of Electrodynamic Quantities in mks and cgs Units.

Quantity	mks units	cgs units	To convert mks to cgs Multiply by
capacitance	farad	centimeter	9×10^{11}
charge	coulomb	statcoulomb	$c \times 10^9$
current	ampere	statampere	$c \times 10^9$
displacement	coulomb/m^2	statcoul./cm^2	$12\pi \times 10^5$
electric field	volt/meter	statvolt/cm	$1/c \times 10^{-4}$
energy	joule	erg	10^7
force	newton	dyne	10^5
H	ampere/m	oersted	$4\pi \times 10^{-3}$
inductance	henry	s^2/cm	$1/9 \times 10^{-11}$
length	meter	centimeter	10^2
magnetic field	tesla	gauss	10^4
magnetic flux	weber	maxwell	10^8
mass	kilogram	gram	10^3
potential	volt	statvolt	$1/c \times 10^{-2}$
resistance	ohm	sec/cm	$1/9 \times 10^{-11}$
time	second	second	1

$c = 2.99792456$

3.7 Bibliography

Griffiths, D. J. *Introduction to Electrodynamics*. Englewood Cliffs, N.J.: Prentice Hall, 1981.

Jackson, J. D. *Classical Electrodynamics*. 2d ed. New York: John Wiley & Sons, 1975.

Konopinski, E. J. *Electromagnetic Fields and Relativistic Particles*. New York: McGraw-Hill, 1981.

Lorrain, P., and Corson, D. R. *Electromagnetic Fields and Waves*. 2d ed. San Francisco: W. H. Freeman & Co., 1970.

4 Optics

4.1 Reflection and Refraction

Consider an electromagnetic wave defined by

$$\vec{E}_i = \vec{E}_{o,i} e^{i\left(\vec{k}_i \cdot \vec{x} - \omega_i t\right)}, \tag{4.1}$$

and

$$\vec{B}_i = \frac{n_1}{k_i} \vec{k}_i \times \vec{E}_i. \tag{4.2}$$

If the wave is incident on the interface between two media, some fraction of the wave will be reflected and some fraction will be transmitted, as shown in figure 4.1.

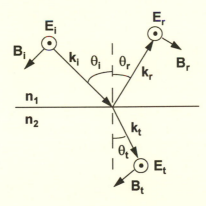

Figure 4.1 Incident, reflected, and transmitted electromagnetic waves.

By inspection, we will define the reflected wave by

$$\vec{E}_r = \vec{E}_{o,r} e^{i\left(\vec{k}_r \cdot \vec{x} - \omega_r t + \phi_r\right)},$$ (4.3)

and

$$\vec{B}_r = \frac{n_1}{k_r} \vec{k}_r \times \vec{E}_r,$$ (4.4)

while the transmitted, or refracted, wave will be defined by

$$\vec{E}_t = \vec{E}_{o,t} e^{i\left(\vec{k}_t \cdot \vec{x} - \omega_t t + \phi_t\right)},$$ (4.5)

and

$$\vec{B}_t = \frac{n_2}{k_t} \vec{k}_t \times \vec{E}_t.$$ (4.6)

We have boundary conditions at the interface that must be satisfied at all times for all points along the interface. Specifically, we require that $\Delta E_{/\!/} = 0$ (eq. 3.161), and $\Delta H_{/\!/} = 0$ (eq. 3.191). By taking the cross product between the electric field and the normal to the interface, the requirement $\Delta E_{/\!/} = 0$ reduces to

$$\hat{n}_1 \times \vec{E}_i + \hat{n}_1 \times \vec{E}_r = \hat{n}_2 \times \vec{E}_t,$$ (4.7)

or

$$E_{o,i} e^{i\left(\vec{k}_i \cdot \vec{x} - \omega_i t\right)} + E_{o,r} e^{i\left(\vec{k}_r \cdot \vec{x} - \omega_r t + \varphi_r\right)} = E_{o,t} e^{i\left(\vec{k}_t' \cdot \vec{x} - \omega_t t + \phi_t\right)}.$$ (4.8)

Because this relation must hold for any value of time and position on the interface, all three waves must have the same functional dependence on the variables t and x. In other words, we also require

$$\left(\vec{k}_i \cdot \vec{x} - \omega_i t\right)\Big|_{y=0} = \left(\vec{k}_r \cdot \vec{x} - \omega_r t + \varphi_r\right)\Big|_{y=0} = \left(\vec{k}_t' \cdot \vec{x} - \omega_t t + \phi_t\right)\Big|_{y=0}.$$ (4.9)

Because this must be true for all values of time, we see that

$$\omega_i = \omega_r = \omega_t. \qquad (4.10)$$

Thus, the scattering of light from an interface, or its transmission into another medium, does not affect its frequency. Equation 4.9 reduces to

$$\left(\vec{k}_i \cdot \vec{x}\right)\Big|_{y=0} = \left(\vec{k}_r \cdot \vec{x} + \varphi_r\right)\Big|_{y=0} = \left(\vec{k}_t' \cdot \vec{x} + \phi_t\right)\Big|_{y=0}. \qquad (4.11)$$

The first two terms give the relation

$$\left(\vec{k}_i - \vec{k}_r\right) \cdot \vec{x}\Big|_{y=0} = \varphi_r. \qquad (4.12)$$

Taking the cross product of the electric field with the normal to the interface would produce a similar result:

$$\left(\vec{k}_i - \vec{k}_r\right) \times \vec{x}\Big|_{y=0} = \varphi_r. \qquad (4.13)$$

However, since there is, by definition, no electric field perpendicular to the plane of incidence, we must conclude that $\varphi_r = 0$. Equation 4.12 then reduces to

$$k_i \sin\theta_i = k_r \sin\theta_r. \qquad (4.14)$$

Because the waves are in the same medium, we see that $k_i = k_r$, and we have deduced the requirement

$$\theta_i = \theta_r, \qquad (4.15)$$

so the angle of incidence must equal the angle of reflection.

Applying a similar line of reasoning to the relationship between the incident and transmitted waves (eq. 4.11) we would conclude that $\phi_t = 0$ and deduce

$$k_i \sin\theta_i = k_t \sin\theta_t. \qquad (4.16)$$

Multiplying both sides by the constant factor c/ω produces Snell's law,

$$n_i \sin\theta_i = n_t \sin\theta_t. \qquad (4.17)$$

Furthermore, since the argument of the exponential terms in equation 4.8 are all identical, we can also deduce that

$$E_{o,i} + E_{o,r} = E_{o,t}. \tag{4.18}$$

Recall that the magnetic field, being perpendicular to the electric field, is parallel to the plane of the interface. The requirement $\Delta H_{\parallel} = 0$ can be obtained by taking the cross product of the magnetic field with the normal to the interface. That is,

$$-\frac{B_i}{\mu_i}\cos\theta_i + \frac{B_r}{\mu_r}\cos\theta_r = -\frac{B_t}{\mu_t}\cos\theta_t. \tag{4.19}$$

Because $\theta_i = \theta_r$ and $B = E/v = nE/c$, this expression reduces to

$$\frac{n_i}{\mu_i}\left(E_{o,i} - E_{o,r}\right)\cos\theta_i = \frac{n_t}{\mu_t}E_{o,t}\cos\theta_t. \tag{4.20}$$

Combining this expression with equation 4.18 yields

$$R \equiv \left(\frac{E_{o,r}}{E_{o,i}}\right) = \frac{\dfrac{n_i}{\mu_i}\cos\theta_i - \dfrac{n_t}{\mu_t}\cos\theta_t}{\dfrac{n_i}{\mu_i}\cos\theta_i + \dfrac{n_t}{\mu_t}\cos\theta_t}, \tag{4.21}$$

and

$$T \equiv \left(\frac{E_{o,t}}{E_{o,i}}\right) = \frac{2\dfrac{n_i}{\mu_i}\cos\theta_i}{\dfrac{n_i}{\mu_i}\cos\theta_i + \dfrac{n_t}{\mu_t}\cos\theta_t}. \tag{4.22}$$

These relations, which are known as the *Fresnel equations*, define the reflection and transmission coefficients R and T.

4.2 Interference

If two electric fields, E_1 and E_2, overlap in space, the total irradiance (eq. 3.81) will be given by

$$E^2 = \left(\vec{E}_1 + \vec{E}_2\right)\cdot\left(\vec{E}_1 + \vec{E}_2\right) = E_1^2 + E_2^2 + 2\vec{E}_1 \cdot \vec{E}_2. \qquad (4.23)$$

The last term is called the interference term and is given by

$$\vec{E}_1 \cdot \vec{E}_2 = E_{1,o}E_{2,o}e^{i\left(\vec{k}_1 \cdot \vec{x}_1 - \omega_1 t + \phi_1\right)}e^{i\left(\vec{k}_2 \cdot \vec{x}_2 - \omega_2 t + \phi_2\right)}, \qquad (4.24)$$

or

$$\vec{E}_1 \cdot \vec{E}_2 = E_{1,o}E_{2,o}e^{i\left[\left(\vec{k}_1 \cdot \vec{x}_1 + \vec{k}_2 \cdot \vec{x}_2\right) - \left(\omega_1 + \omega_2\right)t + \left(\phi_1 + \phi_2\right)\right]}. \qquad (4.25)$$

Obviously, the real part of this term will vary according to the phase difference

$$\delta = \left[\left(\vec{k}_1 \cdot \vec{x}_1 + \vec{k}_2 \cdot \vec{x}_2\right) - \left(\omega_1 + \omega_2\right)t + \left(\phi_1 + \phi_2\right)\right]. \qquad (4.26)$$

Consequently, when $\delta = 0, \pm 2\pi, \pm 4\pi \ldots$, the amplitude of the interference term is a maximum. When $\delta = \pm\pi, \pm 3\pi \ldots$, the amplitude of the interference term is a minimum. These are the conditions for constructive and destructive interference, respectively.

4.3 Diffraction

Closely related to interference is the concept of diffraction. Diffraction is a general characteristic of waves that arises whenever a portion of the wavefront is physically obstructed in some way. Huygens concluded that every unobstructed point of a wavefront, at any given instant in time, serves as a source of spherical secondary wavelets, and the amplitude of the optical field at any further point was the superposition of all these wavelets. This principle, together with interference, describes the characteristic diffraction patterns that are observed in certain applications.

Consider a single slit with plane waves incident on it from below as shown in figure 4.2. Above the slit spherical waves will propagate. A diffraction pattern will be noticed at those points where the path difference is a half-integral number of wavelengths because the opening of the slit can be treated as a superposition of independent wavelet sources.

The total electric field strength reaching an arbitrary point to the right of the slit is, in one dimension, given by

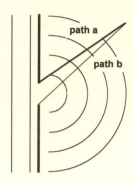

Figure 4.2 Single slit diffraction.

$$E = E_o e^{i(kr_1 - \omega t)} + E_o e^{i(kr_2 - \omega t)} + \ldots + E_o e^{i(kr_N - \omega t)}. \qquad (4.27)$$

This equation simplifies to

$$E = E_o e^{-i\omega t} e^{ikr_1} \left[1 + e^{ik(r_2 - r_1)} + e^{ik(r_3 - r_1)} + \ldots + e^{ik(r_N - r_1)} \right]. \qquad (4.28)$$

From the figure it is easily seen that

$$E = E_o e^{-i\omega t} e^{ikr_1} \left[1 + e^{i\delta} + e^{i2\delta} + \ldots + e^{i(N-1)\delta} \right], \qquad (4.29)$$

where

$$\delta = k(r_2 - r_1) = kd \sin\theta. \qquad (4.30)$$

Obviously, equation 4.29 will have a maximum if $\delta = 2m\pi$, where $m = 0, \pm1, \pm2 \ldots$. Similarly, equation 4.29 will have a minimum if $\delta = m\pi$.

For a circular aperture, rather than a slit, applying a similar analysis would deduce an expression for intensity, as a function of radial position from the center, that depended on Bessel's functions. Sir George Airy showed that the distance from the center of the circle to the center of the first interference minimum (dark ring) is given by

$$q_1 = 1.22 \frac{R\lambda}{2a}, \qquad (4.31)$$

where R is the distance from the aperture to the point of project, a is the radius of the aperture, and λ is the wavelength. If all defects in an optical system could be removed, the image sharpness would ultimately be limited by diffraction.

4.4 Bibliography

Hecht, E., and Zajac, A. *Optics*. Reading, Mass.: Addison-Wesley, 1974.

5 Fluid Dynamics

5.1 The Boltzmann Equation

We define f to be a distribution function (probability density) in six dimensional phase space. That is, f is the probability that a particle has position x to $x + \Delta x$, v_x to $v_x + \Delta v_x$, and so on. During the time δt, the change in probability is seen to be given by

$$\frac{\partial f}{\partial t}\delta t + \frac{\partial f}{\partial x}\delta x + \frac{\partial f}{\partial y}\delta y + \frac{\partial f}{\partial z}\delta z + \frac{\partial f}{\partial v_x}\delta v_x + \frac{\partial f}{\partial v_y}\delta v_y + \frac{\partial f}{\partial v_z}\delta v_z. \quad (5.1)$$

Dividing by δt gives

$$\frac{\delta f}{\delta t} = \frac{\partial f}{\partial t} + \frac{\delta x}{\delta t}\frac{\partial f}{\partial x} + \frac{\delta y}{\delta t}\frac{\partial f}{\partial y} + \frac{\delta z}{\delta t}\frac{\partial f}{\partial z} + \frac{\delta v_x}{\delta t}\frac{\partial f}{\partial v_x} + \frac{\delta v_y}{\delta t}\frac{\partial f}{\partial v_y} + \frac{\delta v_z}{\delta t}\frac{\partial f}{\partial v_z}. \quad (5.2)$$

As $\delta t \to 0$, $\delta x / \delta t \to v_x$, and $\delta v_x / \delta t \to a_x$, so that

$$\frac{\delta f}{\delta t} \to \frac{Df}{Dt} = \frac{\partial f}{\partial t} + v_x\frac{\partial f}{\partial x} + v_y\frac{\partial f}{\partial y} + v_z\frac{\partial f}{\partial z} + a_x\frac{\partial f}{\partial v_x} + a_y\frac{\partial f}{\partial v_y} + a_z\frac{\partial f}{\partial v_z}. \quad (5.3)$$

Consequently, the total change in f as seen by an observer who is following the fluid and watching a particular mass of fluid is given by

$$\frac{Df}{Dt} = \frac{\partial f}{\partial t} + \left(\vec{v} \cdot \vec{\nabla}\right)f + \left(\frac{1}{m}\vec{F} \cdot \vec{\nabla}_v\right)f, \quad (5.4)$$

with $\vec{\nabla}_v = \frac{\partial}{\partial v_x}\hat{i} + \frac{\partial}{\partial v_y}\hat{j} + \frac{\partial}{\partial v_z}\hat{k}$. Note that we have actually defined

$$\frac{D}{Dt} = \frac{\partial}{\partial t} + \left(\vec{v} \cdot \vec{\nabla}\right) + \left(\frac{1}{m}\vec{F} \cdot \vec{\nabla}_v\right) \quad (5.5)$$

to be the total derivative. If particles are neither created nor destroyed, so that probability is conserved, the change in f is zero and equation 5.4 simplifies to

$$\frac{\partial f}{\partial t} + \left(\vec{v} \cdot \vec{\nabla}\right) f + \left(\frac{1}{m}\vec{F} \cdot \vec{\nabla}_v\right) f = 0. \tag{5.6}$$

This expression is called the *Vlasov equation* or the *collisionless Boltzmann equation*.

5.2 The Fluid Equations

5.2.1 The Equation of Continuity

5.2.1.1 Macroscopic View

Consider a mass of fluid with volume V and surface area S. The rate at which mass flows into V is given by

$$-\int_S \rho\vec{v} \cdot d\vec{S}, \tag{5.7}$$

where the minus sign occurs because $d\vec{S}$ points out of V. The mass flow would result in a mass increase given by

$$\frac{d}{dt}\int_V \rho dV = \int_V \frac{\partial \rho}{\partial t} dV. \tag{5.8}$$

Because equations 5.8 and 5.9 must be equal, we have

$$\int_V \left\{\frac{\partial \rho}{\partial t} + \vec{\nabla} \cdot (\rho\vec{v})\right\} dV = 0. \tag{5.9}$$

Because the volume element V can be chosen arbitrarily, we also require

$$\frac{\partial \rho}{\partial t} + \vec{\nabla} \cdot (\rho\vec{v}) = 0, \tag{5.10}$$

which is the equation of mass conservation.

5.2.1.2 Microscopic View

The fluid equations are also the moments of the collisionless Vlasov equation. The first moment is obtained by performing an integration over dV That is, because $\dfrac{Df}{Dt} = 0$, we also require that

$$\int\left[\frac{Df}{Dt}\right]d^3x = \int\left[\frac{\partial f}{\partial t} + \left(\vec{v}\cdot\vec{\nabla}\right)f + \left(\frac{1}{m}\vec{F}\cdot\vec{\nabla}_v\right)f\right]d^3x = 0. \quad (5.11)$$

The first term is

$$\int\frac{\partial f}{\partial t}d\vec{v} = \frac{\partial}{\partial t}\int fd\vec{v} = \frac{\partial n}{\partial t}. \quad (5.12)$$

The second term is

$$\int\vec{v}\cdot\vec{\nabla}fd\vec{v} = \vec{\nabla}\cdot\int\vec{v}fd\vec{v} = \vec{\nabla}\cdot\left(n\langle\vec{v}\rangle\right), \quad (5.13)$$

and the third term is

$$\int\frac{1}{m}\left(\vec{F}\cdot\vec{\nabla}_v\right)fd\vec{v} = \frac{1}{m}\int\left(\frac{\partial}{\partial\vec{v}}\right)\cdot\left(\vec{F}f\right)d\vec{v} = \frac{1}{m}\oint\left(\vec{F}f\right)d\vec{v} = 0. \quad (5.14)$$

The equation of continuity is, from equations 5.12 and 5.13,

$$\frac{\partial n}{\partial t} + \vec{\nabla}\cdot\left(n\langle\vec{v}\rangle\right) = 0. \quad (5.15)$$

If we had used mass density in our integration instead of number density, we would have obtained the equation for mass conservation,

$$\frac{\partial\rho}{\partial t} + \vec{\nabla}\cdot\left(\rho\langle\vec{v}\rangle\right) = 0. \quad (5.16)$$

5.2.2 Conservation of Momentum

5.2.2.1 Macroscopic View

The rate of momentum flow into a volume element of fluid is given by

$$-\int_S (\rho\vec{v})\vec{v}\cdot d\vec{S}.$$ (5.17)

From the definition

$$\vec{F} = \frac{d\vec{p}}{dt},$$ (5.18)

we see that the external forces acting on the fluid may also induce a change in momentum. In general, the external forces may be separated into volume (or body) forces and surface (or pressure) forces. The rate of change in momentum due to these forces is

$$\int_V \rho\vec{f}dV + \int_S \vec{P}\cdot d\vec{S},$$ (5.19)

where f is the force per unit mass and P is the pressure. Consequently, the change in momentum of the fluid is given by

$$\frac{d}{dt}\int_V (\rho\vec{v})dV = \int_V \rho\vec{f}dV + \int_S \vec{P}\cdot d\vec{S} - \int_S (\rho\vec{v})\vec{v}\cdot d\vec{S},$$ (5.20)

or

$$\int_V \frac{\partial}{\partial t}(\rho\vec{v})dV = \int_V \left\{\rho\vec{f} + \vec{\nabla}\cdot\vec{P} - \vec{v}\vec{\nabla}\cdot(\rho\vec{v})\right\}dV.$$ (5.21)

Again, because the volume element V is arbitrary, we have

$$\frac{\partial}{\partial t}(\rho\vec{v}) + \vec{v}\vec{\nabla}\cdot(\rho\vec{v}) = \rho\vec{f} + \vec{\nabla}\cdot\vec{P},$$ (5.22)

which is the equation for conservation of momentum.

5.2.2.2 Microscopic View

To obtain the second moment of the Vlasov equation, we multiply by mv and integrate over dv. The first term is

$$\int m\vec{v}\,\frac{\partial f}{\partial t}dv = m\frac{\partial}{\partial t}\int \vec{v}f dv = m\frac{\partial}{\partial t}(n\vec{v}).$$ (5.23)

The second term is

$$\int m\vec{v}\left(\vec{v}\cdot\vec{\nabla}\right)f dv = m\vec{\nabla}\int f\vec{v}\vec{v}\,dv = m\vec{\nabla}\cdot\left(n\langle\vec{v}\vec{v}\rangle\right),$$ (5.24)

and the third term is

$$m\int \vec{v}\left(\vec{F}\cdot\vec{\nabla}_v\right)f dv.$$ (5.25)

Equation 5.25 is equivalent to

$$\int \vec{\nabla}_v\cdot\left(\vec{v}\vec{F}f\right)dv - \int f\vec{F}\left(\vec{\nabla}_v\cdot\vec{v}\right)dv - \int \vec{v}\left(\vec{F}\cdot\vec{\nabla}_v f\right)dv.$$ (5.26)

The first term is equivalent to a surface term that is zero, and the last term is identically zero. Consequently, equation 5.25 reduces to

$$-\int f\vec{F}\cdot\vec{\nabla}_v\vec{v}\,dv = -n\vec{F}.$$ (5.27)

The momentum equation is, from equations 5.23, 5.24, and 5.27,

$$m\left[\frac{\partial}{\partial t}(n\vec{v}) + \vec{\nabla}\cdot\left(n\langle\vec{v}\vec{v}\rangle\right)\right] = n\vec{F}.$$ (5.28)

Because the force appears explicitly in equation 5.28, this is also referred to as the *force equation*.

Equation 5.28 can be further deconvolved if we separate v into the average fluid velocity u and a thermal velocity ω to obtain

$$m\vec{\nabla}\cdot\left(n\langle\vec{v}\vec{v}\rangle\right) = \vec{\nabla}\cdot\left(mn\langle\vec{u}\vec{u}\rangle\right) + \vec{\nabla}\cdot\left(mn\langle\vec{w}\vec{w}\rangle\right) + 2\vec{\nabla}\cdot\left(mn\langle\vec{u}\vec{w}\rangle\right).$$ (5.29)

The expression $mn\langle\vec{w}\vec{w}\rangle$ is seen to be the pressure tensor because

$$n\frac{1}{2}m\langle\vec{w}\vec{w}\rangle = n\frac{3}{2}k_BT = P. \tag{5.30}$$

We also see that the time average $\langle\vec{u}\vec{w}\rangle$ will be zero. Consequently, equation 5.29 reduces to

$$m\vec{\nabla}\cdot\left(n\langle\vec{v}\vec{v}\rangle\right) = \vec{\nabla}\cdot\left(mn\langle\vec{u}\vec{u}\rangle\right) + \vec{\nabla}\cdot\vec{P}, \tag{5.31}$$

and the momentum equation becomes

$$m\left[\frac{\partial}{\partial t}\left(n(\vec{u}+\vec{w})\right) + \vec{\nabla}\cdot\left(n\langle\vec{u}\vec{u}\rangle\right) + \frac{1}{m}\vec{\nabla}\cdot\vec{P}\right] = n\vec{F}, \tag{5.32}$$

or

$$m\left[\frac{\partial}{\partial t}\left(n\vec{u}\right) + \vec{\nabla}\cdot\left(n\langle\vec{u}\vec{u}\rangle\right)\right] = n\vec{F} - \vec{\nabla}\cdot\vec{P}. \tag{5.33}$$

This last expression is also known as *Euler's equation*. This is explicitly rewritten as

$$m\left[\vec{u}\frac{\partial n}{\partial t} + n\frac{\partial\vec{u}}{\partial t} + \vec{u}\vec{\nabla}(n\vec{u}) + n\vec{u}\left(\vec{\nabla}\cdot\vec{u}\right)\right] = n\vec{F} - \vec{\nabla}\cdot\vec{P}. \tag{5.34}$$

It is seen that the first and third terms are simply the continuity equation, so equation 5.34 reduces to

$$mn\left[\frac{\partial\vec{u}}{\partial t} + \vec{u}\left(\vec{\nabla}\cdot\vec{u}\right)\right] = n\vec{F} - \vec{\nabla}\cdot\vec{P}, \tag{5.35}$$

or

$$\rho\left[\frac{\partial\vec{u}}{\partial t} + \vec{u}\left(\vec{\nabla}\cdot\vec{u}\right)\right] = \rho\vec{f} - \vec{\nabla}\cdot\vec{P}, \tag{5.36}$$

which is identical to equation 5.22.

5.2.3 Conservation of Energy

5.2.3.1 Macroscopic View

We define the total energy of an element of fluid to be

$$\int_V \left\{ \rho e + \frac{\rho}{2} \vec{v} \cdot \vec{v} \right\} dV, \tag{5.37}$$

where the first term is the potential energy and the second term is kinetic energy. The rate of energy flow into the volume element is given by

$$-\int_S \left\{ \rho e + \frac{\rho}{2} \vec{v} \cdot \vec{v} \right\} \vec{v} \cdot d\vec{S}. \tag{5.38}$$

The energy will be constant unless work is done by, or on, the system, or unless heat is added. The work done by the forces acting on the element is

$$\int_V \left(\vec{v} \cdot \rho \vec{f} \right) dV + \int_S \left(\vec{v} \cdot \bar{P} \right) \cdot d\vec{S}, \tag{5.39}$$

while the heat leaving the volume element is

$$\int_S \vec{q} \cdot d\vec{S}. \tag{5.40}$$

The fact that the rate of change of energy is equal to the rate at which work is done, plus the rate at which heat is added, is

$$\frac{d}{dt} \int_V \left\{ \rho e + \frac{\rho}{2} \vec{v} \cdot \vec{v} \right\} dV$$

$$= -\int_S \left\{ \rho e + \frac{\rho}{2} \vec{v} \cdot \vec{v} \right\} \vec{v} \cdot d\vec{S} + \int_V \left(\vec{v} \cdot \rho \vec{f} \right) dV + \int_S \left(\vec{v} \cdot \bar{P} \right) \cdot d\vec{S} + \int_S \vec{q} \cdot d\vec{S}. \tag{5.41}$$

The first term of equation 5.41 is

$$\int_V \frac{\partial}{\partial t} \left\{ \rho e + \frac{\rho}{2} \vec{v} \cdot \vec{v} \right\} dV, \tag{5.42}$$

while the remaining terms are

$$\int_V \left[-\left\{ \rho e + \frac{\rho}{2} \vec{v} \cdot \vec{v} \right\} \left(\vec{\nabla} \cdot \vec{v} \right) + \left(\vec{v} \cdot \rho \vec{f} \right) + \vec{\nabla} \cdot \left(\vec{v} \cdot \vec{P} \right) + \vec{\nabla} \cdot \vec{q} \right] dV . \quad (5.43)$$

As before, because the volume element is arbitrary we have

$$\frac{\partial}{\partial t} \left\{ \rho e + \frac{\rho}{2} \vec{v} \cdot \vec{v} \right\} + \left(\vec{\nabla} \cdot \vec{v} \right) \left\{ \rho e + \frac{\rho}{2} \vec{v} \cdot \vec{v} \right\} = \left(\vec{v} \cdot \rho \vec{f} \right) + \vec{\nabla} \cdot \left(\vec{v} \cdot \vec{P} \right) + \vec{\nabla} \cdot \vec{q} . \quad (5.44)$$

Noting that

$$\frac{\partial}{\partial t} \left\{ \rho e + \frac{\rho}{2} \vec{v} \cdot \vec{v} \right\} = \rho \frac{\partial e}{\partial t} + e \frac{\partial \rho}{\partial t} + \rho \frac{\partial}{\partial t} \left(\frac{1}{2} \vec{v} \cdot \vec{v} \right) + \left(\frac{1}{2} \vec{v} \cdot \vec{v} \right) \frac{\partial \rho}{\partial t} , \quad (5.45)$$

and

$$\left(\vec{\nabla} \cdot \vec{v} \right) \left\{ \rho e + \frac{\rho}{2} \vec{v} \cdot \vec{v} \right\}$$

$$= e \vec{\nabla} \cdot \left(\rho \vec{v} \right) + \left(\rho \vec{v} \right) \cdot \left(\vec{\nabla} e \right) + \left(\frac{1}{2} \vec{v} \cdot \vec{v} \right) \vec{\nabla} \cdot \left(\rho \vec{v} \right) + \left(\rho \vec{v} \right) \cdot \vec{\nabla} \left(\frac{1}{2} \vec{v} \cdot \vec{v} \right) . \quad (5.46)$$

From the equation of mass conservation, equation 5.10, the right-hand side of equation 5.46 reduces to

$$-e \frac{\partial \rho}{\partial t} + \left(\rho \vec{v} \right) \cdot \left(\vec{\nabla} e \right) - \left(\frac{1}{2} \vec{v} \cdot \vec{v} \right) \frac{\partial \rho}{\partial t} + \left(\rho \vec{v} \right) \cdot \vec{\nabla} \left(\frac{1}{2} \vec{v} \cdot \vec{v} \right) . \quad (5.47)$$

Using equations 5.45 and 5.47, equation 5.44 becomes

$$\rho \frac{\partial e}{\partial t} + \rho \frac{\partial}{\partial t} \left(\frac{1}{2} \vec{v} \cdot \vec{v} \right) + \left(\rho \vec{v} \right) \cdot \left(\vec{\nabla} e \right) + \left(\rho \vec{v} \right) \cdot \vec{\nabla} \left(\frac{1}{2} \vec{v} \cdot \vec{v} \right)$$

$$= \left(\vec{v} \cdot \rho \vec{f} \right) + \vec{\nabla} \cdot \left(\vec{v} \cdot \vec{P} \right) + \vec{\nabla} \cdot \vec{q} , \quad (5.48)$$

which is known as the *energy equation*.

5.2.3.2 Microscopic View

We define the total energy of a given particle to be

$$me + \frac{m}{2}\vec{v}\cdot\vec{v}, \tag{5.49}$$

where the first term is the potential energy and the second term is kinetic energy. Allowing equation 5.6 to operate on equation 5.49, and integrating, gives

$$\int\left[\frac{\partial}{\partial t}+\left(\vec{v}\cdot\vec{\nabla}\right)+\left(\frac{1}{m}\vec{F}\cdot\vec{\nabla}_v\right)\right]\left(me + \frac{m}{2}\vec{v}\cdot\vec{v}\right)f\,dv = 0. \tag{5.50}$$

The first term is

$$\int\frac{\partial}{\partial t}\left(me + \frac{m}{2}\vec{v}\cdot\vec{v}\right)f\,dv = m\int\left\{\frac{\partial e}{\partial t}f + e\frac{\partial f}{\partial t}+\frac{\partial\vec{v}}{\partial t}\cdot\vec{v}+\frac{1}{2}\vec{v}\cdot\vec{v}\frac{\partial f}{\partial t}\right\}dv, \tag{5.51}$$

or

$$mn\frac{\partial e}{\partial t}+me\frac{\partial n}{\partial t}+mn\left\langle\frac{\partial\vec{v}}{\partial t}\cdot\vec{v}\right\rangle+\frac{m}{2}\left\langle\vec{v}\cdot\vec{v}\frac{\partial n}{\partial t}\right\rangle. \tag{5.52}$$

The second term of equation 5.50 is

$$m\int\left\{\vec{v}\cdot\left(\vec{\nabla}e\right)f +\vec{v}\cdot\left(\vec{\nabla}f\right)e+\frac{1}{2}\vec{v}\cdot\left[\vec{\nabla}(\vec{v}\cdot\vec{v})\right]f+\frac{1}{2}\vec{v}\cdot\left[\vec{\nabla}f\right](\vec{v}\cdot\vec{v})\right\}dv, \tag{5.53}$$

or

$$m\left\langle n\vec{v}\cdot\left(\vec{\nabla}e\right)\right\rangle+m\left\langle\vec{v}\cdot\left(\vec{\nabla}n\right)e\right\rangle+\frac{m}{2}\left\langle n\vec{v}\cdot\left[\vec{\nabla}(\vec{v}\cdot\vec{v})\right]\right\rangle+\frac{m}{2}\left\langle\vec{v}\cdot\left[\vec{\nabla}n\right](\vec{v}\cdot\vec{v})\right\rangle. \tag{5.54}$$

Note that the fourth term of equations 5.52 and 5.54 are simply e times the continuity equation and sum to zero. Similarly, the second term of equations 5.52 and 5.54 are $(\vec{v}\cdot\vec{v})$ times the continuity equation, and also sum to zero. The sum of equations 5.52 and 5.54 is then

$$mn\frac{\partial e}{\partial t}+mn\left\langle\frac{\partial\vec{v}}{\partial t}\cdot\vec{v}\right\rangle+m\left\langle n\vec{v}\cdot\left(\vec{\nabla}e\right)\right\rangle+\frac{m}{2}\left\langle n\vec{v}\cdot\left[\vec{\nabla}(\vec{v}\cdot\vec{v})\right]\right\rangle. \tag{5.55}$$

The third term in equation 5.50 is

$$\int \left(\frac{1}{m} \vec{F} \cdot \vec{\nabla}_v \right) \left(me + \frac{m}{2} \vec{v} \cdot \vec{v} \right) f dv$$

$$= \int \left\{ \vec{F} \cdot \left(\vec{\nabla}_v e \right) f + \vec{F} \cdot \left(\vec{\nabla}_v f \right) e + \frac{1}{2} \vec{F} \cdot \left[\vec{\nabla}_v (\vec{v} \cdot \vec{v}) \right] f + \frac{1}{2} \vec{F} \cdot \left[\vec{\nabla}_v f \right] (\vec{v} \cdot \vec{v}) \right\} dv. \quad (5.56)$$

All but the third term are zero, so equation 5.56 is simply

$$\langle \vec{F} \cdot \vec{v} n \rangle. \quad (5.57)$$

Combining equations 5.55 and 5.57, we obtain an equivalent form of the energy equation,

$$mn \frac{\partial e}{\partial t} + mn \left\langle \frac{\partial \vec{v}}{\partial t} \cdot \vec{v} \right\rangle + m \left\langle n\vec{v} \cdot \left(\vec{\nabla} e \right) \right\rangle + \frac{m}{2} \left\langle n\vec{v} \cdot \left[\vec{\nabla} (\vec{v} \cdot \vec{v}) \right] \right\rangle + \left\langle \vec{F} \cdot \vec{v} n \right\rangle = 0. \quad (5.58)$$

5.2.4 Using the Fluid Equations: Sound Waves

An excellent example of the application of the fluid equations to problem solving is the derivation of the dispersion relation for sound waves. When a small pressure disturbance occurs in a gas at rest, we may treat the resulting perturbation by linearizing the equations of motion. That is, we assume that pressure and mass density can be expressed as a nonzero equilibrium value plus a small perturbation,

$$P = P_0 + P_1, \quad (5.59)$$

and

$$\rho = \rho_0 + \rho_1. \quad (5.60)$$

Similarly, we assume that the gas is at rest and not subjected to an external force so that the only values for these variables are the perturbation values

$$v = v_1, \quad (5.61)$$

and

$$F = F_1. \quad (5.62)$$

Using these expressions, the equation of continuity (eq. 5.10) becomes (in one dimension)

$$\frac{\partial}{\partial t}(\rho_0 + \rho_1) + \frac{\partial}{\partial x}\left[(\rho_0 + \rho_1)(v_1)\right] = 0. \tag{5.63}$$

If we retain only first-order terms, i.e., terms that depend on at most one perturbation term, the equation simplifies to

$$\frac{\partial \rho_1}{\partial t} + \rho_0 \frac{\partial v_1}{\partial x} = 0. \tag{5.64}$$

Similarly, the momentum equation (eq. 5.22) becomes

$$(\rho_0 + \rho_1)\left[\frac{\partial v_1}{\partial t} + v_1\left(\frac{\partial v_1}{\partial x}\right)\right] = (\rho_0 + \rho_1)\left(\frac{F_1}{m}\right) - \frac{\partial P}{\partial x}, \tag{5.65}$$

or

$$\rho_0 \frac{\partial v_1}{\partial t} = \rho_0\left(\frac{F_1}{m}\right) - \frac{\partial P}{\partial x}. \tag{5.66}$$

Note that we dropped the second term in equation 5.65 because it is of second order with respect to the perturbations. For a polytropic gas, such as air, the pressure satisfies the equation

$$P = p_0\rho^{\gamma} = p_0(\rho_0 + \rho_1)^{\gamma} = p_0\rho_0^{\gamma}\left(1 + \frac{\rho_1}{\rho_0}\right)^{\gamma} \approx p_0\rho_0^{\gamma}\left(1 + \gamma\frac{\rho_1}{\rho_0}\right). \tag{5.67}$$

Consequently, equation 5.66 reduces to

$$\rho_o \frac{\partial v_1}{\partial t} = \rho_o\left(\frac{F_1}{m}\right) - \left(\gamma p_0 \rho_0^{\gamma-1}\right)\frac{\partial \rho_1}{\partial x}. \tag{5.68}$$

Taking the time derivative of equation 5.64 gives

$$\frac{\partial^2 \rho_1}{\partial t^2} + \rho_0 \frac{\partial^2 v_1}{\partial x \partial t} = 0, \tag{5.69}$$

while taking the gradient of equation 5.68 gives

$$\rho_o \frac{\partial^2 v_1}{\partial x \partial t} = \frac{\rho_o}{m}\left(\frac{\partial F_1}{\partial x}\right) - \left(\gamma p_o \rho_o^{\gamma-1}\right)\frac{\partial^2 \rho_1}{\partial x^2}. \tag{5.70}$$

Plugging equation 5.69 into equation 5.70 gives

$$\frac{\partial^2 \rho_1}{\partial t^2} - \left(\gamma p_0 \rho_0^{\gamma-1}\right)\frac{\partial^2 \rho_1}{\partial x^2} = -\frac{\rho_0}{m}\left(\frac{\partial F_1}{\partial x}\right). \tag{5.71}$$

This is the equation for sound waves with velocity

$$c_o = \left(\gamma p_0 \rho_0^{\gamma-1}\right)^{1/2}. \tag{5.72}$$

5.3 Circulation and Vorticity

The circulation contained in a closed contour of a volume element of fluid is defined by

$$\Gamma = \oint \vec{v} \cdot d\vec{l} = \int_S \left(\vec{\nabla} \times \vec{v}\right) \cdot d\vec{S} = \int_S \vec{\omega} \cdot d\vec{S}, \tag{5.73}$$

where

$$\vec{\omega} = \vec{\nabla} \times \vec{v} \tag{5.74}$$

is defined as the *vorticity*. Flows for which $\vec{\omega} = 0$ are called *irrotational*, flows for which $\vec{\omega} \neq 0$ are called *rotational*. The time derivative of the circulation is

$$\frac{d\Gamma}{dt} = \frac{d}{dt}\oint \vec{v} \cdot d\vec{l} = \oint \frac{d\vec{v}}{dt} \cdot d\vec{l} + \oint \vec{v} \cdot \frac{d}{dt}\left(d\vec{l}\right). \tag{5.75}$$

The second term is

$$\vec{v} \cdot \frac{d}{dt}\left(d\vec{l}\right) \rightarrow \vec{v} \cdot \frac{d}{dt}\left(\delta \vec{l}\right) = \vec{v} \cdot \delta\left(\frac{d\vec{l}}{dt}\right) = \vec{v} \cdot \delta\vec{v} = \delta\left(\frac{v^2}{2}\right). \tag{5.76}$$

When this is integrated along a closed contour, it will reduce to zero. If the acceleration is the result of a force that is derivable from a potential, then equation 5.75 reduces to

$$\frac{d\Gamma}{dt} = -\frac{1}{m}\oint \vec{\nabla} V \cdot d\vec{l} = -\frac{1}{m}\int_S \left[\vec{\nabla} \times \left(\vec{\nabla} V\right)\right] \cdot d\vec{S} = 0, \qquad (5.77)$$

and it is seen that circulation is conserved.

5.4 Shear Stress Tensor and the Navier-Stokes Equations

The pressure P is related to the stress tensor σ by

$$P_i = \sigma_{ij} n_j, \qquad (5.78)$$

where n_j is the unit normal. A fluid cannot exert a torque; consequently, the expression for the total moment exerted by the surface forces at the boundary of a volume about an arbitrary point within the volume is

$$\oint \varepsilon_{ijk} r_j P_i dA = \oint \varepsilon_{ijk} r_j \sigma_{kl} n_l dA = 0, \qquad (5.79)$$

where ε_{ijk} is the alternating tensor with the value ± 1. From Gauss's theorem (sec. 1.2.3) we see that the expression reduces to

$$\int \varepsilon_{ijk} \frac{\partial}{\partial r_l}\left(r_j \sigma_{kl}\right) dV = \int \varepsilon_{ijk}\left(\sigma_{kj} + r_j \frac{\partial \sigma_{kl}}{\partial r_l}\right) dV = 0. \qquad (5.80)$$

As $dV \to 0$, the first term is seen to be of higher order in V as follows. Let the average radius of V be l. The first term goes as σl^3, while the second term goes as $\sigma l^4/L$, where L is the length over which σ changes. Divide by l^3 and let $l \to 0$, as $dV \to 0$, and the second term will be identically zero. Consequently, the integral is identically zero only if

$$\varepsilon_{ijk} \sigma_{kj} = 0. \qquad (5.81)$$

Consequently, the stress tensor must be symmetrical. Along the principal axes of the tensor we have

$$\sigma_{ij} = \begin{pmatrix} \sigma_{11} & 0 & 0 \\ 0 & \sigma_{22} & 0 \\ 0 & 0 & \sigma_{33} \end{pmatrix}. \tag{5.82}$$

We define the static fluid pressure as

$$p = -\frac{1}{3}(\sigma_{11} + \sigma_{22} + \sigma_{33}), \tag{5.83}$$

and rewrite equation 5.82 as

$$\sigma_{ij} = \begin{pmatrix} -p & 0 & 0 \\ 0 & -p & 0 \\ 0 & 0 & -p \end{pmatrix} + \begin{pmatrix} \sigma_{11} + p & 0 & 0 \\ 0 & \sigma_{22} + p & 0 \\ 0 & 0 & \sigma_{33} + p \end{pmatrix}, \tag{5.84}$$

or simply

$$\sigma_{ij} = -p\delta_{ij} + \tau_{ij}. \tag{5.85}$$

The first term, $-p\delta_{ij}$, is called the *isotropic stress tensor* and is due to the hydrostatic pressure. The second term is called the *deviatoric stress tensor* or the *shear stress tensor*.

The rate of rotation of a fluid is defined by

$$\frac{1}{2}\left(\frac{\partial u_i}{\partial x_j} - \frac{\partial u_j}{\partial x_i} \right), \tag{5.86}$$

while the rate of shearing is defined by

$$\frac{1}{2}\left(\frac{\partial u_i}{\partial x_j} + \frac{\partial u_j}{\partial x_i} \right). \tag{5.87}$$

These two tensors form the antisymmetric and symmetric parts of the *deformation tensor*, which is defined by

$$e_{ij} = \frac{1}{2}\left(\frac{\partial u_i}{\partial x_j} - \frac{\partial u_j}{\partial x_i} \right) + \frac{1}{2}\left(\frac{\partial u_i}{\partial x_j} + \frac{\partial u_j}{\partial x_i} \right) = \frac{\partial u_i}{\partial x_j}. \tag{5.88}$$

If we require that τ_{ij} be a linear combination of the elements of the deformation tensor, it is seen that the general form of τ_{ij} is

$$\tau_{ij} = \alpha_{ijkl} \frac{\partial v_k}{\partial x_l}. \tag{5.89}$$

However, because τ_{ij} must be symmetrical (eq. 5.81) this relation reduces to

$$\tau_{ij} = \frac{1}{2} \beta_{ijkl} \left(\frac{\partial v_k}{\partial x_l} + \frac{\partial v_l}{\partial x_k} \right). \tag{5.90}$$

We further require that it be isotropic. That is, we require that it not show any preference for the orientation of the coordinate system chosen. The most general isotropic tensor of rank four is

$$\beta_{ijkl} = \lambda \delta_{ij} \delta_{kl} + \mu \left(\delta_{ik} \delta_{jl} + \delta_{il} \delta_{jk} \right) + \gamma \left(\delta_{ik} \delta_{jl} - \delta_{il} \delta_{jk} \right), \tag{5.91}$$

where λ, μ, and γ are constants. By symmetry, we must require $\gamma = 0$, and it is seen that equation 5.91 reduces to

$$\tau_{ij} = \lambda \delta_{ij} \frac{\partial v_k}{\partial x_k} + \mu \left(\frac{\partial v_i}{\partial x_j} + \frac{\partial v_j}{\partial x_i} \right). \tag{5.92}$$

Consequently, the stress tensor is given by

$$\sigma_{ij} = -p \delta_{ij} + \lambda \delta_{ij} \frac{\partial u_k}{\partial x_k} + \mu \left(\frac{\partial u_i}{\partial x_j} + \frac{\partial u_j}{\partial x_i} \right). \tag{5.93}$$

Knowing the form of the stress tensor, we see that

$$\frac{\partial \sigma_{ij}}{\partial x_i} = -\frac{\partial p}{\partial x_i} + \frac{\partial}{\partial x_i} \left(\lambda \frac{\partial u_k}{\partial x_k} \right) + \frac{\partial}{\partial x_i} \left[\mu \left(\frac{\partial u_i}{\partial x_j} + \frac{\partial u_j}{\partial x_i} \right) \right]. \tag{5.94}$$

Substituting this into the momentum equation (eq. 5.35) gives the Navier-Stokes equations,

$$m\left[\frac{\partial}{\partial t}(nu_i) + \frac{\partial}{\partial x_i}\left(n\langle u_j u_k\rangle\right)\right]$$

$$= nF_i - \frac{\partial p}{\partial x_i} + \frac{\partial}{\partial x_i}\left(\lambda \frac{\partial u_k}{\partial x_k}\right) + \frac{\partial}{\partial x_i}\left[\mu\left(\frac{\partial u_i}{\partial x_j} + \frac{\partial u_j}{\partial x_i}\right)\right]. \tag{5.95}$$

5.4.1 Flow between Parallel Surfaces

Consider the case of a viscous liquid flowing between two plates whose separation distance is assumed to be small in comparison to its other dimensions. It is assumed that the flow is aligned with the x-axis. Because there is no flow along the y or z axes, the pressure will be a function of x only. Similarly, the velocity will be a function of y only. The Navier-Stokes equation (eq. 5.95) reduces to

$$0 = -\frac{\partial p}{\partial x} + \frac{\partial}{\partial y}\left(\lambda \frac{\partial u}{\partial y}\right), \tag{5.96}$$

or

$$\frac{\partial p}{\partial x} = \lambda \frac{\partial^2 u}{\partial y^2}. \tag{5.97}$$

Because of the independence of the variables, each side of equation 5.97 must be equal to a constant, and we have

$$\frac{\partial p}{\partial x} = -G, \tag{5.98}$$

which has solution

$$p = p_o - Gx. \tag{5.99}$$

Note that the minus sign is a choice of convention and is included to indicate that a higher pressure on the left will push the fluid to the right. Equation 5.97 also reduces to

$$\frac{\partial^2 u}{\partial y^2} = -\frac{G}{\lambda}, \tag{5.100}$$

which has solution

$$u(y) = -\frac{G}{2\lambda}\left(y^2 + Ay + B\right). \tag{5.101}$$

We impose the boundary conditions $u(0) = u(h) = 0$, which imply that $B = 0$ and $A = -h$. The solution is

$$u(y) = -\frac{G}{2\lambda}(y - h)y. \tag{5.102}$$

5.4.2 Flow in a Pipe

Consider a pipe of circular cross section, of which the velocity is only a function of the radial position. The Navier-Stokes equation (eq. 5.95) in cylindrical coordinates, reduces to

$$0 = -\frac{\partial p}{\partial z} + \frac{1}{r}\frac{\partial}{\partial r}\left(\lambda r \frac{\partial u}{\partial r}\right). \tag{5.103}$$

Again, because of the independence of the variables, we see immediately that

$$p = p_o - Gz. \tag{5.104}$$

Equation 5.103 reduces to

$$\frac{1}{r}\frac{\partial}{\partial r}\left(\lambda r \frac{\partial u}{\partial r}\right) = -G, \tag{5.105}$$

which has solution

$$u(r) = -\frac{G}{4\lambda}r^2 + A\ln r + B. \tag{5.106}$$

Because the velocity must be finite at $r = 0$, we require $A = 0$. Similarly, the boundary condition $u(a) = 0$, where a is the radius of the pipe, requires $B = \frac{G}{4\lambda}a^2$. The solution is

$$u(r) = \frac{G}{4\lambda}\left(a^2 - r^2\right). \tag{5.107}$$

5.5 Bernoulli's Equation

If the force $\vec{F}$ is derivable from a potential function of the form

$$\vec{F} = -\vec{\nabla}U, \tag{5.108}$$

the equation for conservation of momentum (eq. 5.35) may be rewritten as

$$mn\left[\frac{\partial \vec{u}}{\partial t} + \vec{u}\left(\vec{\nabla} \cdot \vec{u}\right)\right] = -n\vec{\nabla}U - \vec{\nabla} \cdot \vec{P}. \tag{5.109}$$

From the vector identities, it is seen that

$$\vec{u}\left(\vec{\nabla} \cdot \vec{u}\right) = \vec{\nabla}\left(\frac{1}{2}\vec{u} \cdot \vec{u}\right) - \vec{u} \times \left(\vec{\nabla} \times \vec{u}\right), \tag{5.110}$$

so that equation 5.109 is

$$\frac{\partial \vec{u}}{\partial t} + \vec{\nabla}\left(\frac{1}{2}\vec{u} \cdot \vec{u}\right) - \vec{u} \times \left(\vec{\nabla} \times \vec{u}\right) = -\frac{1}{m}\vec{\nabla}U - \frac{1}{mn}\vec{\nabla} \cdot \vec{P}. \tag{5.111}$$

Consider the pressure term. We see that

$$d\vec{x} \cdot \left(\frac{1}{mn}\vec{\nabla} \cdot \vec{P}\right) = \frac{1}{mn}\left(d\vec{x} \cdot \left(\vec{\nabla} \cdot \vec{P}\right)\right) = \frac{dP}{mn} = d\int \frac{dP}{mn} = d\vec{x} \cdot \vec{\nabla}\left(\int \frac{dP}{mn}\right), \tag{5.112}$$

so that

$$\frac{\vec{\nabla} \cdot \vec{P}}{mn} = \vec{\nabla}\left(\int \frac{dP}{mn}\right). \tag{5.113}$$

Using this result equation 5.109 becomes

$$\frac{\partial \vec{u}}{\partial t} + \vec{\nabla}\left(\frac{1}{2}\vec{u} \cdot \vec{u}\right) - \vec{u} \times \left(\vec{\nabla} \times \vec{u}\right) = -\frac{1}{m}\vec{\nabla}U - \vec{\nabla}\left(\int \frac{dP}{mn}\right), \tag{5.114}$$

or

$$\frac{1}{m}\vec{\nabla}\left[\left(\frac{m}{2}\vec{u} \cdot \vec{u}\right) + U + \left(\int \frac{dP}{n}\right)\right] = \frac{\partial \vec{u}}{\partial t} + \vec{u} \times \left(\vec{\nabla} \times \vec{u}\right). \tag{5.115}$$

Consider steady state flow so that $\frac{\partial \vec{u}}{\partial t} = 0$. If we take the vector product of $\vec{u}$ with the previous expression, we see that the right-hand side will be zero. The left-hand side will involve the term $\vec{u} \cdot \vec{\nabla}$ of the expression in brackets. This is the steady-state form of the derivative from the continuity equation, so we conclude that

$$\left(\frac{m}{2} \vec{u} \cdot \vec{u}\right) + U + \left(\int \frac{dP}{n}\right) = \text{constant}. \tag{5.116}$$

This expression is known as *Bernoulli's equation* and is usually written in the form

$$\frac{1}{2} \rho v^2 + \rho g h + P = \text{constant}. \tag{5.117}$$

5.6 Bibliography

Currie, I. G. *Fundamental Mechanics of Fluids.* New York: McGraw-Hill, 1974.

Paterson, A. R. *A First Course in Fluid Dynamics.* Cambridge, U.K.: Cambridge University Press, 1983).

Batchelor, G. K., *An Introduction to Fluid Dynamics*, Cambridge, U.K.: Cambridge University Press, 1967.

6   Plasma Physics

6.1   Basic Plasma Properties

6.1.1   Debye Shielding

Consider a test particle of charge $q_t > 0$, with near infinite mass, located at the origin. The charge will repel ions and attract electrons. That is, it will gather a shielding cloud that will screen its own charge. Poisson's equation (in cgs units) is

$$\vec{\nabla} \cdot \vec{E} = \nabla^2 V = -4\pi\rho = 4\pi\left[e(n_e - n_i) - q_t\delta(\vec{r})\right]. \qquad (6.1)$$

In the chapter on thermodynamics, it is shown (eq. 13.70) that the ratio of states with energy E_2 to energy E_1 is given by

$$\frac{n(E_2)}{n(E_1)} = \exp\left(\frac{-(E_2 - E_1)}{kT}\right), \qquad (6.2)$$

which gives

$$n_e(r) = n_o\exp\left(\frac{eV(r)}{kT_e}\right), \qquad (6.3a)$$

and

$$n_i(r) = n_o\exp\left(-\frac{eV(r)}{kT_i}\right), \qquad (6.3b)$$

where $E = qV$ and n_o is the equilibrium density far from the test charge. If $qV/T_{i,e} \ll 1$, we may expand the exponential in a Taylor series to find that, away from the origin,

$$\nabla^2 V = \frac{4\pi e^2 n_o}{k}\left(\frac{1}{T_e}+\frac{1}{T_i}\right)V(r).$$ (6.4)

We define the Debye length for electrons and ions by the relation

$$\lambda_{e,i} = \left(\frac{kT_{e,i}}{4\pi n_o e^2}\right)^{1/2},$$ (6.5)

and the total Debye length by

$$\frac{1}{\lambda_D^2} = \frac{1}{\lambda_e^2}+\frac{1}{\lambda_i^2}.$$ (6.6)

Substituting this expression into equation 6.4 gives

$$\frac{1}{r^2}\frac{d}{dr}\left(r^2\frac{dV}{dr}\right) = \frac{V(r)}{\lambda_D^2}.$$ (6.7)

The solution for $V(r)$ is seen to be

$$V(r) = \frac{q_t}{r}\exp\left(-\frac{r}{\lambda_D}\right),$$ (6.8)

so that the effect of the plasma is to reduce the potential due to a charged object exponentially.

6.1.2 The Plasma Parameter

In a plasma of density n_0 the average distance between particles is $n_0^{-1/3}$. The average potential energy is

$$|V| \approx \frac{q^2}{r} \approx n_o^{1/3}e^2.$$ (6.9)

The kinetic energy of a particle is given by

$$\frac{1}{2}m_s\langle v\rangle^2 = \frac{3}{2}kT = \frac{3}{2}m_s v_s^2.$$ (6.10)

For a plasma to exist, we require that the kinetic energy be much greater than the potential energy, that is,

$$n_o^{1/3} e^2 \ll kT. \tag{6.11}$$

Rearranging terms gives

$$\frac{n_o^{1/3} e^2}{kT} \ll 1, \tag{6.12}$$

or

$$\frac{kT}{n_o^{1/3} e^2} = 4 \pi n_o^{2/3} \left(\frac{kT}{4 \pi n_o e^2} \right) \gg 1. \tag{6.13}$$

We define

$$\wedge_s = n_o \left(\frac{kT}{4 \pi n_o e^2} \right)^{3/2} = n_o \lambda_D^3 \gg 1, \tag{6.14}$$

where $\wedge_s$ is the plasma parameter, as a necessary condition for a plasma to exist.

6.1.3 The Plasma Frequency

Consider a slab of plasma of thickness l, where $n_i = n_e$, and $m_i \sim \infty$. If the electrons in the slab are displaced from the background of ions, the charge displacement will establish an electric field which will pull the electrons back toward the ions. However, once they reach the equilibrium position, the electrons will have momentum and will overshoot. This situation will be repeated in the other direction, and it is seen that the end result will be an oscillation of the electrons. If the electrons are displaced a distance d, then in one dimension we have

$$\vec{\nabla} \cdot \vec{E} = \frac{\partial E}{\partial x} = 4 \pi \rho. \tag{6.15}$$

The electric field over most of the slab of plasma is dependent upon the displacement distance δ and is given by

$$E \approx -4\pi n_o e \delta. \tag{6.16}$$

The force/unit area is simply the electric field multiplied by the charge per unit area, or

$$\frac{F}{A} = \left(-4\pi n_o e \delta\right)\left(e n_o L\right). \tag{6.17}$$

By definition, force equals mass times acceleration, so

$$\left(-4\pi n_o e \delta\right)\left(e n_o L\right) = \left(n_o m_e L\right)\ddot{\delta}, \tag{6.18}$$

which reduces to

$$\left(\frac{4\pi n_o e^2}{m_e}\right)\delta + \ddot{\delta} = 0. \tag{6.19}$$

This is the equation for a simple harmonic oscillator having a fundamental frequency

$$f_{p,e} = \frac{\omega_{p,e}}{2\pi} = \frac{1}{2\pi}\left(\frac{4\pi n_o e^2}{m_e}\right)^{1/2}, \tag{6.20}$$

which is called the *plasma frequency*.

6.1.4 Collisions

Consider the situation shown in figure 6.1. A particle with charge q, mass m is incident upon q_0, $m_0 = \infty$ with a speed v_0. If the particle were undeflected, it would have position $x = v_0 t$ along the upper dashed line. If the scattering angle is small, the final parallel speed is approximately v_0, and $v_\perp$ is found from

$$mv_\perp = \int_{-\infty}^{+\infty} F_\perp(t)\, dt. \tag{6.21}$$

We have

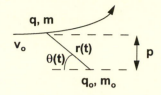

Figure 6.1 Collision geometry.

$$F = \frac{qq_o}{r^2}\hat{r},$$

(6.22)

or

$$F_\perp = \frac{qq_o}{r^2} \sin\theta.$$

(6.23)

In terms of the impact parameter p,

$$F_\perp = \frac{qq_o}{(p/\sin\theta)^2} \sin\theta = \frac{qq_o}{p^2} \sin^3\theta,$$

(6.24)

so that

$$v_\perp = \frac{qq_o}{mp^2} \int_{-\infty}^{+\infty} \sin^3\theta(t)\,dt.$$

(6.25)

We define

$$x = -r\cos\theta = -\frac{p\cos\theta}{\sin\theta} = v_o t,$$

(6.26)

so that

$$dt = \frac{p}{v_o \sin^2\theta}\,d\theta.$$

(6.27)

Equation 6.25 reduces to

$$v_\perp = \frac{qq_o}{mv_o P} \int_0^\pi \sin\theta d\theta = \frac{2qq_o}{mv_o P}. \tag{6.28}$$

We define

$$p_o = \frac{2qq_o}{mv_o}. \tag{6.29}$$

If $p < p_o$, then $v_\perp > v_o$. If this is the case, we say that a large scattering angle collision has occurred. The cross section for large scattering is

$$\sigma_{LA} = \pi p_o^2. \tag{6.30}$$

A large-angle collision occurs after the first particle has traversed an average distance

$$l = v_o t = \frac{1}{n\pi d^2}. \tag{6.31}$$

If $d = p_o$ the frequency between collisions is

$$v_L = \pi v_o n p_o^2 = \frac{4\pi n_o e^4}{m_e^2 v_o^3}. \tag{6.32}$$

6.2 Single Particle Motion

6.2.1 Cyclotron Motion

A charged particle moving with velocity v in the presence of either an electric or magnetic field experiences a force given by

$$\vec{F} = m\vec{a} = q(\vec{E} + \vec{v} \times \vec{B}). \tag{6.33}$$

In the absence of an electric field, if any component of v is perpendicular to B the result will be a force that acts at right angles to both v and B. Because the force acts perpendicular to the line of travel, it cannot change the magnitude of v, only its direction. Consequently, the force will be constant and will cause the particle to gyrate around the magnetic field. If B is aligned with the z-axis and v is entirely perpendicular to B, equation 6.33 reduces to

$$
m\dot{v}_x = qBv_y,
$$
$$
m\dot{v}_y = -qBv_x,
$$
$$
m\dot{v}_z = 0. \tag{6.34}
$$

Differentiating the first expression and substituting it into the second expression gives

$$
\ddot{v}_x = -\left(\frac{qB}{m}\right)^2 v_x. \tag{6.35}
$$

This expression defines the cyclotron gyration frequency, or simply the cyclotron frequency, which is given by

$$
\Omega = \left(\frac{qB}{m}\right). \tag{6.36}
$$

The implication is that the particle must gyrate around the magnetic field lines with frequency given by equation 6.36. The constraint on circular motion is

$$
a = \frac{v^2}{r}. \tag{6.37}
$$

Consequently, the gyroradius (sometimes called the Larmour radius) of a charged particle is given by

$$
r = \frac{mv}{qB}. \tag{6.38}
$$

6.2.2 The E cross B Drift

Consider a particle with $v_z = 0$ in a magnetic field $\vec{B} = B_o\hat{z}$, with an electric field $\vec{E} = E_o\hat{y}$. The magnetic field will turn the particle; therefore the electric field will not be able to accelerate the particle indefinitely. The result is a drift to the right, as shown in figure 6.2.

If we average over many gyroperiods, the average acceleration will be zero; therefore

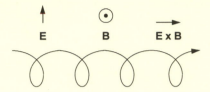

Figure 6.2 The E cross B drift.

$$\langle m\dot{v} \rangle = 0 = q_s \vec{E}_o + \frac{q_s}{c} \vec{v}_d \times \vec{B}. \tag{6.39}$$

Taking the cross product of this expression with B gives

$$q_s \vec{E}_o \times \vec{B} + \frac{q_s}{c}\left(\vec{v}_d \times \vec{B}\right) \times \vec{B} = 0, \tag{6.40}$$

and solving for v_d gives

$$\vec{v}_d = c\frac{\vec{E}_o \times \vec{B}_o}{B_o^2}, \tag{6.41}$$

which is the $E \times B$ drift velocity. If a general force $F_\perp$ were used in place of the electric force $q_s E_o$, we would have

$$\vec{v}_d = \frac{1}{\Omega_s}\left(\frac{\vec{F}_\perp}{m_s} \times \hat{B}_o\right), \tag{6.42}$$

where Ω_s is the cyclotron frequency for the species in question.

6.2.3 The Grad B Drift

If we have a particle in a magnetic field $B_o\hat{z}$ that increases in strength in the y-direction, the gyroradius will be smaller at large y and larger at small y. As a result, positively charged particles will drift, as shown in figure 6.3.

The force on a charged particle is

$$m\dot{\vec{v}} = \frac{q_s}{c}\left(\vec{v} \times \vec{B}\right). \tag{6.43}$$

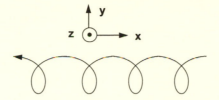

Figure 6.3 The grad B drift.

If we perform a Taylor series expansion on B about the guiding center of the particle, we get

$$\vec{B} = \vec{B}_0 + \left(\vec{r} \cdot \vec{\nabla}\right)\vec{B}_0. \tag{6.44}$$

If we also let

$$\vec{v} = \vec{v}_0 + \vec{v}_1, \tag{6.45}$$

we find

$$m_s\left(\dot{\vec{v}}_0 + \dot{\vec{v}}_1\right) = \frac{q_s}{c}\left[\left(\vec{v}_0 + \vec{v}_1\right) \times \vec{B}_0 + \left(\vec{v}_0 + \vec{v}_1\right) \times \left(\vec{r} \cdot \vec{\nabla}\right)\vec{B}_0\right]. \tag{6.46}$$

The terms of order zero and one are

$$m_s\dot{\vec{v}}_0 = \frac{q_s}{c}\vec{v}_0 \times \vec{B}_0, \tag{6.47}$$

and

$$m_s\dot{\vec{v}}_1 = \frac{q_s}{c}\left(\vec{v}_1 \times \vec{B}_0\right) + \frac{q_s}{c}\left[\vec{v}_0 \times \left(\vec{r} \cdot \vec{\nabla}\right)\vec{B}_0\right], \tag{6.48}$$

respectively. Assuming v_1 is $\perp$ to B_0, the v cross B force will cause the particle to move essentially in a circle, so that $\left\langle \dot{\vec{v}}_1 \right\rangle = 0$. Equation 6.48 reduces to

$$\frac{q_s}{c}\left\langle \vec{v}_1 \times \vec{B}_0 \right\rangle + \frac{q_s}{c}\left\langle \vec{v}_0 \times \left(\vec{r} \cdot \vec{\nabla}\right)\vec{B}_0 \right\rangle = 0, \tag{6.49}$$

or

$$\vec{v}_1 = \frac{1}{B_o^2}\left\langle \left[\vec{v}_0 \times \left(\vec{r}\cdot\vec{\nabla}\right)\vec{B}_0\right] \times \vec{B}_0 \right\rangle. \tag{6.50}$$

Using

$$\vec{r} = \left(-\frac{v_0}{\Omega}\cos\Omega t, \frac{v_0}{\Omega}\sin\Omega t, 0\right), \tag{6.51}$$

and

$$\vec{v}_0 = v_0\left(\sin\Omega t, \cos\Omega t, 0\right), \tag{6.52}$$

where Ω is the cyclotron frequency. We find that

$$\vec{v}_1 = \frac{1}{2B_0^2}\frac{v_0^2}{\Omega_s}\left(\vec{B}_0 \times \nabla\vec{B}_0\right). \tag{6.53}$$

Charged particles also experience curvature drifts and polarization drifts.

6.3 The Ponderomotive Force

Consider a charged particle oscillating in a high frequency electric field $E(t) = E_o \cos\omega t$ with an amplitude that varies smoothly in space, i.e., $E(x,t) = E_0(x) \cos\omega t$. If the region of strong field is to the right, the first oscillation will bring the particle into regions of strong field, where it can be given a strong push to the left. When the particle turns around, the particle is in a region of weaker field and the push to the right is not as strong. The result is acceleration away from the region of strong field. We have

$$m_s\ddot{x} = q_s E_0(x)\cos\omega t. \tag{6.54}$$

We let $x = x_0 + x_1$, where 0 is the oscillation center, and expand E_0 about x_0. This gives

$$m_s\left(\ddot{x}_0 + \ddot{x}_1\right) = q_s\left[E_0(x_0) + x_1\frac{dE_0}{dx}\bigg|_{x_0} + \ldots\right]\cos\omega t. \tag{6.55}$$

If we average over time, so that $\langle \ddot{x}_1 \rangle = 0$, we find that

$$m_s \langle \ddot{x}_0 \rangle = q_s \left. \frac{dE_0}{dx} \right|_{x_0} \langle x_1 \cos \omega t \rangle. \tag{6.56}$$

However, since $\ddot{x}_1 \gg \ddot{x}_0$ and $E_0 \gg x_1 \dfrac{dE_0}{dx}$, we have

$$m_s \langle \ddot{x}_1 \rangle = q_s E_0 \cos \omega t. \tag{6.57}$$

Using this expression in equation 6.49 and solving for $\ddot{x}_0$, we obtain the ponderomotive force

$$F_p = -\frac{q_s^2}{4 m_s \omega^2} \frac{d}{dx} \left(E_0^2 \right). \tag{6.58}$$

6.4 Plasma Waves

6.4.1 Plasma Oscillations

To derive the relation for plasma oscillations from the fluid equations, we make the following assumptions: (1) no magnetic fields, (2) no thermal motions,(3) fixed ions, (4) infinite plasma, and (5) one-dimensional. Because the ions don't move the equation of continuity, the equation of motion and Poisson's equation are

$$\frac{\partial n_e}{\partial t} + \vec{\nabla} \cdot \left(n_e \vec{v}_e \right) = 0, \tag{6.59}$$

$$m n_e \left[\frac{\partial \vec{v}_e}{\partial t} + \left(\vec{v}_e \cdot \vec{\nabla} \right) \vec{v}_e \right] = -e n_e \vec{E}, \tag{6.60}$$

and

$$\vec{\nabla} \cdot \vec{E} = 4 \pi e \left(n_i - n_e \right). \tag{6.61}$$

STEP 1. Linearize. Let $n_e = n_0 + n_1$, $v_e = v_0 + v_1$, $E = E_0 + E_1$. The equation of continuity is

$$\frac{\partial(n_0 + n_1)}{\partial t} + \bar{\nabla} \cdot \left((n_0 + n_1)(\bar{v}_0 + \bar{v}_1) \right) = 0. \tag{6.62}$$

STEP 2. The initial conditions require $\nabla n_0 = v_0 = E_0 = 0$. Applying these relations and dropping second-order terms, the previous expression reduces to

$$\frac{\partial n_1}{\partial t} + \bar{\nabla} \cdot \left(n_0 \bar{v}_1 \right) = 0. \tag{6.63}$$

STEP 3. Keeping only second-order terms, the equation of motion reduces to

$$m \frac{\partial v_1}{\partial t} = -eE_1, \tag{6.64}$$

while Poisson's equation gives

$$\nabla \cdot E_1 = -en_1. \tag{6.65}$$

If we are looking for plane wave solutions, we may let $\dfrac{\partial}{\partial t} \to -i\omega$ and $\dfrac{\partial}{\partial x} \to ik$. Equations 6.63, 6.64, and 6.65 reduce to

$$-i\omega n_1 + in_0 k v_1 = 0, \tag{6.66}$$

$$-im\omega v_1 = -eE_1, \tag{6.67}$$

and

$$ikE_1 = -4\pi en_1. \tag{6.68}$$

Solving for the three unknowns (n_1, v_1, and E_1) gives

$$f = \frac{\omega}{2\pi} = \frac{1}{2\pi} \left(\frac{4\pi n_0 e^2}{m} \right)^{1/2}, \tag{6.69}$$

which is identical to equation 6.20.

6.4.2 Ion Plasma Waves

We wish to consider low-frequency plasma waves, where ion motions are not negligible. The equation of continuity and equation of motion for the electrons and ions are

$$\frac{\partial n_e}{\partial t} + \frac{\partial}{\partial x}(n_e v_e) = 0, \tag{6.70}$$

$$m_e n_e \frac{\partial v_e}{\partial t} + m_e n_e v_e \frac{\partial v_e}{\partial x} = -\frac{\partial P_e}{\partial x} - e n_e E, \tag{6.71}$$

$$\frac{\partial n_i}{\partial t} + \frac{\partial}{\partial x}(n_i v_i) = 0, \tag{6.72}$$

and

$$m_i n_i \frac{\partial v_i}{\partial t} + m_i n_i v_i \frac{\partial v_i}{\partial x} = -\frac{\partial P_i}{\partial x} + e n_i E, \tag{6.74}$$

respectively. Poisson's equation is now

$$\frac{\partial E}{\partial x} = 4\pi e (n_i - n_e). \tag{6.74}$$

When we linearize all terms that depend on $v \frac{\partial v}{\partial x}$ will be of second order and can be ignored. The pressure terms are simplified by

$$P_s = n_s m_s \langle v^2 \rangle = n_s k_B T_s, \tag{6.75}$$

so that

$$\nabla P_s = \nabla(n_s T_s) = n_s \nabla T_s + T_s \nabla n_s = \gamma_s T_s \nabla n_s, \tag{6.76}$$

where γ_s is the ratio of the specific heats. Linearizing gives

$$\frac{\partial}{\partial t} n_{e,1} + n_{e,0} \frac{\partial}{\partial x}(v_e) = 0, \tag{6.77}$$

$$m_e n_{e,0} \frac{\partial}{\partial t} v_e = -\gamma_e T_e \frac{\partial}{\partial x} n_{e,1} - e n_{e,0} E, \tag{6.78}$$

$$\frac{\partial}{\partial t} n_{i,1} + n_{i,0} \frac{\partial}{\partial x}(v_i) = 0, \tag{6.79}$$

$$m_i n_{i,0} \frac{\partial}{\partial t} v_i = -\gamma_i T_i \frac{\partial}{\partial x} n_{i,1} + e n_{i,0} E, \tag{6.80}$$

$$\frac{\partial}{\partial x} E = 4\pi e \left(n_{i,1} - n_{e,1} \right). \tag{6.81}$$

Based on the initial assumptions, $n_{i,1} \sim n_{e,1}$ and $v_e \sim v_i$. Combining equations 6.78 and 6.80 gives

$$(m_e + m_i) n_0 \frac{\partial v_e}{\partial t} = -(\gamma_e T_e + \gamma_i T_i) \frac{\partial n_{e,1}}{\partial x}. \tag{6.82}$$

We eliminate v_e by taking the time derivative of the continuity equations, 6.77 and 6.79, to obtain

$$(m_e + m_i) \frac{\partial^2 n_{e,1}}{\partial t^2} = (\gamma_e T_e + \gamma_i T_i) \frac{\partial^2 n_{e,1}}{\partial x^2}. \tag{6.83}$$

Because $m_i \gg m_e$, this reduces to

$$\frac{\partial^2 n_{e,1}}{\partial t^2} = \left(\frac{\gamma_e T_e + \gamma_i T_i}{m_i} \right) \frac{\partial^2 n_{e,1}}{\partial x^2}. \tag{6.84}$$

This is the ion acoustic dispersion relation,

$$\omega^2 = k^2 C_s^2, \tag{6.85}$$

with

$$C_s = \left(\frac{\gamma_e T_e + \gamma_i T_i}{m_i} \right)^{1/2}. \tag{6.86}$$

6.4.3 Electrostatic Waves

If electrons undergoing plasma oscillations have thermal velocities, they may carry information about what is happening in the oscillating region. We assume that each species has a distribution function,

$$f_s = f_{s,0} + f_{s,1}, \tag{6.87}$$

where

$$n_0 = \int f_{s,0}(v)\,dv. \tag{6.88}$$

Choosing the E-field to be aligned with the x-axis and assuming spatial oscillations only in this direction, the Vlasov equation is

$$\frac{\partial f_s}{\partial t} + v_x \frac{\partial f_s}{\partial x} + \frac{q_s}{m_s} E \frac{\partial f_s}{\partial x} = 0. \tag{6.89}$$

The zero-order terms are

$$\frac{\partial f_{s,0}}{\partial t} + v_x \frac{\partial f_{s,0}}{\partial x} = 0, \tag{6.90}$$

and the first-order terms are

$$\frac{\partial f_{s,1}}{\partial t} + v_x \frac{\partial f_{s,1}}{\partial x} + \frac{q_s}{m_s} E \frac{\partial f_{s,0}}{\partial x} = 0. \tag{6.91}$$

Because we seek phase wave solutions, this reduces to

$$-i\omega f_{s,1} + ikv_x f_{s,1} = -\frac{q_s}{m_s} E \frac{\partial f_{s,0}}{\partial x}, \tag{6.92}$$

which gives

$$f_{s,1}(\vec{x},\vec{v},t) = -\frac{i}{\omega - kv_x} \frac{q_s}{m_s} E \frac{\partial f_{s,0}}{\partial x}. \tag{6.93}$$

From Poisson's equation,

$$\vec{\nabla} \cdot \vec{E} = ikE = 4\pi e(n_i - n_e) = 4\pi e \int (f_{i,1} - f_{e,1}) dv, \qquad (6.94)$$

or

$$ikE = 4\pi e(-ieE) \int \left(\frac{1}{m_i} \frac{1}{\omega - kv_x} \frac{\partial f_{i,0}}{\partial v_x} + \frac{1}{m_e} \frac{1}{\omega - kv_x} \frac{\partial f_{e,0}}{\partial v_x} \right) dv. \qquad (6.95)$$

This gives the dispersion relation for electrostatic waves,

$$1 + \frac{\omega^2}{k^2} \int \frac{\partial_u g(u)}{\omega/k - u} du = 0, \qquad (6.96)$$

where

$$g(v_x) = \frac{m_e}{n_0 m_i} \int f_{i,0}(\vec{v}) dv_y dv_z + \frac{1}{n_0} \int f_{e,0}(\vec{v}) dv_y dv_z. \qquad (6.97)$$

If the ions and electrons are Maxwellian,

$$f_{s,0} = \frac{n_0}{(2\pi)^{3/2} v_s^3} \exp \left(\frac{-\left(v_x^2 + v_y^2 + v_z^2\right)}{2v_s^2} \right), \qquad (6.98)$$

whereupon

$$g(v_x) = \frac{1}{(2\pi)^{1/2} v_e} \exp \left(\frac{-v_x^2}{2v_e^2} \right) + \frac{m_e}{m_i} \frac{1}{(2\pi)^{1/2} v_i} \exp \left(\frac{-v_x^2}{2v_i^2} \right). \qquad (6.99)$$

If we consider only high-frequency waves, called *Langmuir waves*, we may assume that $\omega/k \gg u$ for all u for which $g(u)$ is applicable, so that $d_u g(u) = 0$ at $u = \omega/k$. If we also assume that m_e/m_i is small enough to ignore, we can integrate by parts to find, because the boundary terms go to zero,

$$1 - \frac{\omega_e^2}{k^2} \int_{-\infty}^{+\infty} \frac{g(u)}{\left(\frac{\omega}{k} - u \right)^2} du = 0, \qquad (6.100)$$

or

$$1 - \frac{\omega_e^2}{\omega^2} \int_{-\infty}^{+\infty} g(u)\left(1 + \frac{2uk}{\omega} + \frac{3u^2k^2}{\omega^2} + \ldots\right) du = 0. \qquad (6.101)$$

Truncating the series after the third term, we find

$$\int g(u)\,du = 1, \qquad (6.102a)$$

$$\int g(u)u\,du = 0, \qquad (6.102b)$$

$$\int g(u)u^2\,du = v_e^2. \qquad (6.102c)$$

We have

$$1 - \frac{\omega_e^2}{\omega^2} - \frac{3k^2 v_e^2 \omega_e^2}{\omega^4} = 0, \qquad (6.103)$$

or, assuming $k^2 v_e^2 \ll \omega^2$,

$$\omega^2 = \omega_e^2 + 3k^2 v_e^2. \qquad (6.104)$$

This is the Langmuir wave dispersion relation.

6.4.4 Electromagnetic Waves

We wish to find the dispersion relation for transverse electromagnetic (EM) waves. Transverse implies that $\vec{k} \cdot \vec{E} = \vec{k} \cdot \vec{B} = 0$. From Maxwell's equations,

$$\vec{\nabla} \times \vec{E} = -\frac{1}{c}\frac{\partial \vec{B}}{\partial t}, \qquad (6.105)$$

$$\vec{\nabla} \times \vec{B} = \frac{4\pi}{c}\vec{J} + \frac{1}{c}\frac{\partial \vec{E}}{\partial t}, \qquad (6.106)$$

with

$$\vec{J} = -en_e\vec{v}_e. \tag{6.107}$$

From the fluid equations,

$$\frac{\partial n_e}{\partial t} + \vec{\nabla}\cdot(n_e\vec{v}_e) = 0, \tag{6.108}$$

$$m_e n_e \frac{\partial \vec{v}_e}{\partial t} + m_e n_e(\vec{v}_e\cdot\vec{\nabla}) = -\vec{P}_e - en_e\vec{E} - \frac{en_e}{c}\vec{v}_e\times\vec{B}. \tag{6.109}$$

If we assume that $\vec{v}_e\times\vec{B}$ can be neglected and that $\vec{k}\cdot\vec{v}_e = 0$, then

$$\vec{\nabla}\times\vec{E} = -\frac{1}{c}\frac{\partial\vec{B}}{\partial t}, \tag{6.110}$$

$$\vec{\nabla}\times\vec{B} = -\frac{4\pi en_e}{c}\vec{v}_e + \frac{1}{c}\frac{\partial\vec{E}}{\partial t}, \tag{6.111}$$

and

$$m_e n_e \frac{\partial\vec{v}_e}{\partial t} = -en_e\vec{E}. \tag{6.112}$$

We find that

$$-c\vec{\nabla}\times(\vec{\nabla}\times\vec{E}) = \frac{4\pi n_e e^2}{m_e c^2}\vec{E} + \frac{1}{c}\frac{\partial^2\vec{E}}{\partial t^t}. \tag{6.113}$$

Letting $\vec{\nabla}\times(\vec{\nabla}\times) \to k^2$, we get

$$(\omega^2 - k^2 c^2 - \omega_e^2)\vec{E} = 0, \tag{6.114}$$

or

$$\omega^2 = k^2 c^2 + \omega_e^2, \tag{6.115}$$

which is the dispersion relation for electromagnetic waves.

Example 6.1
Consider a plasma consisting of two homogeneous streams of electrons with equal densities and equal but opposite velocities of magnitude v'.

Show that waves with wave number $k < k_c = \sqrt{2}\,\dfrac{w_p}{v'}$, where ω_p is the plasma frequency, are unstable.

This problem will be easier to solve in a frame of reference where one stream of electrons has velocity zero and the other has velocity $v_0 = 2v'$. We may then linearize about equilibrium with the assumptions

$$n = n_0 + n_1, \tag{6.116}$$

$$v = v_0 + v_1, \tag{6.117}$$

where n_0 and v_0 do not vary in time or space. The equation of continuity for the stream with velocity $2v'$ is

$$\frac{\partial}{\partial t}(n_{01} + n_{11}) + \vec{\nabla} \cdot \left[(n_{01} + n_{11})(\vec{v}_{01} + \vec{v}_{11})\right] = 0, \tag{6.118}$$

which reduces to

$$\frac{\partial}{\partial t} n_{11} + v_{01} \nabla n_{11} + n_{01} \vec{\nabla} \cdot \vec{v}_{11} = 0. \tag{6.119}$$

Similarly, the equation of motion is

$$m_e(n_{01} + n_{11})\left[\frac{\partial}{\partial t}(v_{01} + v_{11}) + \left\{(\vec{v}_{01} + \vec{v}_{11}) \cdot \vec{\nabla}\right\}(\vec{v}_{01} + \vec{v}_{11})\right]$$
$$= (n_{01} + n_{11})qE_1, \tag{6.120}$$

or

$$m_e n_{01} \frac{\partial}{\partial t} v_{11} + m_1 n_{01} v_{01} \vec{\nabla} \cdot \vec{v}_{11} = n_{01} qE_1, \tag{6.121}$$

where we have assumed $E_0 = 0$. By comparison, the equation of continuity for the stream of electrons at rest is

$$\frac{\partial}{\partial t} n_{12} + n_{02} \vec{\nabla} \cdot \vec{v}_{12} = 0, \tag{6.122}$$

and the equation of motion is

$$m_e n_{02} \frac{\partial}{\partial t} v_{12} = n_{02} q E_1. \tag{6.123}$$

Since we are looking for plane wave solutions, we let $\nabla \to ik$, and $\frac{\partial}{\partial t} \to -i\omega$. Equations 6.122 and 6.123 reduce to

$$-i\omega m_e n_{01} v_{11} + ik m_e n_{01} v_{01} v_{11} = n_{01} q E_1, \tag{6.124}$$

and

$$-i\omega m_e n_{02} v_{12} = n_{02} q E_1. \tag{6.125}$$

Making use of the fact that $v_{01} = 2v'$ and $q = -e$, it can be seen that

$$v_{12} = \frac{e E_1}{i\omega m_e}, \tag{6.126}$$

and

$$v_{22} = \frac{e E_1}{-i\omega m_e + ik 2 m_e v'}. \tag{6.127}$$

Equations 6.124 and 6.125 reduce to

$$-i\omega n_{11} + ik v_{01} n_{11} + ik n_{01} v_{11} = 0, \tag{6.128}$$

and

$$-i\omega n_{12} + ik n_{02} v_{12} = 0. \tag{6.129}$$

Using the values for v_{11} and v_{12} given in equations 6.126 and 6.127 it can be seen that

$$n_{11} = n_{01} \frac{-ik e E_1}{m_e (\omega - 2v' k)^2}, \tag{6.130}$$

and

$$n_{12} = n_{02} \frac{-ikeE_1}{m_e \omega^2}. \tag{6.131}$$

Substituting these expressions into Poisson's equation

$$ikE_1 = \vec{\nabla} \cdot \vec{E} = 4\pi\rho = -4\pi e(n_{11} + n_{12}), \tag{6.132}$$

gives

$$1 = \frac{4\pi e^2 n_0}{m_e} \left[\frac{1}{\omega^2} + \frac{1}{(\omega - 2v'k)^2} \right] = \omega_{p,e}^2 \left[\frac{1}{(\omega + v'k)^2} + \frac{1}{(\omega - v'k)^2} \right]. \tag{6.133}$$

This in turn reduces to

$$(\omega + v'k)^2 (\omega - v'k)^2 = \omega_{p,e}^2 \left[(\omega + v'k)^2 + (\omega - v'k)^2 \right], \tag{6.134}$$

or simply

$$\omega^4 - \omega^2 \left(2\omega_{p,e}^2 \right) + \left(v'^4 k^4 - 2v'^2 k^2 \omega_{p,e}^2 \right) = 0. \tag{6.135}$$

Solving this expression for ω gives

$$\omega^2 = \omega_{p,e}^2 \pm \sqrt{\omega_{p,e}^4 + 2\omega_{p,e}^2 v'^2 k^2 - v'^4 k^4}. \tag{6.136}$$

It is seen that if $k < \sqrt{2}\, \dfrac{\omega_{p,e}}{v'}$, then $\omega^2 < 0$, ω is imaginary, and the wave is unstable.

Example 6.2
Consider a plasma in a uniform static magnetic field B_0. The plasma consists of singly charged ions of mass m_i and electrons of mass m_e, with equilibrium plasma density n_0. Assume that the ions are very cold in comparison to the electrons, $T_i \sim 0$, and that the electrons have a finite temperature T_e. Derive the dispersion relation for small-amplitude electrostatic ion-cyclotron waves propagating nearly perpendicular to B_0. Assume that the electrons move along the magnetic field and are in Boltzmann equilibrium.

When we linearize about equilibrium, we obtain

$$n_e \rightarrow n_{0e} + n_{1e}, \tag{6.137}$$

$$n_i \rightarrow n_{0i} + n_{1i}, \tag{6.138}$$

$$v_e \rightarrow v_{0e} + v_{1e}, \tag{6.130}$$

and

$$v_i \rightarrow v_{1i}. \tag{6.140}$$

The equations of continuity for the electrons and ions are

$$\frac{\partial}{\partial t} n_{e1} + n_{e0} \vec{\nabla} \cdot \vec{v}_{e1} = 0, \tag{6.141}$$

and

$$\frac{\partial}{\partial t} n_{i1} + n_{i0} \vec{\nabla} \cdot \vec{v}_{i1} = 0, \tag{6.142}$$

respectively. The fact that the electrons are in Boltzmann equilibrium implies that $n_e = n_{e0} \exp^{-\frac{eV}{kT_e}}$ or

$$n_{e1} = n_{e0} \frac{e\phi_1}{kT_e}, \tag{6.143}$$

where $\vec{E}_1 = -\nabla \phi_1$. Similarly, because we are looking for plane wave solutions, equation 6.142 implies that

$$n_{i1} = n_{i0} \frac{k}{\omega} v_{i1}. \tag{6.144}$$

From the problem definition we may assume that both n_0 and B_0 are constant and uniform, and that $v_0 = E_0 = 0$. The equations of motion for electrons and ions are

$$m_e n_{0e} \frac{\partial}{\partial t} \vec{v}_{e1} = -e n_{e0} \left[\vec{E}_1 + \vec{v}_{e1} \times \vec{B}_0 \right] + \vec{\nabla} \cdot \vec{P}_e, \tag{6.145}$$

and

$$m_i n_{0i} \frac{\partial}{\partial t} \vec{v}_{i1} = +en_{i0}\left[\vec{E}_1 + \vec{v}_{i1} \times \vec{B}_0\right], \qquad (6.146)$$

respectively. Because $m_e \ll m_i$ we assume $m_e = 0$. With the proper choice of geometry, B_o in the z-direction with k, E in the xy-plane, the vector equation of motion for the electrons reduces to

$$en_{e0}\left[E_{1,x} + v_{e1,y} B_0\right] - ikk_B T_e n_{e1} = 0, \qquad (6.147)$$

where we have used the definition of pressure provided by equation 6.67. The vector equation of motion for the ions gives

$$-i\omega m_i n_{0i} v_{i1,x} = +en_{i0}\left[-ik\phi_1 + v_{i1,y} B_0\right], \qquad (6.148)$$

and

$$-i\omega m_i n_{0i} v_{i1,y} = -en_{i0}\left[v_{i1,x} B_0\right]. \qquad (6.149)$$

Solving these last two equations for $v_{i1,x}$ gives

$$v_{i1,x} = \frac{ek\phi_1}{\omega m_i}\left(1 - \frac{e^2 B_0^2}{\omega^2 m_i^2}\right)^{-1}. \qquad (6.150)$$

By definition, $\Omega_i = \frac{eB_0}{m_i}$ is the ion cyclotron frequency, so

$$v_{i1,x} = \frac{ek\phi_1}{\omega m_i}\left(1 - \frac{\Omega_i^2}{\omega^2}\right)^{-1}. \qquad (6.151)$$

From equation 6.142,

$$\phi_1 = \frac{n_{e1}}{n_{e0}} \frac{kT_e}{e}, \qquad (6.152)$$

so

$$v_{i1,x}\left(1 - \frac{\Omega_i^2}{\omega^2}\right) = \frac{ek}{\omega m_i}\frac{n_{e1}}{n_{e0}}\frac{k_B T_e}{e}. \tag{6.153}$$

In the plasma approximation, $n_e \sim n_i$, so

$$\frac{n_{e1}}{n_{e0}} \approx \frac{n_{i1}}{n_{i0}} = \frac{k}{\omega}v_{i1,x}, \tag{6.154}$$

Using this in the preceding equation gives

$$v_{i1,x}\left(1 - \frac{\Omega_i^2}{\omega^2}\right) = \frac{ek}{\omega m_i}\frac{k_B T_e}{e}\frac{k}{\omega}v_{i1,x}, \tag{6.155}$$

or

$$\left(1 - \frac{\Omega_i^2}{\omega^2}\right) = \frac{k^2}{\omega^2}\frac{k_B T_e}{m_i}. \tag{6.156}$$

Rearranging terms gives

$$\omega^2 = \Omega_i^2 + k^2\left(\frac{k_B T_e}{m_i}\right) = \Omega_i^2 + k^2 v_s^2, \tag{6.157}$$

which is the dispersion relation for electrostatic ion cyclotron waves.

6.6 Magnetohydrodynamics

If we allow for the possibility of collisions in the plasma and reexamine the Vlasov equation, we would find that integrating over v would still provide the same equation of continuity for electrons and ions (eq. 5.12). However, the momentum equation would pick up a collisional term of the form

$$m_s n_s \frac{\partial \vec{v}_s}{\partial t} + m_s n_s \left(\vec{v}_s \cdot \vec{\nabla}\right)\vec{v}_s = -\nabla \vec{P}_s + q_s n_s \left(\vec{E} + \frac{1}{c}\vec{v}_s \times \vec{B}\right) + \vec{k}_s(z), \tag{6.158}$$

where

$$\vec{k}_s = m_s \int \vec{v} \left(\frac{\partial f_s}{\partial t} \right)_c d\vec{v}. \tag{6.159}$$

We can combine the electron and the ion equations to obtain a one-fluid model. This fluid is characterized by the following:

Mass density

$$\rho_m(\vec{x}) = m_e n_e(\vec{x}) + m_i n_i(\vec{x}) \cong m_i n_i(\vec{x}) \tag{6.160}$$

Charge density

$$\rho_c(\vec{x}) = q_e n_e(\vec{x}) + q_i n_i(\vec{x}) = e(n_i(\vec{x}) - n_e(\vec{x})) \tag{6.161}$$

Center of mass fluid flow velocity

$$\vec{v} = \frac{1}{\rho_m} \left(m_i n_i \vec{v}_i + m_e n_e \vec{v}_e \right) \tag{6.162}$$

Current density

$$\vec{J} = q_i n_i \vec{v}_i + q_e n_e \vec{v}_e \tag{6.163}$$

Total pressure

$$P = P_e + P_i. \tag{6.164}$$

If we multiply the ion continuity equation by m_i and the electron continuity equation by m_e and add the two, we obtain the mass conservation law,

$$\frac{\partial \rho_m}{\partial t} + \vec{\nabla} \cdot (\rho_m \vec{v}) = 0. \tag{6.165}$$

If we repeat this procedure with q_s instead of m_s, we obtain the charge conservation law

$$\frac{\partial \rho_c}{\partial t} + \vec{\nabla} \cdot (\vec{J}) = 0. \tag{6.166}$$

Consider next the equations of motion. If we regard v_s and $\dfrac{\partial n_s}{\partial t}$ as small quantities, neglect their products, and recall that $k_e = -k_i$, we can add the two equations of motion to obtain force equations

$$\rho_m \frac{\partial \vec{v}}{\partial t} = -\vec{\nabla} \cdot \vec{P} + \rho_c \vec{E} + \frac{1}{c} \vec{J} \times \vec{B}. \tag{6.167}$$

If we now multiply the force equations by q_s / m_s, add the ion version to the electron version, and neglect quadratic terms in v_s and $\dfrac{\partial n_s}{\partial t}$, we obtain

$$\frac{\partial \vec{J}}{\partial t} = -\frac{e}{m_i} \vec{\nabla} \cdot \vec{P}_i + \frac{e}{m_e} \vec{\nabla} \cdot \vec{P}_e + \left(\frac{e^2 n_e}{m_e} + \frac{e^2 n_i}{m_i} \right) \vec{E}$$

$$+ \frac{e^2 n_e}{m_e c} \vec{v}_e \times \vec{B} + \frac{e^2 n_i}{m_i c} \vec{v}_i \times \vec{B} + \left(\frac{e}{m_i} + \frac{e}{m_e} \right) \vec{k}_i. \tag{6.168}$$

We substitute

$$\frac{e^2 n_e}{m_e c} \vec{v}_e = \frac{e}{m_e c} \left(n_e e \vec{v}_e - n_i e \vec{v}_i \right) + \frac{e^2}{m_e m_i c} \left(m_i n_i \vec{v}_i \right), \tag{6.169}$$

which is by definition also equal to

$$\frac{e^2 n_e}{m_e c} \vec{v}_e = -\frac{e}{m_e c} \vec{J} + \frac{e^2}{m_e m_i c} \left(\rho_m \vec{v} \right). \tag{6.170}$$

If we neglect m_i^{-1} as being much smaller than m_e^{-1} and assume that $P_e \sim P_i = P/2$ and $n_i \sim n_e$, we find that equation 6.168 reduces to

$$\frac{\partial \vec{J}}{\partial t} = \frac{e}{2 m_e} \nabla P + \frac{e^2 \rho_m}{m_e m_i} \left(\vec{E} + \frac{1}{c} \vec{v} \times \vec{B} \right) + \frac{e}{m_e c} \vec{J} \times \vec{B} + \frac{e}{m_e} \vec{k}_i. \tag{6.171}$$

Recall that k_i is the change in ion momentum due to collisions with electrons. It is reasonable to assume that k_i is a function of the relative velocity, $v_i - v_e$, between the two species. That is,

$$\bar{k}_i = C_1(\bar{v}_i + \bar{v}_e) = C_2 \bar{J} = -\frac{\rho_m e}{m_i \sigma} \bar{J}. \tag{6.172}$$

Using this relation we obtain the generalized form of Ohm's law,

$$\frac{m_e m_i}{e^2 \rho_m} \frac{\partial \bar{J}}{\partial t} = \frac{m_i}{2\rho_m e} \nabla P + \left(\bar{E} + \frac{1}{c} \bar{v} \times \bar{B} \right) - \frac{m_i}{\rho_m e c} \bar{J} \times \bar{B} + \frac{1}{\sigma} \bar{J}. \tag{6.173}$$

Equations 6.165, 6.166, 6.167, and 6.173 are the magnetohydrodynamics (MHD) equations. For low frequencies we may ignore $\dfrac{\partial \bar{J}}{\partial t}$; for low temperatures, ignore $\vec{\nabla} \cdot \vec{P}$; when the current is small, ignore $\bar{J} \times \bar{B}$ in comparison to $\bar{v} \times \bar{B}$. Ohm's law becomes

$$\left(\bar{E} + \frac{1}{c} \bar{v} \times \bar{B} \right) - \frac{1}{\sigma} \bar{J} = 0. \tag{6.174}$$

When $n_i \sim n_e$, we obtain the ideal MHD equations,

$$\frac{\partial \rho_m}{\partial t} + \vec{\nabla} \cdot (\rho_m \bar{v}) = 0, \tag{6.175}$$

$$\rho_m \frac{\partial \bar{v}}{\partial t} = \nabla P + \frac{1}{c} \bar{J} \times \bar{B}, \tag{6.176}$$

$$\vec{\nabla} \times (\bar{v} \times \bar{B}) = \frac{\partial \vec{B}}{\partial t}, \tag{6.177}$$

$$\vec{\nabla} \times \bar{B} = \frac{4\pi}{c} \bar{J}. \tag{6.178}$$

6.7 Bibliography

Chen, F. F. *Introduction to Plasma Physics and Controlled Fusion*. vol. 1: *Plasma Physics*. 2d ed. New York: Plenum Press, 1984.

Nicholson, D. *Introduction to Plasma Theory*. New York: John Wiley & Sons, 1983.

7 Relativity

7.1 Special Relativity

7.1.1 The Postulates of Special Relativity

Postulate 1:
Valid physical laws should apply equally well for any two observers in uniform relative motion, each being equally entitled to consider himself at rest, and neither being able to detect within his own system of reference any effects of motion ascribed to it by the other.

Postulate 2:
All observers, even when in uniform relative motion, will find the same value c for the speed of light in empty space.

7.1.2 The Lorentz Transformations

Consider two arbitrary inertial frames S and S' in which Cartesian coordinate lattices have been set up. It is assumed that, relative to frame S, the frame S' is moving to the right with velocity v, as shown in figure 7.1. The transformation between these coordinates must be linear and S and S' assign equal and opposite velocities to each other. By linearity

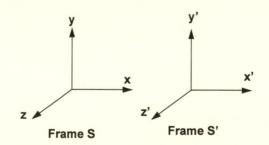

Figure 7.1 Two arbitrary inertial frames S and S'.

$$y' = \alpha y, \tag{7.1}$$

and

$$z' = \beta z. \tag{7.2}$$

To determine α, reverse the direction of the x and z axes in S and S' by performing a rotation about the y-axis. This interchanges the role of S and S' (an observer in S' would conclude that S is moving with velocity v), so that the primed and unprimed symbols switch positions. Therefore, equation 7.1 is equivalent to

$$y = By', \tag{7.3}$$

and $\alpha = \pm 1$. We choose $\alpha = 1$ because as $v \to 0$, $y \to y'$. A similar argument shows $\beta = 1$. Because x' must be linear with respect to x, and since $x = vt$ implies $x' = 0$, we require

$$x' = \gamma(x - vt). \tag{7.4}$$

Since $x' = -vt'$ gives $x = 0$ we also require

$$x = \gamma'(x' + vt'). \tag{7.5}$$

If we reverse the x and z axes, so that $x \to -x$ and $x' \to -x'$, equation 7.4 becomes

$$x' = \gamma(x + vt). \tag{7.6}$$

However, since the roles of x and x' have been reversed, equation 7.5 becomes

$$x' = \gamma'(x + vt), \tag{7.7}$$

thus $\gamma = \gamma'$. By the second postulate, $x = ct$ implies $x' = ct'$. In other words,

$$ct' = \gamma t(c - v), \tag{7.8}$$

must also equal

$$ct = \gamma t'(c + v). \tag{7.9}$$

From these two equations we find that

$$\gamma = \gamma(v) = \frac{1}{\left(1 - \dfrac{v^2}{c^2}\right)^{1/2}}. \tag{7.10}$$

The Lorentz transformations are then given by

$$x' = \gamma(x - vt), \tag{7.11}$$

$$y' = y, \tag{7.12}$$

$$z' = z, \tag{7.13}$$

and

$$t' = \gamma\left(t - \frac{vx}{c^2}\right). \tag{7.14}$$

If we define

$$\frac{v}{c} = \beta = \tanh\xi, \tag{7.15}$$

and

$$\gamma = \cosh\xi, \tag{7.16}$$

then

$$x' = -t\sinh\xi + x\cosh\xi, \tag{7.17}$$

and

$$t' = t\cosh\xi - x\sinh\xi. \tag{7.18}$$

Therefore, the Lorentz transformations are simply a rotation in "spacetime" because

$$\begin{pmatrix} t' & x' & y' & z' \end{pmatrix} = \begin{pmatrix} \cosh\xi & -\sinh\xi & 0 & 0 \\ -\sinh\xi & \cosh\xi & 0 & 0 \\ 0 & 0 & 1 & 0 \\ 0 & 0 & 0 & 1 \end{pmatrix} \begin{pmatrix} t \\ x \\ y \\ z \end{pmatrix}. \qquad (7.19)$$

7.1.2.1 Length Contraction

If we consider the example of a rod of length $\Delta x'$ in frame S' (its rest frame) that is moving uniformly with respect to frame S, equation 7.11 indicates that the length of the rod in frame S is given by

$$\Delta x' = \gamma \Delta x, \qquad (7.20)$$

or

$$L = \frac{L_0}{\gamma}, \qquad (7.21)$$

where L_0 is the length of the rod in its rest frame, S', and L is the length of the rod in the moving frame (as perceived by the rod). As a result, the length of a body moving in uniform motion is reduced by a factor of γ in the direction of its motion. Other dimensions of the object remain unaffected.

The interpretation of equation 7.21 is that a man running at the speed 0.866 c, so that $\gamma = 2$, could fit a 20 m pole into a 10 m garage. This result gives rise to a well-known paradox. Couldn't the man holding the pole claim that he is at rest and that he sees a 5 m garage approaching? The answer is yes, but the pole would still fit into the garage. After the garage had completely overlapped the pole, there would still be 15 m of pole sticking out the door. At this point the end of the pole in the garage would start to be compressed toward the other end. Because of the finite speed at which this compression wave must travel, the opposite end of the pole would not know that the pole was being compressed until the compression wave reached it. Because the garage must travel only 15 m to reach the far end of the pole, while the signal from the back end must travel 20 m, the garage will win the race provided that its velocity is at least (15 m)/(20 m) = 0.75 c or greater. Because v is 0.866 c, the garage will be able to cover the pole.

7.1.2.2 Time Dilation

Consider two events that occur at a fixed point in the moving frame S' a time interval $\Delta t'$ apart. Because the events occur at the same point, equation

7.11 requires that $\Delta x = v\Delta t$. The time difference between the events in the moving frame, S', and the rest frame, S, is then given by equation 7.14 as

$$\Delta t' = \gamma\left(\Delta t - \frac{v\Delta x}{c^2}\right) = \gamma\left(\Delta t - \frac{v^2\Delta t}{c^2}\right) = \frac{\Delta t}{\gamma}, \tag{7.22}$$

or

$$T_0 = \frac{T}{\gamma}, \tag{7.23}$$

where T_0 is the time interval in the moving frame and T is the time interval in the rest frame. Because the time interval in the moving frame is a factor of γ smaller than in the rest frame, we say that time in the moving frame slows down by a factor of γ with respect to the rest frame.

This leads to a second paradox often called the *twin paradox*. One twin gets on a rocket ship and accelerates to 0.866 c, so that $\gamma = 2$. If this twin travels for one year, as measured by his moving clock, then turns around and heads back at the same speed, the moving twin will report that the trip required two years. However, her nonmoving twin on Earth would report that the time required for the journey was four years. Can the twin on the spaceship argue that she was at rest, and it was the twin on Earth who was moving? In this case the answer is no. In order to make a transition from a rest frame to a moving frame, there must be an acceleration. Whichever twin feels an acceleration can no longer argue that her frame is the rest frame. In other words, the twin on the spaceship cannot argue that it was the other twin who was moving, so there is no paradox.

7.1.3 Velocity Transformation and Proper Acceleration

In the two frames S and S' we define

$$u = \left(\frac{\Delta x}{\Delta t}, \frac{\Delta y}{\Delta t}, \frac{\Delta z}{\Delta t}\right) = (u_1, u_2, u_3), \tag{7.24}$$

and

$$u' = \left(\frac{\Delta x'}{\Delta t}, \frac{\Delta y'}{\Delta t}, \frac{\Delta z'}{\Delta t}\right) = (u'_1, u'_2, u'_3). \tag{7.25}$$

From the Lorentz transformations we have

$$\Delta x' = \gamma(\Delta x - v\Delta t), \tag{7.26}$$

$$\Delta y' = \Delta y, \tag{7.27}$$

$$\Delta z' = \Delta z, \tag{7.28}$$

and

$$\Delta t' = \gamma\left(\Delta t - \frac{v\Delta x}{c^2}\right). \tag{7.29}$$

Substituting equations 7.26–7.29 into equation 7.25 gives

$$u'_1 = \frac{\Delta x'}{\Delta t'} = \frac{\gamma(\Delta x - v\Delta t)}{\gamma\left(\Delta t - \frac{v\Delta x}{c^2}\right)} = \frac{\frac{\Delta x}{\Delta t} - v}{1 - \frac{v\Delta x}{c^2\Delta t}} = \frac{u_1 - v}{1 - \frac{vu_1}{c^2}}, \tag{7.30}$$

$$u'_2 = \frac{\Delta y'}{\Delta t'} = \frac{\Delta y}{\gamma\left(\Delta t - \frac{v\Delta x}{c^2}\right)} = \frac{1}{\gamma}\frac{\frac{\Delta y}{\Delta t}}{\left(1 - \frac{v\Delta x}{c^2\Delta t}\right)} = \frac{u_2}{\gamma\left(1 - \frac{vu_1}{c^2}\right)}, \tag{7.31}$$

and

$$u'_3 = \frac{\Delta z'}{\Delta t'} = \frac{\Delta z}{\gamma\left(\Delta t - \frac{v\Delta x}{c^2}\right)} = \frac{1}{\gamma}\frac{\frac{\Delta z}{\Delta t}}{\left(1 - \frac{v\Delta x}{c^2\Delta t}\right)} = \frac{u_3}{\gamma\left(1 - \frac{vu_1}{c^2}\right)}. \tag{7.32}$$

We define proper acceleration to be the acceleration of a particle relative to its instantaneous rest frame. We consider one-dimensional motion in the x-direction. The inverse of equation 7.30 is

$$u_1 = \frac{u'_1 + v}{1 - \frac{vu'_1}{c^2}}. \tag{7.33}$$

Differentiating this expression gives

$$\Delta u_1 = \frac{1 - \dfrac{v^2}{c^2}}{\left(1 + \dfrac{vu'_1}{c^2}\right)^2} \Delta u'_1. \tag{7.34}$$

If S' is the instantaneous rest frame of the particle, then $u' = 0$ and $u = v$ momentarily, and equation 7.34 reduces to

$$\Delta u_1 = 1 - \frac{v^2}{c^2} \Delta u'_1 = \frac{\Delta u'_1}{\gamma^2}. \tag{7.35}$$

Applying the same conditions to equation 7.14 gives

$$\Delta t' = \frac{\Delta t}{\gamma}. \tag{7.36}$$

Equations 7.35 and 7.36 can be combined to obtain the proper acceleration,

$$\alpha = \frac{du'}{dt'} = \gamma^3 \frac{du}{dt} = \frac{d}{dt}(\gamma u). \tag{7.37}$$

Example 7.1
A rocket has a constant acceleration g in its instantaneous inertial frame. If it starts from rest near Earth, how far from Earth will the rocket be in forty years as measured on Earth? How far will the rocket be after forty years as measured in the rocket?

The velocity in the rocket's rest frame is

$$v = \frac{dx}{dt} = \alpha\tau. \tag{7.38}$$

From equations 7.37, 7.15, and 7.16 we have

$$\alpha = \frac{d}{dt}(\gamma u) = \frac{d}{dt}(c \cosh\xi \tanh\xi) = c \cosh\xi \frac{d\xi}{dt} = c\gamma \frac{d\xi}{dt}, \tag{7.39}$$

or

$$\xi = \frac{\alpha\tau}{c}. \tag{7.40}$$

Also from equation 7.15 we have

$$\beta\gamma = \frac{v}{c}\gamma = \sinh\xi = \sinh\left(\frac{\alpha\tau}{c}\right). \tag{7.41}$$

Because $dt = \gamma d\tau$, equation 7.40 also gives

$$\frac{\gamma}{c}\frac{dx}{dt} = \sinh\left(\frac{\alpha\tau}{c}\right) = \frac{1}{c}\frac{dx}{d\tau}, \tag{7.42}$$

which reduces to

$$dx = c\sinh\left(\frac{\alpha\tau}{c}\right)d\tau, \tag{7.43}$$

or

$$x = \frac{c^2}{\alpha}\cosh\left(\frac{\alpha\tau}{c}\right) + \text{const.} \tag{7.44}$$

If we require $x = 0$ when $t = 0$, then

$$x = \frac{c^2}{\alpha}\left[\cosh\left(\frac{\alpha\tau}{c}\right) - 1\right] \tag{7.45}$$

is the distance traveled in Earth's frame. By definition, the distance traveled in the rocket's frame is simply

$$x' = \frac{1}{2}\alpha\tau^2. \tag{7.46}$$

Using the constants defined in the problem, we see that $x' \sim 7.8 \times 10^{18}$ m, while $x \sim 3.44 \times 10^{33}$ m.

7.1.4 Relativistic Energy and Momentum

For a particle with speed $v \ll c$, the classical definitions of momentum and energy apply. That is,

$$\vec{p} = m\vec{v}, \tag{7.47}$$

and

$$E = E(0) + \frac{1}{2}mv^2. \tag{7.48}$$

Expressions for the relativistic momentum and energy of a particle must be consistent with the Lorentz transformations and reduce to the above for $v \ll c$. We define the relativistic momentum,

$$\vec{p} = m(v)\vec{v}, \tag{7.49}$$

where $m(0) = m$.

To determine the form of $m(v)$, we consider a glancing collision between two identical particles in frames S and S', which are defined to be the *precollision rest frames* of the two particles, labeled A and A', respectively. By symmetry, for each particle the postcollision component of velocity that is perpendicular to the original direction of travel will be identical in magnitude, but opposite in direction. This constraint holds true in each frame. Equation 7.31 provides the final transverse velocity of particle A' relative to S as

$$u'_\perp = \frac{u_\perp}{\gamma\left(1 + \dfrac{vu'}{c^2}\right)}, \tag{7.50}$$

where u' is the postcollision component of velocity that is parallel to the original direction of travel of A' in S'. Transverse momentum conservation in frame S then requires

$$Mu = M'u'_\perp = \frac{M'u_\perp}{\gamma\left(1 + \dfrac{vu'}{c^2}\right)}, \tag{7.51}$$

where M and M' are the postcollision masses of A and A', respectively, as measured in S. As u and $u'_\perp \to 0$, then $M \to m(0)$ and $M' \to m(v)$, so that equation 7.51 reduces to

$$m(v) = m(0)\gamma(v). \tag{7.52}$$

Similarly, from the definition of kinetic energy we have

$$KE = \int_0^u \sum F ds = \int_0^u \frac{d(\gamma m u)}{dt} ds = \int_0^u u \, d(\gamma m u).\tag{7.53}$$

It is straightforward to see that

$$d(\gamma m u) = m \gamma^3 du,\tag{7.54}$$

so that

$$KE = \int_0^u m \gamma^3 u \, du = mc^2(\gamma - 1).\tag{7.55}$$

The quantity mc^2 is defined as the rest energy of the particle. The total energy is the sum of the kinetic energy and the rest energy and is given by

$$E = \gamma mc^2 = \left(p^2 c^2 + m^2 c^4\right)^{1/2}.\tag{7.56}$$

It is convenient to define the energy-momentum four-vector,

$$p^0 = m\eta^0 = \frac{E}{c},\tag{7.57}$$

and

$$p^i = m\eta^i,\tag{7.58}$$

which transform properly under the Lorentz transformations as

$$p^{0\prime} = \gamma p^0 - \gamma\beta p^1,\tag{7.59}$$

$$p^{1\prime} = \gamma p^1 - \gamma\beta p^0,\tag{7.60}$$

$$p^{2\prime} = p^2,\tag{7.61}$$

and

$$p^{3\prime} = p^3.\tag{7.62}$$

7.1.5 Relativistic Electrodynamics

If we define the current density in terms of a four-vector,

$$J^\mu = \left(c\rho, J_x, J_y, J_z\right),$$ (7.63)

we find that the continuity equation

$$\nabla \cdot \vec{J} = -\frac{\partial \rho}{\partial t}$$ (7.64)

reduces to

$$\partial_\mu J^\mu = 0.$$ (7.65)

Similarly, we note that the E and B fields are expressed in terms of potentials

$$\vec{E} = -\frac{1}{c}\frac{\partial \vec{A}}{\partial t} - \nabla\phi,$$ (7.66)

and

$$\vec{B} = \nabla \times \vec{A}.$$ (7.67)

We have

$$E_x = -\frac{1}{c}\frac{\partial A_x}{\partial t} - \frac{\partial \phi}{\partial x} = -\left(\partial^0 A^1 - \partial^1 A^0\right),$$ (7.68)

and

$$B_x = -\left(\partial^y A^z - \partial^z A^y\right).$$ (7.69)

This suggests defining

$$F^{\alpha\beta} = \partial^\alpha A^\beta - \partial^\beta A^\alpha,$$ (7.70)

where

$$F^{\alpha\beta} = \begin{pmatrix} 0 & -E_x & -E_y & -E_z \\ E_x & 0 & -B_z & B_y \\ E_y & B_z & 0 & -B_x \\ E_z & -B_y & B_x & 0 \end{pmatrix}. \tag{7.71}$$

is termed the field strength tensor. The dual field strength tensor is defined by

$$\tilde{F}^{\alpha\beta} = \frac{1}{2}\varepsilon^{\alpha\beta\chi\delta}F_{\chi\delta} = \frac{1}{2}\varepsilon^{\alpha\beta\chi\delta}g_{\chi\kappa}F^{\kappa\lambda}g_{\lambda\delta}, \tag{7.72}$$

or

$$\tilde{F}^{\alpha\beta} = \begin{pmatrix} 0 & -B_x & -B_y & -B_z \\ B_x & 0 & E_z & -E_y \\ B_y & -E_z & 0 & E_x \\ B_z & E_y & -E_x & 0 \end{pmatrix}. \tag{7.73}$$

The homogeneous Maxwell equations are

$$\partial_\nu F^{\mu\nu} = J^\mu, \tag{7.74}$$

while the inhomogeneous Maxwell equations are

$$\partial_\nu \tilde{F}^{\mu\nu} = 0. \tag{7.75}$$

7.2 General Relativity

7.2.1 The Structure of Spacetime

Newtonian mechanics separated events in "space" from events in "time." That is, two spatial events would be said to occur some distance (dx, dy, dz) apart, while two temporal events would be separated by some time interval dt. In relativity, there is no distinction between space and time, and we speak only of "spacetime." In spacetime, events are separated by the spacetime interval

$$ds^2 = c^2 dt^2 - dx^2 - dy^2 - dz^2, \tag{7.76}$$

which is seen to be invariant under the Lorentz transformations. In tensor notation, equation 7.76 is written as

$$ds^2 = g_{\mu\nu} dx^\mu dx^\nu, \tag{7.77}$$

with

$$g_{\mu\nu} = \begin{pmatrix} 1 & 0 & 0 & 0 \\ 0 & -1 & 0 & 0 \\ 0 & 0 & -1 & 0 \\ 0 & 0 & 0 & -1 \end{pmatrix} \tag{7.78}$$

being the metric, or Minkowski, tensor. Equations 7.76 and 7.78 are specific to Cartesian coordinates, while equation 7.77 is general to any curvilinear coordinates. From the definition of the metric tensor (eqs. 1.144 and 7.77), we note that

$$g_{\mu\nu} \frac{dx^\mu}{d\tau} \frac{dx^\nu}{d\tau} = 1, \tag{7.79}$$

and is therefore a constant of the motion. A "geodesic" is defined as the line of shortest length between any two points in spacetime. The motion of particles in free fall will be along a geodesic. We define the Lagrangian function

$$L = \sqrt{g_{\mu\nu} \frac{dx^\mu}{d\tau} \frac{dx^\nu}{d\tau}}. \tag{7.80}$$

The condition for the stationary value (eq. 2.68) is

$$\delta \int L \, ds = 0, \tag{7.81}$$

which has solution

$$\frac{d}{d\tau}\frac{\partial L}{\partial\left(\dfrac{dx^\gamma}{d\tau}\right)} - \frac{\partial L}{\partial x^\gamma} = 0. \tag{7.82}$$

These terms are

$$\frac{d}{d\tau}\frac{\partial L}{\partial\left(\dfrac{dx^\gamma}{d\tau}\right)} = \frac{d}{d\tau}\frac{\partial}{\partial\left(\dfrac{dx^\gamma}{d\tau}\right)}\left[\left(g_{\mu\nu}\frac{dx^\mu}{d\tau}\frac{dx^\nu}{d\tau}\right)^{1/2}\right], \tag{7.83}$$

and

$$\frac{\partial L}{\partial x^\gamma} = \frac{\partial}{\partial x^\gamma}\left[\left(g_{\mu\nu}\frac{dx^\mu}{d\tau}\frac{dx^\nu}{d\tau}\right)^{1/2}\right]. \tag{7.84}$$

Making use of equation 7.79, we can verify that equation 7.82 reduces to

$$\frac{d}{d\tau}\left(g_{\mu\nu}\frac{dx^\mu}{d\tau}\right) - \frac{1}{2}g_{\alpha\beta,\mu}\frac{dx^\alpha}{d\tau}\frac{dx^\beta}{d\tau} = 0, \tag{7.85}$$

which is known as the *geodesic equation*. If we use the definition of the four-velocity

$$u^\alpha = \frac{dx^\alpha}{ds}, \tag{7.86}$$

equation 7.85 can be written as

$$du^\nu + \Gamma^\nu_{\alpha\beta}u^\alpha dx^\beta = 0, \tag{7.87}$$

using the definition of the Christoffel symbol provided in equation 1.151. It is always possible to find coordinates such that any single given point of the metric is that of flat spacetime. The coordinates which, at a given point, make the Christoffel symbols and the first derivative of the metric zero are called geodesic coordinates. A particle placed at the "flat spacetime" point will move with constant velocity or remain at rest as seen in the *geodesic coordinates*. Note that the derivatives of the metric are zero only at one point for one instant of time.

If we have a vector field, A_β, and take covariant derivatives (eq. 1.151) with respect to μ and ν, we would find that

$$A_{\beta;\mu;\nu} \neq A_{\beta;\nu;\mu}. \tag{7.88}$$

The difference is given by

$$A_{\beta;\mu;\nu} - A_{\beta;\nu;\mu} = R^\alpha_{\beta\mu\nu} A_\alpha, \tag{7.89}$$

where

$$R^\alpha_{\beta\mu\nu} = -\Gamma^\nu_{\alpha\beta,\mu} + \Gamma^\alpha_{\beta\nu,\mu} + \Gamma^\sigma_{\beta\nu}\Gamma^\alpha_{\sigma\mu} - \Gamma^\sigma_{\beta\mu}\Gamma^\alpha_{\sigma\nu}, \tag{7.90}$$

is defined to be the *Riemann curvature tensor*, because the tensor carries information about the local curvature of spacetime. A contraction of the Riemann tensor

$$R_{\beta\mu} = R^\alpha_{\beta\mu\alpha} \tag{7.91}$$

is called the *Ricci tensor*. A further contraction,

$$R = R^\alpha_\alpha = R^{\alpha\beta}_{\beta\alpha}, \tag{7.92}$$

produces the curvature scalar.

7.2.2 The Vacuum Field Equations

By comparison with Poisson's equation, the standard Newtonian gravitational field equation can be written in the form

$$\nabla^2\varphi = 4\pi G\rho, \tag{7.93}$$

which can be integrated to give the usual relation

$$\varphi = \frac{GM}{r}. \tag{7.94}$$

These equations predict the acceleration due to gravity as

$$\vec{g} = -\frac{\partial \varphi}{\partial r} \hat{r} . \tag{7.95}$$

Equation 7.93 may be generalized to tensor form in order to obtain the field equations of general relativity, which would relate the curvature of spacetime to the presence of masses. By inspection, the most general second-order tensor that contains information on mass distributions is the stress-energy tensor, $T_{\mu\nu}$. Consequently, the source term (the right-hand side of 7.93) will be a function of $T_{\mu\nu}$. In more general terms, equation 7.95 may be written as

$$g_i = -\frac{\partial \varphi}{\partial x_i} = -\varphi_i . \tag{7.96}$$

The relative acceleration of two particles separated by a small three-vector x^i is

$$\frac{d^2 x^i}{dt^2} = \Delta g_i = -\varphi_{ij} \eta^j . \tag{7.97}$$

Consequently, the second derivatives of the potential relate to the intrinsic field. This indicates that the left-hand side of the proper field equations must contain some form of the Riemann curvature tensor. The most general relation that satisfies our requirements and places no unwarranted restrictions on $T_{\mu\nu}$ is

$$R_{\mu\nu} - \frac{1}{2} g_{\mu\nu} R = -8\pi G T_{\mu\nu}, \tag{7.98}$$

which is the Einstein field equation.

7.2.3 The Schwarzschild Solution

The solution to the Einstein field equation for a spherically symmetric mass distribution at rest is called the *Schwarzschild solution*. To meet the requirement for spherical symmetry, any solution to equation 7.98 must involve only those combinations of x, y, z, dx, dy, and dz that are left invariant by spatial rotations. We can see that the only such combinations are

$$r = \left(x^2 + y^2 + z^2 \right)^{1/2} , \tag{7.99}$$

$$dr^2 + r^2 d\theta^2 + r^2 \sin^2 \theta d\phi^2 = dx^2 + dy^2 + dz^2, \qquad (7.100)$$

and

$$r dr = x dx + y dy + z dz. \qquad (7.101)$$

Consequently, in spherical coordinates the spacetime interval (eq. 7.76) must be of the form

$$ds^2 = A(r)dt^2 - B(r)\left(dr^2 + r^2 d\theta^2 + r^2 \sin^2 \theta d\phi^2\right) - C(r)dr^2, (7.102)$$

where $A(r)$, $B(r)$, and $C(r)$ are unknown functions of r. A change of coordinates, where $r\sqrt{B(r)} \to r'$ with all other variables remaining unchanged, gives

$$ds^2 = A'(r')dt'^2 - B'(r')dr'^2 r'^2 d\theta'^2 - r'^2 \sin^2 \theta' d\phi'^2. \qquad (7.103)$$

By convention the metric tensor is written in the form

$$g_{\mu\nu} = \begin{pmatrix} e^N & 0 & 0 & 0 \\ 0 & -e^L & 0 & 0 \\ 0 & 0 & -r^2 & 0 \\ 0 & 0 & 0 & -r^2 \sin^2 \theta \end{pmatrix}. \qquad (7.104)$$

In a vacuum (no sources) the Einstein field equation becomes

$$R_\mu^\nu - \frac{1}{2}\delta_\mu^\nu R = 0. \qquad (7.105)$$

Plugging in all nonzero Christoffel symbols gives

$$R_0^0 - \frac{1}{2}R = -e^{-L}\left(\frac{L'}{r} - \frac{1}{r^2}\right) - \frac{1}{r^2} = 0, \qquad (7.106)$$

$$R_1^1 - \frac{1}{2}R = e^{-L}\left(\frac{N'}{r} + \frac{1}{r^2}\right) - \frac{1}{r^2} = 0, \qquad (7.107)$$

$$R_2^2 - \frac{1}{2}R = e^{-L}\left(\frac{N''}{2} - \frac{L'N'}{4} + \frac{N'^2}{4} + \frac{N'-L'}{2r}\right) = 0, \quad (7.108)$$

and

$$R_3^3 - \frac{1}{2}R = e^{-L}\left(\frac{N''}{2} - \frac{L'N'}{4} + \frac{N'^2}{4} + \frac{N'-L'}{2r}\right) = 0. \quad (7.109)$$

Equation 7.106 has the unique solution,

$$e^L = \frac{1}{1-\dfrac{C}{r}}, \quad (7.110)$$

where C is a constant. Using this relation, it can be shown that

$$e^N = 1 - \frac{C}{r}. \quad (7.111)$$

Comparison with linear field theory indicates that $C = 2GM$, so the Schwarzschild solution is

$$ds^2 = \left(1 - \frac{2GM}{r}\right)dt^2 - \frac{dr^2}{\left(1 - \dfrac{2GM}{r}\right)} - r^2 d\theta^2 - r^2 \sin^2\theta d\phi^2. \,(7.112)$$

7.2.4 Predictions of General Relativity

7.2.4.1 Gravitational Red Shift

From the definition of the metric tensor, it follows that for a light pulse

$$ds = c\sqrt{-g_{00}(r)}\,dt. \quad (7.113)$$

Consequently, the frequency of the light pulse is given by

$$\nu_i = \frac{1}{\Delta t_i} = \frac{c\sqrt{-g_{00}(r_i)}}{\Delta s}. \quad (7.114)$$

Because Δs is an invariant, we have

$$\Delta s = \frac{c\sqrt{-g_{00}(r_f)}}{v_f} = \frac{c\sqrt{-g_{00}(r_i)}}{v_i}, \tag{7.115}$$

where r_i and r_f are different locations along the same geodesic. From equation 7.115 we see that

$$\frac{v_i}{v_f} = \sqrt{\frac{-g_{00}(r_i)}{-g_{00}(r_f)}}. \tag{7.116}$$

The frequency shift is therefore

$$\Delta v = v_f - v_i = v_i \left(\sqrt{\frac{-g_{00}(r_i)}{-g_{00}(r_f)}} - 1 \right). \tag{7.117}$$

From equation 7.104 and the definition $C = 2GM$, we see that

$$g_{00}(r) = 1 - \frac{2GM}{r}, \tag{7.118}$$

so

$$\Delta v \approx v_i(GM) \left(\frac{1}{r_i} - \frac{1}{r_f} \right). \tag{7.119}$$

7.2.4.2 Precession of Perihelion of Orbits

The equation of motion for a planet in a gravitational field must satisfy the geodesic equation (eq. 7.85). Using equation 7.104 in equation 7.85 gives

$$\frac{d^2r}{d\tau^2} + \frac{1}{2}\frac{dL}{d\tau}\left(\frac{dt}{d\tau}\right)^2 - re^{-L}\left(\frac{d\theta}{d\tau}\right)^2 + \frac{1}{2}e^{N-L}\frac{dN}{d\tau}\left(c\frac{dt}{d\tau}\right)^2 = 0, \tag{7.120}$$

$$\frac{d^2\theta}{d\tau^2} + \frac{2}{r}\frac{dr}{d\tau}\frac{d\theta}{d\tau} - \cos\theta\sin\theta\left(\frac{d\phi}{d\tau}\right)^2 = 0, \qquad (7.121)$$

$$\frac{d^2\phi}{d\tau^2} + \frac{2}{r}\frac{dr}{d\tau}\frac{d\phi}{d\tau} + 2\cot\theta\frac{d\phi}{d\tau}\frac{d\theta}{d\tau} = 0, \qquad (7.122)$$

and

$$\frac{d^2t}{d\tau^2} + \frac{dN}{dr}\frac{dr}{d\tau}\frac{dt}{d\tau} = 0. \qquad (7.123)$$

If $\theta = \pi/2$ and $d\theta/dt = 0$ initially, then $d^2\theta/dt^2 = 0$ and the motion is confined to the $\theta = \pi/2$ plane for all time. With this constraint, equation 7.122 simplifies to

$$\frac{d^2\phi}{d\tau^2} + \frac{2}{r}\frac{dr}{d\tau}\frac{d\phi}{d\tau} = 0, \qquad (7.124)$$

which has solution

$$\frac{d\phi}{d\tau} = \frac{A}{r^2}, \qquad (7.125)$$

with A being a constant. Similarly, equation 7.123 has solution

$$\frac{dt}{d\tau} = Be^{-N}, \qquad (7.126)$$

with B being a constant. If we explicitly write out equation 7.79, we obtain

$$e^N\left(\frac{dt}{d\tau}\right)^2 - e^L\left(\frac{dr}{d\tau}\right)^2 - r^2\left(\frac{d\theta}{d\tau}\right)^2 - r^2\sin^2\theta\left(\frac{d\phi}{d\tau}\right)^2 = 1. \quad (7.127)$$

Noting that

$$e^N = e^{-L} = 1 - \frac{2MG}{r}, \qquad (7.128)$$

and applying the constraint $\theta = \pi/2$, equation 7.127 reduces to

$$\frac{B^2}{1-\frac{2mG}{r}} - \frac{\left(\frac{dr}{d\tau}\right)^2}{1-\frac{2mG}{r}} - \frac{A^2}{r^2} = 1. \tag{7.129}$$

If we make the substitution $u = 1/r$, we have

$$\dot{r} = \frac{dr}{d\phi}\dot{\phi} = -\frac{1}{u^2}\frac{du}{d\phi}\frac{A}{r^2} = -\frac{du}{d\phi}A, \tag{7.130}$$

which, when combined with equation 7.129, produces

$$B^2 - A^2\left(\frac{du}{d\phi}\right)^2 - A^2u^2(1-2GMu) = (1-2GMu). \tag{7.131}$$

Taking a second derivative of this equation, and simplifying, gives

$$\frac{d^2u}{d\phi^2} + u - \frac{GM}{A^2} - 3GMu^2 = 0. \tag{7.132}$$

This last term is the relativistic correction to the precession of the perihelion. For the case of the planet Mercury, it can be shown that it is equivalent to 0.01 arcseconds per revolution or 40 arcseconds per century.

7.2.4.3 Deflection of Light

A solution to equation 7.132 can be obtained using an approximation method. A general solution to

$$\frac{d^2u}{d\phi^2} + u - \frac{GM}{A^2} = 0 \tag{7.133}$$

is

$$u = \frac{\sin\phi}{R} + \frac{GM}{A^2}, \tag{7.134}$$

where R is a constant (assumed to be the radius of the Sun). Shifting the origin by a constant term reduces equation 7.132 to the form

$$\frac{d^2u}{d\phi^2} + u = 3GMu^2.$$ (7.135)

If the left-hand side of equation 7.135 is set to zero, it is seen to have solution

$$u = \frac{\sin\phi}{R}.$$ (7.136)

Substituting this expression into the right-hand side of equation 7.135 gives

$$3GMu^2 = 3GM\left(\frac{\sin\phi}{R}\right)^2 = \frac{3GM}{R^2}\left(1 - \cos^2\phi\right).$$ (7.137)

A solution to

$$\frac{d^2u}{d\phi^2} + u = \frac{3GM}{R^2}\left(1 - \cos^2\phi\right)$$ (7.138)

is

$$u = \frac{\sin\phi}{R} + \frac{3GM}{2R^2}\left(1 + \frac{1}{3}\cos2\phi\right).$$ (7.139)

For large R, ϕ is small and $\cos2\phi \to 1$, so that the second term in equation 7.139, the relativistic correction, is

$$\delta = \frac{2GM}{R}.$$ (7.140)

This is the magnitude of the deflection of the ray between the Sun and infinity. By symmetry, the deflection of the ray before the Sun is equal in magnitude so that the total deflection is twice equation 7.140. For the case of a light ray grazing the surface of the Sun, the deflection is 1.75 arc seconds.

7.3 Bibliography

Misner, C. W., Thorne, K. S., and Wheeler, J. A. *Gravitation*. San Francisco: W. H. Freeman, 1970.
Ohanian, H. C. *Gravitation and Spacetime*. New York: W. W. Norton, 1976.

Rindler, W. *Essential Relativity*. 2d ed. New York: Springer-Verlag, 1976.

Weber, J. *General Relativity and Gravitational Waves*. New York: Interscience Publishers, 1961.

8 Quantum Mechanics

8.1 Fundamental Postulates of Quantum Mechanics

Like Newton's laws or Maxwell's equations, the laws of quantum mechanics cannot be derived. The justification for quantum mechanics is its agreement with observation. The mathematical justification for quantum mechanics evolve from a number of fundamental assumptions that are summarized, without elaboration, as follows.

8.1.1 State Functions

We shall make the assumption that the state of a particle at time t is completely describable by some function ψ that we shall call the *state function* (wave function) of the particle or system. Only those state functions that are physically admissible correspond to realizable physical states. We also make the plausible and physically necessary assumptions that the probability of finding a particle is large where ψ is large and small where ψ is small. If $P(x,t)dx$ is the relative probability of finding the particle at time t within a volume dx centered about x, then

$$P(x,t) = |\psi(x,t)|^2 = \psi^*(x,t)\psi(x,t), \qquad (8.1)$$

and the absolute probability is

$$\rho(x,t)dx = \frac{P(x,t)dx}{\int P(x,t)dx}. \qquad (8.2)$$

For normalized state functions,

$$\int \psi^* \psi dx = 1, \qquad (8.3)$$

and

$$\rho(x,t) = \psi^*(x,t)\psi(x,t) \tag{8.4}$$

is the probability amplitude. An important property of the state function is that if ϕ_1 describes one possible state and ϕ_2 describes a second possible state. A third state can be formed from a linear combination of ϕ_1 and ϕ_2. In general,

$$\psi = \sum_\mu c_\mu v_\mu, \tag{8.5}$$

where c_μ are arbitrary constants and v_μ are independent state functions. This principle of superposition is necessary to support experimental observations of interference.

8.1.2 Operators

Each dynamic variable that relates to the motion of the particle can be represented by a linear operator. We assume that the eigenvalues of a physical operator form a complete set. One of the eigenvalues ω_μ is the only possible value of a precise measurement of the dynamical variable represented by Ω. That is,

$$\Omega v_\mu = \omega_\mu v_\mu. \tag{8.6}$$

The number of measurements that result in the eigenvalue ω_μ are proportional to the square of the magnitude of the coefficient of v_μ in the expansion of ψ. It will be shown that $E = -\dfrac{\hbar}{i}\dfrac{\partial}{\partial t}$, $\vec{p} \sim \dfrac{\hbar}{i}\vec{\nabla}$, and $\vec{r} \sim \vec{r}$ are the energy, momentum, and position operator, respectively.

The act of measurement forces the system into an eigenstate of the dynamic variable which is being measured. If operator A represents a physically observable quantity, then its expectation value, defined by

$$\langle A \rangle = \langle \psi | A | \psi \rangle = \langle A\psi | \psi \rangle = \langle \psi | A\psi \rangle, \tag{8.7}$$

must always be real. If $\langle A \rangle$ is to be real, the last two expressions must also be real.

Example 8.1
A two-component system is in the state

$$|\psi(t)\rangle = \frac{1}{\sqrt{2}}\begin{pmatrix}1\\0\end{pmatrix}e^{-i\alpha t} + \frac{1}{\sqrt{2}}\begin{pmatrix}0\\1\end{pmatrix}e^{-i\beta t}. \tag{8.8}$$

Find (1) the density matrix, $\rho(t)$, for the state, (2) the expectation value of the Hamiltonian in this state, and (3) the probability that the system is in the state $\frac{1}{\sqrt{2}}\begin{pmatrix}1\\1\end{pmatrix}$.

(1) By definition, $\rho(x,t) = \psi^*(x,t)\psi(x,t)$, so

$$\rho_{mn} = \psi^*\psi = \frac{1}{2}\begin{pmatrix}1 & e^{-i(\alpha-\beta)t}\\ e^{+i(\alpha-\beta)t} & 1\end{pmatrix}. \tag{8.9}$$

(2) Again by definition,

$$\langle\psi|H|\psi\rangle = -\frac{\hbar}{i}\langle\psi|\frac{\partial}{\partial t}|\psi\rangle$$

$$= -\frac{\hbar}{i}\left[\frac{1}{\sqrt{2}}\left(e^{-i\alpha t} \quad e^{-i\beta t}\right)\right]\left[\frac{1}{\sqrt{2}}\begin{pmatrix}-i\alpha e^{-i\alpha t}\\ -i\beta e^{-i\beta t}\end{pmatrix}\right]$$

$$= \frac{\hbar}{2}(\alpha+\beta). \tag{8.10}$$

(3) The probability of being found in a state ψ' is

$$\langle\psi'|\psi_0\rangle^2 = \left|\left[\frac{1}{\sqrt{2}}(1 \quad 1)\right]\left[\frac{1}{\sqrt{2}}\begin{pmatrix}e^{-i\alpha t}\\ e^{-i\beta t}\end{pmatrix}\right]\right|^2 = \frac{1}{4}\left(e^{-i\alpha t} + e^{-i\beta t}\right)^2. \tag{8.11}$$

8.1.3 Summary

1. Any physical system is completely described by a normalized vector $|\psi\rangle$ in a Hilbert space. All possible information about the system can be derived from this state vector by the rules that follow.

2. To every physical observable there corresponds a Hermitian operator on the Hilbert space.

3. The only allowed physical results of measurements of the observable corresponding to the operator A are the eigenvalues of A.

4. If, on a system in the state $|\psi\rangle$, we make a measurement of the observable A, then the probability that the measured value is λ_n, that is, $A|\phi_n\rangle = \lambda_n|\phi_n\rangle$, is given by $|\langle\phi_n|\psi\rangle|^2$.

5. For every system there exists a Hermitian operator H, the Hamiltonian or energy operator, which determines the time development of $|\psi\rangle$ by

$$i\hbar\frac{\partial}{\partial t}|\psi\rangle = H|\psi\rangle.$$

6. If one measures the observable corresponding to the operator A and finds the value λ_n, then immediately after the measurement the state vector satisfies $i\hbar\frac{\partial}{\partial t}|\psi\rangle = H|\psi\rangle$. We say that the measurement projects the original state vector onto the eigenvectors of A.

8.2 Mathematical Interpretation

8.2.1 The Heisenberg Uncertainty Principle

If we have a positive definite operator A and a wave function χ, then the Schwartz inequality states

$$\int \chi^* A\chi d^3r \geq 0. \tag{8.12}$$

If we let $\chi = f + \lambda g$, the inequality is

$$\int (f+\lambda g)^* A(f+\lambda g)d^3r$$
$$\geq \int f^* Afd^3r + \lambda^* \int g^* Afd^3r + \lambda\int f^* Agd^3r + |\lambda|^2\int g^* Agd^3r. \tag{8.13}$$

This is true for any λ. If we let

$$\lambda = -\frac{\int g^* A f d^3 r}{\int g^* A g d^3 r} \quad , \tag{8.14}$$

the inequality is

$$\int f^* A f d^3 \vec{r} - \frac{\int g A^* f^* d^3 r}{\int g A^* g^* d^3 r} \int g^* A f d^3 r$$

$$-\frac{\int g^* A f d^3 r}{\int g^* A g d^3 r} \int f^* A g d^3 r + \frac{\left|\int g^* A f d^3 r\right|^2}{\left|\int g^* A g d^3 r\right|} \int g^* A g d^3 r. \tag{8.15}$$

Because A is a positive definite operator, $A = A^*$. The inequality is

$$\int \chi^* A \chi d^3 r = \int f^* A f d^3 r - \frac{\left|\int f^* A g d^3 r\right|^2}{\int g^* A g d^3 r} \geq 0, \tag{8.16}$$

or

$$\int f^* A f d^3 r \int g^* A g d^3 r \geq \left|\int f^* A g d^3 r\right|^2. \tag{8.17}$$

If we let $A = 1$ and choose

$$f = \alpha \psi, \tag{8.18a}$$

$$\alpha = x - \langle x \rangle, \tag{8.18b}$$

$$g = \beta \psi, \tag{8.18c}$$

$$\beta = p_x - \langle p_x \rangle, \tag{8.18d}$$

then

$$\int f^* A f d^3 r = \int \psi^* (x - \langle x \rangle)^2 \psi d^3 r = (\Delta x)^2, \tag{8.19}$$

and

$$\int g^* A\, g d^3 r = \int \psi^*\left(p_x - \langle p_x \rangle\right)^2 \psi d^3 r = (\Delta p_x)^2. \qquad (8.20)$$

Therefore,

$$\Delta x^2 \Delta \rho_x^{\ 2} \ge \left| \int f^* A g d^3 r \right|^2 = \left| \int \psi^* \alpha \beta \psi d^3 r \right|^2. \qquad (8.21)$$

If we rewrite

$$\alpha\beta = \tfrac{1}{2}(\alpha\beta - \beta\alpha) + \tfrac{1}{2}(\alpha\beta + \beta\alpha), \qquad (8.22)$$

then

$$\left| \int \psi^* \alpha \beta \psi d^3 r \right|^2 = \tfrac{1}{4} \left| \int \psi^*(\alpha\beta - \beta\alpha)\psi d^3 r + \int \psi^*(\alpha\beta + \beta\alpha)\psi d^3 r \right|$$

$$= \tfrac{1}{4}\left\{ \left| \int \psi^*(\alpha\beta - \beta\alpha)\psi d^3 r \right|^2 + \left| \int \psi^*(\alpha\beta + \beta\alpha)\psi d^3 r \right|^2 \right\}$$

$$+ \tfrac{1}{4}\left(\int \psi^*(\alpha\beta - \beta\alpha)\psi d^3 r \right)^* \left(\int \psi^*(\alpha\beta + \beta\alpha)\psi d^3 r \right)$$

$$+ \tfrac{1}{4}\left(\int \psi^*(\alpha\beta - \beta\alpha)\psi d^3 r \right)\left(\int \psi^*(\alpha\beta + \beta\alpha)\psi d^3 r \right)^*. \quad (8.23)$$

But

$$\int \psi^* \alpha \beta \psi d^3 r = \int (\alpha\psi)^* (\beta\psi) d^3 r$$

$$= \left[\int (\beta\psi)^* (\alpha\psi) d^3 r \right]^*$$

$$= \left[\int \psi^* \beta\alpha\psi d^3 r \right]^*. \qquad (8.24)$$

The last two terms in the previous expression are

$$\left(\langle \alpha\beta \rangle - \langle \beta\alpha \rangle\right)^* \left(\langle \alpha\beta \rangle + \langle \beta\alpha \rangle\right) + \left(\langle \alpha\beta \rangle - \langle \beta\alpha \rangle\right)\left(\langle \alpha\beta \rangle + \langle \beta\alpha \rangle\right)^*, \quad (8.25)$$

which is

$$(\langle\beta\alpha\rangle - \langle\alpha\beta\rangle)(\langle\alpha\beta\rangle + \langle\beta\alpha\rangle) + (\langle\alpha\beta\rangle - \langle\beta\alpha\rangle)(\langle\beta\alpha\rangle + \langle\alpha\beta\rangle), \quad (8.26)$$

or

$$\langle\beta\alpha\rangle\langle\alpha\beta\rangle - \langle\alpha\beta\rangle\langle\alpha\beta\rangle + \langle\beta\alpha\rangle\langle\beta\alpha\rangle - \langle\alpha\beta\rangle\langle\beta\alpha\rangle$$
$$+\langle\alpha\beta\rangle\langle\beta\alpha\rangle - \langle\beta\alpha\rangle\langle\beta\alpha\rangle + \langle\alpha\beta\rangle\langle\alpha\beta\rangle - \langle\beta\alpha\rangle\langle\alpha\beta\rangle. \quad (8.27)$$

All these terms cancel, therefore

$$(\Delta x)^2 (\Delta p_x)^2 \geq \frac{1}{4}\left|\int \psi^*(\alpha\beta - \beta\alpha)\psi d^3 r\right|^2 + \frac{1}{4}\left|\int \psi^*(\alpha\beta + \beta\alpha)\psi d^3 r\right|^2. \quad (8.28)$$

Expanding the terms in the first integrand, we obtain

$$(\alpha\beta - \beta\alpha)\psi = (x - \langle x\rangle)(p_x - \langle p_x\rangle)\psi - (p_x - \langle p_x\rangle)(x - \langle x\rangle)\psi$$

$$= (xp_x - \langle x\rangle p_x - \langle p_x\rangle x + \langle x\rangle\langle p_x\rangle)\psi$$

$$- (p_x x - \langle p_x\rangle x - \langle x\rangle p_x + \langle p_x\rangle\langle x\rangle)$$

$$= (xp_x - p_x x)\psi$$

$$= \frac{\hbar}{i}\left[x\frac{\partial\psi}{\partial x} - \frac{\partial}{\partial x}(x\psi)\right]$$

$$= -\frac{\hbar}{i}\psi. \quad (8.29)$$

What about the second integrand? If this integral is zero, the uncertainty is a minimum. If so, then

$$\int \psi^*(\alpha\beta + \beta\alpha)\psi d^3 r = 0, \quad (8.30)$$

which is true only if $\alpha\psi = \gamma\beta\psi$. Therefore,

$$(x + \langle x\rangle)\psi = \gamma\left(\frac{\hbar}{i}\frac{\partial}{\partial x} + \langle p_x\rangle\right)\psi, \quad (8.31)$$

$$\frac{\partial \psi}{\partial x} = \frac{i}{\gamma \hbar}\left(x + \langle x \rangle\right)\psi - \frac{i\langle p_x \rangle \psi}{\hbar}, \tag{8.32}$$

and

$$\psi = N \exp\left(\frac{i}{2\gamma \hbar}(x+\langle x \rangle)^2 + \frac{i\langle p_x \rangle x}{\hbar}\right), \tag{8.33}$$

with N being a constant. If this is the form of ψ, then

$$\int \psi^*(\alpha\beta + \beta\alpha)\psi d^3 r = \left(\frac{1}{\gamma} + \frac{1}{\gamma^*}\right)\int \psi^* \alpha^2 \psi d^3 r. \tag{8.34}$$

This implies that γ^* is imaginary. We want $\psi^*\psi$ to be nonnegative, therefore γ^* is negative. If we require the state function to be normalized and

$$\int \left(x - \langle x \rangle\right)|\psi|^2 d^3 r = (\Delta x)^2, \tag{8.35}$$

then

$$\psi(x) = \left[2\pi(\Delta x)^2\right]^{-\frac{1}{4}} \exp\left(-\frac{(x+\langle x \rangle)^2}{4(\Delta x)^2} + \frac{i\langle p_x \rangle x}{\hbar}\right). \tag{8.36}$$

If $\psi(x)$ has this form, then

$$(\Delta x)^2(\Delta p_x)^2 \ge \frac{1}{4}\left|(i\hbar\psi)\right|^2, \tag{8.37}$$

or

$$(\Delta x)(\Delta p_x) \ge \frac{\hbar}{2}. \tag{8.38}$$

8.2.2 The Schrödinger Equation

A consequence of the Heisenberg uncertainty principle (eq. 8.38) is that quanta of matter or radiation must be represented by wave packets. For this reason the state function ψ is also called a wave function. That is, ψ must have properties of plane waves, $\psi \propto e^{i(kx - \omega t)}$. This form also satisfies the empirical requirement that wave functions may be superimposed to produce interference. In agreement with experimental observations, a continuous traveling harmonic wave will carry de Broglie momentum, and energy, according to the relations $p = \hbar k$ and $E = \hbar \omega$, with $k = \dfrac{2\pi}{\lambda}$ and $\omega = 2\pi\nu$. The classical energy equation is $E = p^2/2m$. We define the energy operator by

$$E \rightarrow i\hbar \frac{\partial}{\partial t}, \tag{8.39}$$

because

$$E\psi = \hbar\omega\psi = i\hbar(-i\omega)\psi = i\hbar \frac{\partial}{\partial t}\psi. \tag{8.40}$$

Similarly, the momentum operator is defined by

$$\vec{p} \rightarrow -i\hbar\vec{\nabla}, \tag{8.41}$$

because

$$p\psi = \hbar k\psi = -i\hbar(ik)\psi = -i\hbar\vec{\nabla}\psi. \tag{8.42}$$

A quantitative description of the motion of a particle can be obtained if we can develop an equation of motion that has harmonic waves, and more complicated waves, as its solutions. The equation must be linear, to allow for superposition. The equation must also involve constants such as $\hbar$, m, e as coefficients. If we try

$$\frac{\partial\psi}{\partial t} = \gamma \frac{\partial^2 \psi}{\partial x^2}, \tag{8.43}$$

where

$$\gamma = \frac{i\omega}{k^2} = \frac{i\hbar E}{p^2} = \frac{i\hbar}{2m}, \tag{8.44}$$

this equation meets our requirements and may be rewritten in the form

$$ i\hbar \frac{\partial \psi}{\partial t} = -\frac{\hbar^2}{2m} \frac{\partial^2 \psi}{\partial x^2}. \tag{8.45} $$

With the definition of the energy and momentum operators given on the previous page, this expression is equivalent to $E = p^2/2m$ and is known as the *one-dimensional Schrödinger equation*.

In the presence of forces $\vec{F} = -\nabla V$, we have

$$ E = \frac{p^2}{2m} + V. \tag{8.46} $$

If V is independent of E, p, then

$$ i\hbar \frac{\partial \psi}{\partial t} = -\frac{\hbar^2}{2m} \frac{\partial^2 \psi}{\partial x^2} + V(\vec{r}, t)\psi. \tag{8.47} $$

8.2.2.1 Time Dependence of y

If we make the assumption that $\psi(x,t)$ can be written as $\psi(x)T(t)$, then for a free particle the Schrödinger equation is

$$ i\hbar\psi \frac{\partial T}{\partial t} = -\frac{\hbar^2}{2m} \frac{\partial^2 \psi}{\partial x^2} T, \tag{8.48} $$

or equivalently

$$ (i\hbar)\frac{1}{T} \frac{\partial T}{\partial t} = \frac{(i\hbar)^2}{2m} \frac{1}{\psi} \frac{\partial^2 \psi}{\partial x^2}. \tag{8.49} $$

This can only be true if both sides of the equation are equal to a constant, α. The left-hand side of the equation then has the solution

$$ T(t) = \exp\left(\frac{-i\alpha t}{\hbar}\right), \tag{8.50} $$

while the right-hand side has solution

$$ E\psi = \frac{p^2}{2m}\psi = \frac{(i\hbar)^2}{2m} \frac{\partial^2 \psi}{\partial x^2} = \alpha\psi. \tag{8.51} $$

Thus, we have

$$\psi(x,t) = \psi(x)\exp^{\left(-\frac{iEt}{\hbar}\right)}. \qquad (8.52)$$

8.2.2.2 Center of Mass Coordinates

If we are dealing with the motion of two particles, many types of potentials will depend only on the motion between the particles. The Schrödinger equation has the explicit form

$$\left[-\frac{\hbar^2}{2m_1}\left(\frac{\partial^2}{\partial x_{1,i}^2}\right) - \frac{\hbar^2}{2m_2}\left(\frac{\partial^2}{\partial x_{2,i}^2}\right) + V(\bar{x}_1,\bar{x}_2)\right]\psi(\bar{x}_1,\bar{x}_2,t) = i\hbar\frac{\partial}{\partial t}\psi(\bar{x}_1,\bar{x}_2,t). \quad (8.53)$$

We define the relative coordinates:

$$x = x_1 - x_2,$$

$$y = y_1 - y_2,$$

$$z = z_1 - z_2, \qquad (8.54)$$

and

$$MX = m_1 x_1 + m_2 x_2,$$

$$MY = m_1 y_1 + m_2 y_2,$$

$$MZ = m_1 z_1 + m_2 z_2, \qquad (8.55)$$

where

$$M = m_1 + m_2. \qquad (8.56)$$

We have

$$m_1 x = m_1 x_1 - m_1 x_2, \qquad (8.57)$$

and

$$MX - m_1 x = (m_1 + m_2) x_2. \tag{8.58}$$

Solving for x_2 gives

$$x_2 = X - \frac{m_1}{m_1 + m_2} x. \tag{8.59}$$

Similarly, it can be shown that

$$x_1 = X + \frac{m_2}{m_1 + m_2} x. \tag{8.60}$$

Differentiating these expressions gives

$$\frac{\partial}{\partial x_1} = \frac{\partial X}{\partial x_1} \frac{\partial}{\partial X} + \frac{\partial x}{\partial x_1} \frac{\partial}{\partial x} = \frac{m_1}{M} \frac{\partial}{\partial X} + \frac{\partial}{\partial x},$$

$$\frac{\partial^2}{\partial x_1^2} = \frac{m_1^2}{M^2} \frac{\partial^2}{\partial X^2} + \frac{2m_1}{M} \frac{\partial^2}{\partial X \partial x} + \frac{\partial^2}{\partial x^2}, \tag{8.61}$$

and

$$\frac{\partial}{\partial x_2} = \frac{m_2}{M} \frac{\partial}{\partial X} - \frac{\partial}{\partial x},$$

$$\frac{\partial^2}{\partial x_2^2} = \frac{m_2^2}{M^2} \frac{\partial^2}{\partial X^2} - \frac{2m_2}{M} \frac{\partial^2}{\partial X \partial x} + \frac{\partial^2}{\partial x^2}. \tag{8.62}$$

Therefore,

$$\frac{1}{m_1} \frac{\partial^2}{\partial x_1^2} + \frac{1}{m_2} \frac{\partial^2}{\partial x_2^2}$$

$$= \left[\frac{m_1}{M^2} + \frac{m_2}{M^2} \right] \frac{\partial^2}{\partial X^2} + \left[\frac{2}{M} - \frac{2}{M} \right] \frac{\partial^2}{\partial X \partial x} + \left[\frac{1}{m_1} + \frac{1}{m_2} \right] \frac{\partial^2}{\partial x^2}, \tag{8.63}$$

or simply

$$\frac{1}{m_1}\frac{\partial^2}{\partial x_1^2} + \frac{1}{m_2}\frac{\partial^2}{\partial x_2^2} = \frac{1}{M}\frac{\partial^2}{\partial X^2} + \frac{1}{\mu}\frac{\partial^2}{\partial x^2}, \qquad (8.64)$$

where

$$\mu = \frac{m_1 m_2}{m_1 + m_2}. \qquad (8.65)$$

The Schrödinger equation becomes

$$\left[-\frac{\hbar^2}{2M}\left(\frac{\partial^2}{\partial X_i^2}\right) - \frac{\hbar^2}{2\mu}\left(\frac{\partial^2}{\partial x_i^2}\right) + V(\vec{x},t) \right]\psi(\vec{X},\vec{x},t) = i\hbar\frac{\partial}{\partial t}\psi(\vec{X},\vec{x},t). \qquad (8.66)$$

If

$$\psi(\vec{X},\vec{x},t) = u(\vec{x},t)U(\vec{X},t)\exp\left(-\frac{i(E-E')t}{\hbar}\right), \qquad (8.67)$$

then the Schrödinger equation reduces to two separable equations,

$$-\frac{\hbar^2}{2\mu}\nabla^2 u + Vu = Eu, \qquad (8.68)$$

and

$$-\frac{\hbar^2}{2M}\nabla^2 U = E'u. \qquad (8.69)$$

8.2.3 Ehrenfest's Theorem

We can show that the expectation values of the position and momentum operators agree in the classical and quantum limits. By definition,

$$\frac{d}{dt}\langle x \rangle = \frac{d}{dt}\int \psi^* x\psi d^3r = \int \frac{\partial \psi^*}{\partial t}x\psi d^3r + \int \psi^* x\frac{\partial \psi}{\partial t}d^3r. \qquad (8.70)$$

From the Schrödinger equation, equation 8.47, we obtain

$$\frac{d}{dt}\langle x\rangle = \frac{1}{i\hbar}\left\{-\int\left(-\frac{\hbar^2}{2m}\nabla^2\psi^* +V\psi^*\right)x\psi d^3r + \int \psi^* x\left(-\frac{\hbar^2}{2m}\nabla^2\psi +V\psi\right)d^3r\right\}$$

$$= \frac{i\hbar}{2m}\left\{\int\left(-\nabla^2\psi^*\right)x\psi d^3r + \int \psi^* x\left(\nabla^2\psi\right)d^3r\right\} \tag{8.71}$$

The second integral above may be integrated by parts

$$\int\left(\nabla^2\psi^*\right)x\psi d^3r = -\int\left(\nabla\psi^*\right)\cdot\nabla(x\psi)d^3r + \oint\left(x\psi\nabla\psi^*\right)_n dS. \tag{8.72}$$

The surface integral goes to zero at infinity. A second partial integration of the remaining term gives

$$\int\left(\nabla^2\psi^*\right)x\psi d^3r = -\int \psi^*\nabla^2(x\psi)d^3r. \tag{8.73}$$

As a check,

$$\nabla\cdot\left(x\psi\nabla\psi^*\right) = \left(\nabla\psi^*\right)\cdot\nabla(x\psi) + \left(\nabla^2\psi^*\right)(x\psi), \tag{8.74}$$

and

$$\left(\nabla\psi^*\right)\cdot\nabla(x\psi) = \nabla\cdot\left(\psi^*\nabla(x\psi)\right) - \psi^*\nabla^2(x\psi). \tag{8.75}$$

Thus, from equation 8.71

$$\frac{d}{dt}\langle x\rangle = \frac{i\hbar}{2m}\int \psi^*\left[x\nabla^2\psi - \nabla^2(x\psi)\right]d^3r = -\frac{i\hbar}{2m}\int \psi^*\frac{\partial\psi}{\partial x}d^3r = \frac{1}{m}\langle p_x\rangle, \tag{8.76}$$

which is the classical definition.

For the momentum,

$$\frac{d}{dt}\langle p_x\rangle = -i\hbar\frac{d}{dt}\int \psi^*\frac{\partial\psi}{\partial x}d^3r = -i\hbar\left(\int \psi^*\frac{\partial}{\partial x}\frac{\partial\psi}{\partial t}d^3r + \int\frac{\partial\psi^*}{\partial t}\frac{\partial\psi}{\partial x}d^3r\right). \tag{8.77}$$

From the Schrödinger equation (eq. 8.47) this is equal to

$$\left\{\int\Psi^*\frac{\partial}{\partial x}\left(-\frac{\hbar^2}{2m}\nabla^2\psi + V\psi\right)d^3r + \int\left(-\frac{\hbar^2}{2m}\nabla^2\psi^* + V\psi^*\right)\frac{\partial\psi}{\partial x}d^3r\right\}, \tag{8.78}$$

so that equation 8.77 reduces to

$$\frac{d}{dt}\langle p_x \rangle = -\int \psi^* \left[\frac{\partial}{\partial x}(V\psi) - V\frac{\partial \psi}{\partial x} \right] d^3r = -\int \psi^* \frac{\partial V}{\partial x}\psi d^3r = \left\langle -\frac{\partial V}{\partial x} \right\rangle, \quad (8.79)$$

which is in agreement with the classical definition.

8.2.4 The Virial Theorem

The quantum mechanical Virial theorem is also similar to the classical result. By comparison with section 2.1.1,

$$\frac{d}{dt}\langle \vec{r} \cdot \vec{p} \rangle = \frac{1}{i\hbar}\langle [(\vec{r} \cdot \vec{p}), H] \rangle, \quad (8.80)$$

where

$$[(\vec{r} \cdot \vec{p}), H] = \left[(xp_x + yp_y + zp_z), \frac{p_x^2 + p_y^2 + p_z^2}{2m} + V(x,y,z) \right]. \quad (8.81)$$

When dealing with the term $p^2/2m$, we need only consider the terms $\left[xp_x, p_x^2 \right]$ cyclically. By definition,

$$\left[xp_x, p_x^2 \right] = xp_x^3 - p_x^2 xp_x = \left(xp_x^2 - p_x^2 x \right)p_x. \quad (8.82)$$

This reduces to

$$\left(xp_x^2 - p_x xp_x + p_x xp_x - p_x^2 x \right)p_x \quad (8.83)$$

or

$$(xp_x - p_x x)p_x^2 + p_x(xp_x - p_x x)p_x, \quad (8.84)$$

which is seen to be equal to $2i\hbar p_x^2$.

For the potential term,

$$[xp_x, V] = xp_x V - Vxp_x. \quad (8.85)$$

If this acts on a wave function ψ, we obtain

$$[xp_x, V]\psi = xp_x V\psi - Vxp_x\psi, \tag{8.86}$$

which is equal to

$$-i\hbar\left\{x\frac{\partial}{\partial x}(V\psi) - Vx\frac{\partial}{\partial x}\psi\right\} = -i\hbar x\frac{\partial V}{\partial x}\psi. \tag{8.87}$$

Therefore,

$$[xp_x, V] = -i\hbar x\frac{\partial V}{\partial x}. \tag{8.88}$$

The result is

$$[(\vec{r}\cdot\vec{p}), H] = \frac{i\hbar}{m}\left(p_x^2 + p_y^2 + p_z^2\right) - i\hbar\left(x\frac{\partial V}{\partial x} + y\frac{\partial V}{\partial y} + z\frac{\partial V}{\partial z}\right), \tag{8.89}$$

or

$$[(\vec{r}\cdot\vec{p}), H] = 2i\hbar T - i\hbar\left(\vec{r}\cdot\vec{\nabla}V\right). \tag{8.90}$$

The conclusion is

$$2\langle T\rangle = \langle\vec{r}\cdot\vec{\nabla}V\rangle, \tag{8.91}$$

in agreement with the classical result.

8.2.5 Conservation of Probability

From the previous definition of probability (eq. 8.4) we have

$$\int P(x,t)dx = \int_{-\infty}^{+\infty}\psi^*(x,t)\psi(x,t)dx. \tag{8.92}$$

Thus

$$\frac{d}{dt}\int P(x,t)dx = \int_{-\infty}^{+\infty}\left[\frac{\partial \psi^*}{\partial t}\psi + \psi^*\frac{\partial \psi}{\partial t}\right]dx. \qquad (8.93)$$

From the Schrödinger equation $\dfrac{\partial \psi}{\partial t} = \dfrac{i\hbar}{2m}\nabla^2\psi$, so

$$\frac{d}{dt}\int P(x,t)dx = \left(\frac{i\hbar}{2m}\right)\int_{-\infty}^{+\infty}\left[\psi^*\nabla^2\psi - \psi\nabla^2\psi^*\right]dx, \qquad (8.94)$$

or

$$\frac{d}{dt}\int P(x,t)dx = \left(\frac{i\hbar}{2m}\right)\int_{-\infty}^{+\infty}\frac{\partial}{\partial x}\left[\psi^*\frac{\partial \psi}{\partial x} - \psi\frac{\partial \psi^*}{\partial x}\right]dx. \qquad (8.95)$$

If we define the vector

$$S(\vec{r},t) = \left(\frac{i\hbar}{2m}\right)\frac{\partial}{\partial x}\left[\psi^*\nabla\psi - \psi\nabla\psi^*\right], \qquad (8.96)$$

then

$$\frac{d}{dt}\int P(\vec{r},t)d^3r = -\int_{-\infty}^{+\infty}\vec{\nabla}\cdot S(\vec{r},t)d^3r, \qquad (8.97)$$

or

$$\frac{dP(\vec{r},t)}{dt} + \vec{\nabla}\cdot S(\vec{r},t) = 0. \qquad (8.98)$$

Therefore, $S(\vec{r},t)$ is interpreted as the probability current density, and probability is conserved.

8.3 Applications: One-Dimensional Systems

8.3.1 Square Well—Perfectly Rigid Walls

If we have the potential shown in figure 8.1, then in the regions $|x| < a$ the Schrödinger equation is

$$-\frac{\hbar^2}{2m}\frac{d^2u}{dx^2} = Eu. \qquad (8.99)$$

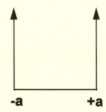

Figure 8.1 Infinite square well.

This has solution

$$u(x) = A\sin\alpha x + B\cos\alpha x, \qquad (8.100)$$

with

$$\alpha = \left(\frac{2mE}{\hbar^2}\right)^{1/2}. \qquad (8.101)$$

However, $u(x) = 0$ if $x = \pm a$, therefore

$$A\sin\alpha a + B\cos\alpha a = 0, \qquad (8.102)$$

and

$$-A\sin\alpha a + B\cos\alpha a = 0. \qquad (8.103)$$

This requires that $A\sin\alpha a = 0 = B\cos\alpha a$. There are two possible classes of solutions. Either $A = 0$ and $\cos\alpha a = 0$, or $B = 0$ $\sin\alpha a = 0$. The two solutions are

$$u_1(x) = B\cos\frac{n\pi x}{2a}, \tag{8.104}$$

and

$$u_2(x) = A\sin\frac{n\pi x}{2a}. \tag{8.105}$$

In either case, it can be shown that a consequence of these solutions is that the energy levels are quantized. That is,

$$E = \frac{\pi^2\hbar^2 n^2}{8ma^2}. \tag{8.106}$$

8.3.2 Square Well—Finite Potential Step

If we have the situation shown in figure 8.2, then the Schrödinger equation is modified as follows.

For the region $|x| < a$, we have the situation described by equations 8.99, 8.100 and 8.101.

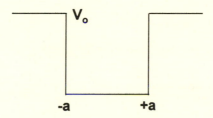

Figure 8.2 Finite square well.

In the region $|x| > a$, the Schrödinger equation is

$$-\frac{\hbar^2}{2m}\frac{d^2u}{dx^2} + V_o u = Eu. \tag{8.107}$$

If $E < V_0$, we have bound states with solutions

$$u(x) = Ce^{-\beta x} + De^{+\beta x}, \tag{8.108}$$

where

$$\beta = \left[\frac{2m(V_o - E)}{\hbar^2} \right]^{1/2}. \tag{8.109}$$

If $x > 0$, then D must equal zero so that $u(\infty) = 0$. Similarly, if $x < 0$, then C must equal zero so that $u(-\infty) = 0$.

We can match the solutions at the boundaries if we require that both u and du/dx are continuous. Continuity of $u(x)$ at $x = \pm a$ requires

$$A \sin \alpha a + B \cos \alpha a = C e^{-\beta a}, \tag{8.110}$$

and

$$-A \sin \alpha a + B \cos \alpha a = D e^{-\beta a}. \tag{8.111}$$

Continuity of du/dx at $x = \pm a$ requires

$$\alpha A \cos \alpha a - \alpha B \cos \alpha a = -\beta C e^{-\beta a}, \tag{8.112}$$

and

$$\alpha A \cos \alpha a + \alpha B \cos \alpha a = \beta D e^{-\beta a}. \tag{8.113}$$

From these relations we obtain

$$2 A \sin \alpha a = (C - D) e^{-\beta a}, \tag{8.114}$$

$$2 \alpha A \cos \alpha a = -\beta (C - D) e^{-\beta a}, \tag{8.115}$$

$$2 B \cos \alpha a = (C + D) e^{-\beta a}, \tag{8.116}$$

and

$$2 \alpha B \sin \alpha a = \beta (C + D) e^{-\beta a}. \tag{8.117}$$

Again, we have two classes of solutions: (i) $A = 0$, $C = D$, $\alpha \tan \alpha a = \beta$, or (ii) $B = 0$, $C = -D$, $\alpha \cot \alpha a = -\beta$.

8.3.3 One-Dimensional Square Potential Barrier

Consider the potential shown in figure 8.3.

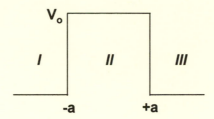

Figure 8.3 Square potential barrier.

The Schrödinger equation and its solutions are as follows:

Region I:

$$-\frac{\hbar^2}{2m}\frac{\partial^2\psi}{\partial x^2} = E\psi,\qquad(8.118)$$

and

$$\psi = Ae^{ikx} + Be^{-ikx},\qquad(8.119)$$

with $k = \sqrt{\dfrac{2mE}{\hbar^2}}$.

Region II:

$$-\frac{\hbar^2}{2m}\frac{\partial^2\psi}{\partial x^2} + V_0\psi = E\psi,\qquad(8.120)$$

with

$$\psi = Ce^{i\alpha x} + De^{-i\alpha x},\qquad(8.121)$$

with $\alpha = \sqrt{\dfrac{2m(E - V_0)}{\hbar^2}}$.

Region III:

$$-\frac{\hbar^2}{2m}\frac{\partial^2 \psi}{\partial x^2} = E\psi, \tag{8.122}$$

and

$$\psi = Ee^{ikx}, \tag{8.123}$$

with $k = \sqrt{\dfrac{2mE}{\hbar^2}}$.

To solve the problem, note that

$$\left|\frac{E}{A}\right|^2 = \text{transmission coefficient}, \tag{8.124}$$

$$\left|\frac{B}{A}\right|^2 = \text{reflection coefficient}. \tag{8.125}$$

We can solve for these coefficients by making use of the boundary conditions. That is, ϕ and $d\phi/dx$ are continuous at 0 and a.

For $x = 0$,

$$\phi \rightarrow A + B = C + D, \tag{8.126}$$

and

$$\phi' \rightarrow ikA - ikB = i\alpha C - i\alpha D. \tag{8.127}$$

For $x = a$,

$$\phi \rightarrow Ce^{i\alpha a} + De^{-i\alpha a} = Ee^{ika}, \tag{8.128}$$

and

$$\phi' \rightarrow i\alpha Ce^{i\alpha a} - i\alpha De^{-i\alpha a} = ikEe^{ika}. \tag{8.129}$$

Solving these equations, we find

$$\frac{B}{A} = \frac{\left(k^2 - \alpha^2\right)\left(1 - e^{2i\alpha a}\right)}{\left(k + \alpha\right)^2 - \left(k - \alpha\right)^2 e^{2i\alpha a}},$$ (8.130)

and

$$\frac{E}{A} = \frac{4k\alpha e^{i(\alpha - k)a}}{\left(k + \alpha\right)^2 - \left(k - \alpha\right)^2 e^{2i\alpha a}},$$ (8.131)

which gives

$$\left|\frac{B}{A}\right|^2 = \left[1 + \frac{4E(E - V_o)}{V_o^2 \sin^2 \alpha a}\right]^{-1},$$ (8.132)

and

$$\left|\frac{E}{A}\right|^2 = \left[1 + \frac{V_o^2 \sin^2 \alpha a}{4E(E - V_o)}\right]^{-1}.$$ (8.133)

Note that $\left|\dfrac{B}{A}\right|^2 + \left|\dfrac{E}{A}\right|^2 = 1$.

8.3.4 Linear Harmonic Oscillator

For a simple harmonic oscillator, $F = -kx$, the Schrödinger equation is

$$-\frac{\hbar^2}{2m}\frac{d^2u}{dx^2} + \frac{1}{2}kx^2 u = Eu.$$ (8.134)

If we make the substitutions $\xi = \alpha x$ and introduce a dimensionless eigenvalue λ, we can rewrite this equation in a simpler form. That is,

$$\frac{d^2u}{dx^2} + \left[\frac{1}{2}kx^2\left(\frac{-2m}{\hbar^2}\right) - E\right]u = 0$$ (8.135)

reduces to

$$\frac{d^2u}{d\xi^2} + \left[\lambda - \xi^2\right]u = 0, \tag{8.136}$$

where $\alpha^4 = \frac{mk}{\hbar^2}$ and $\lambda = \frac{2E}{\hbar}\left(\frac{m}{k}\right)^{1/2} = \frac{2E}{\hbar\omega_c}$. If we look in the limit as $\xi \rightarrow \pm\infty$, the equation has two solutions of the form

$$u(\xi) = e^{\pm\frac{1}{2}\xi^2}. \tag{8.137}$$

As $\xi \rightarrow +\infty$ only the solution with the minus sign is acceptable. We try a solution of the form

$$u(\xi) = H(\xi)e^{-\frac{1}{2}\xi^2}. \tag{8.138}$$

This has derivatives

$$u'(\xi) = H'(\xi)e^{-\frac{1}{2}\xi^2} - \xi H(\xi)e^{-\frac{1}{2}\xi^2}, \tag{8.139}$$

and

$$u''(\xi) = H''(\xi)e^{-\frac{1}{2}\xi^2} - 2\xi H'(\xi)e^{-\frac{1}{2}\xi^2} + \left(\xi^2 - 1\right)H(\xi)e^{-\frac{1}{2}\xi^2}. \tag{8.140}$$

Using these in equation 8.136, we obtain the constraint

$$H'' - 2\xi H' + (\lambda - 1)H = 0. \tag{8.141}$$

Trying a solution for H of the form $H = \xi^s\left(a_0 + a_1\xi + a_2\xi^2 + \ldots\right)$ we obtain a recursion relation for the coefficients, which must be zero for each power of ξ. That is,

$$s(s-1)a_0 = 0, \tag{8.142}$$

$$(s+1)sa_1 = 0, \tag{8.143}$$

$$(s+2)(s+1)a_2 - (2s+1-\lambda)a_0 = 0, \qquad (8.144)$$

$$(s+3)(s+2)a_3 - (2s+3-\lambda)a_1 = 0, \qquad (8.145)$$

which in general reduce to

$$(s+v+2)(s+v+1)a_{v+2} - (2s+2v+1-\lambda)a_v = 0. \qquad (8.146)$$

This first relation has solutions $s = 0$, $s = 1$, or $a_0 = 0$. The second relation has solutions $s = 0$ or $a_1 = 0$. Higher values of a_n can be defined in terms of a_0 and a_1. $H(\xi)$ is said to be even if $s = 0$ and a_0 is nonzero. Similarly, $H(\xi)$ is said to be odd if $s = 1$ and a_1 is nonzero. The series must terminate to preserve the boundary conditions. This occurs if

$$2s+2v+1 = \lambda, \qquad (8.147)$$

which gives

$$\lambda = 2n+1. \qquad (8.148)$$

Because $\lambda = \dfrac{2E}{\hbar}\left(\dfrac{m}{k}\right)^{1/2}$, the solution for E is

$$E = \left(n + \frac{1}{2}\right)\hbar\omega_c. \qquad (8.149)$$

8.3.4.1 Hermite Polynomials

If $\lambda = 2n + 1$, the previous expression for H becomes

$$H_n'' - 2\xi H_n' + 2nH_n = 0, \qquad (8.150)$$

where H_n is the nth Hermite polynomial. We can express H_n in terms of a generating function $S(\xi,s)$, defined by

$$S(\xi,s) = e^{\xi^2-(s-\xi)^2} = e^{-s^2+2s\xi} = \sum_{n=0}^{\infty} \frac{H_n(\xi)}{n!}s^n. \qquad (8.151)$$

To see that H_n satisfies equation 8.134, we first compute $\dfrac{\partial S}{\partial \xi}$ and $\dfrac{\partial S}{\partial s}$. By examination

$$\frac{\partial S}{\partial \xi} = \sum_{n=0}^{\infty} \frac{s^n}{n!} H_n'(\xi)$$

$$= 2se^{-s^2+2s\xi}$$

$$= \sum_{n=0}^{\infty} \frac{2s^{n+1}}{n!} H_n(\xi). \tag{8.152}$$

Equating equal powers of s, we see that this can only be true if

$$\frac{2s^m}{(m-1)!} H_{m-1} = \frac{s^m}{m!} H_m', \tag{8.153}$$

or

$$H_m' = 2mH_{m-1}. \tag{8.154}$$

Similarly,

$$\frac{\partial S}{\partial s} = \sum_{n=0}^{\infty} \frac{s^{n-1}}{(n-1)!} H_n(\xi)$$

$$= (-2s + 2\xi)e^{-s^2+2s\xi}$$

$$= -2\sum_{n=0}^{\infty} \frac{s^{n+1}}{n!} H_n(\xi) + 2\xi \sum_{n=0}^{\infty} \frac{s^n}{n!} H_n(\xi). \tag{8.155}$$

Again, equating equal powers of s requires that

$$\frac{s^m}{m!} H_{m+1} = -2\frac{s^m}{(m-1)!} H_{m-1} + 2\xi H_m, \tag{8.156}$$

or

$$H_{m+1} = -2mH_{m-1}2\xi H_m. \tag{8.157}$$

The lowest order differential equation that can be constructed from equations 8.154 and 8.157 is

$$H_n'' - 2\xi H_n' + 2nH_n = 0. \tag{8.158}$$

If we compute $\dfrac{\partial^n S}{\partial s^n}$, then set $s = 0$, the result is $H_n(\xi)$. That is

$$\left. \frac{\partial^n S}{\partial s^n} \right|_{s=0} = e^{\xi^2} \frac{\partial^n}{\partial s^n} e^{-(s-\xi)^2} \bigg|_{s=0}$$

$$= (-1)^n e^{\xi^2} \frac{\partial^n}{\partial \xi^n} e^{-(s-\xi)^2} \bigg|_{s=0}$$

$$= (-1)^n e^{\xi^2} \frac{\partial^n}{\partial \xi^n} e^{-\xi^2}$$

$$= \frac{\partial^n}{\partial s^n} \left(\sum_{n=0}^{\infty} \frac{H_n(\xi)}{n!} s^n \right) \bigg|_{s=0}$$

$$= H_n(\xi). \tag{8.159}$$

Therefore,

$$H_n(\xi) = (-1)^n e^{\xi^2} \frac{\partial^n}{\partial \xi^n} e^{-\xi^2}, \tag{8.160}$$

so that

$$H_o(\xi) = 1, \tag{8.161}$$

$$H_1(\xi) = 2\xi, \tag{8.162}$$

$$H_2(\xi) = 4\xi^2 - 2, \tag{8.163}$$

and so on.

8.3.4.2 Harmonic Oscillator Wave Functions

Equation 8.138 shows that the harmonic oscillator wave function is of the form

$$u_n(x) = N_n H_n(\alpha x) e^{-\frac{1}{2}\alpha^2 x^2}. \tag{8.164}$$

The condition for normalization is

$$\int_{-\infty}^{+\infty} |u_n(x)|^2 \, dx = \frac{|N_n|^2}{\alpha} \int_{-\infty}^{+\infty} H_n^2(\xi) e^{-\xi^2} \, d\xi = 1. \tag{8.165}$$

From the definition of the Hermite polynomials, equation 8.160, we have

$$\sum_{n=0}^{\infty} \sum_{m=0}^{\infty} \frac{s^n t^m}{n! m!} \int_{-\infty}^{+\infty} H_n(\xi) H_m(\xi) e^{-\xi^2} \, d\xi$$

$$= \int_{-\infty}^{+\infty} e^{-s^2 + 2s\xi} e^{-t^2 + 2t\xi} e^{-\xi^2} \, d\xi. \tag{8.166}$$

By inspection, the last term reduces to

$$e^{-s^2 + 2s\xi} e^{-t^2 + 2t\xi} e^{-\xi^2}$$

$$= e^{+2st} e^{-(s+t)^2 + 2(s+t)\xi - \xi^2} = e^{+2st} e^{-[(s+t)-\xi]^2}, \tag{8.167}$$

so that the integral has solution

$$\pi^{1/2} e^{2st} = \pi^{1/2} \sum_{n=0}^{\infty} \frac{(2st)^n}{n!}. \tag{8.168}$$

By equating powers of s, t, we obtain

$$\int_{-\infty}^{+\infty} H_n^2(\xi) e^{-\xi^2} \, d\xi = \pi^{1/2} 2^n n! \tag{8.169}$$

and

$$\int_{-\infty}^{+\infty} H_n(\xi) H_m(\xi) e^{-\xi^2} d\xi = 0, \qquad (8.170)$$

for $n \neq m$. Therefore,

$$N_n = \left(\frac{\alpha}{\pi^{1/2} 2^n n!} \right)^{1/2}, \qquad (8.171)$$

and

$$u_n(x) = \left(\frac{\alpha}{\pi^{1/2} 2^n n!} \right)^{1/2} H_n(\alpha x) e^{-\frac{1}{2}\alpha^2 x^2}. \qquad (8.172)$$

From this definition of u_n, it follows that

$$\langle x \rangle = \int_{-\infty}^{+\infty} u_n^*(x) x u_m(x) dx = \frac{1}{\alpha} \left(\frac{n+1}{2} \right)^{1/2}, \qquad (8.173)$$

if $m = n + 1$,

$$\langle x \rangle = \int_{-\infty}^{+\infty} u_n^*(x) x u_m(x) dx = \frac{1}{\alpha} \left(\frac{n}{2} \right)^{1/2}, \qquad (8.174)$$

if $m = n - 1$, and

$$\langle x \rangle = \int_{-\infty}^{+\infty} u_n^*(x) x u_m(x) dx = 0, \qquad (8.175)$$

otherwise.

8.3.4.3 Matrix Theory of the Harmonic Oscillator

The Hamiltonian for the linear harmonic oscillator

$$H = \frac{p^2}{2m} + \frac{1}{2}kx^2 \tag{8.176}$$

and the condition

$$[x, p] = xp - px = i\hbar \tag{8.177}$$

are all that is needed to determine the energy eigenvalues of the system. If H is diagonal, we have

$$\langle k|H|l\rangle = E_k\langle k|l\rangle = \frac{1}{2m}\langle k|p|j\rangle\langle j|p|l\rangle + \frac{1}{2}k\langle k|x|j\rangle\langle j|x|l\rangle, \tag{8.178}$$

where

$$\langle j|p|l\rangle = \langle l|p|j\rangle^*, \ \langle j|x|l\rangle = \langle l|x|j\rangle^*. \tag{8.179}$$

We have

$$[x, H] = xH - Hx = x\left(\frac{p^2}{2m} + \frac{1}{2}kx^2\right) - \left(\frac{p^2}{2m} + \frac{1}{2}kx^2\right)$$

$$= \frac{1}{2m}\left(xp^2 - p^2x\right)$$

$$= \frac{1}{2m}\left(xp^2 - pxp + pxp - p^2x\right)$$

$$= \frac{1}{2m}\{[x, p]p + p[x, p]\}$$

$$= \frac{p}{m}i\hbar x. \tag{8.180}$$

Similarly,

$$[p, H] = -i\hbar kx. \tag{8.181}$$

These can be rewritten in matrix notation

$$[x, H] = \langle k|x|j\rangle\langle j|H|l\rangle - \langle k|H|j\rangle\langle j|x|l\rangle$$

$$= (E_l - E_k)\langle k|x|l\rangle$$

$$= \frac{i\hbar}{m}\langle k|p|l\rangle, \tag{8.182}$$

and

$$[p, H] = \langle k|p|j\rangle\langle j|H|l\rangle - \langle k|H|j\rangle\langle j|p|l\rangle$$

$$= (E_l - E_k)\langle k|p|l\rangle$$

$$= -i\hbar k\langle k|x|l\rangle. \tag{8.183}$$

In addition, we have the constraints

$$\langle k|x|l\rangle = \langle k|p|l\rangle = 0 \tag{8.184}$$

and

$$(E_l - E_k) = \pm\hbar\left(\frac{k}{m}\right)^{1/2} = \pm\hbar\omega_c. \tag{8.185}$$

That is, the only possible states are those whose energy eigenvalues differ by $\pm\hbar\omega_c$. Using $k = m\omega_c^2$, we rewrite equations 8.182 and 8.183 as

$$-im\omega_c(E_l - E_k)\langle k|x|l\rangle = \hbar\omega_c\langle k|p|l\rangle, \tag{8.186}$$

and

$$-i\hbar k\langle k|x|l\rangle = (E_l - E_k)\langle k|p|l\rangle. \tag{8.187}$$

Adding these two gives

$$(E_l - E_k)\langle k|-im\omega_c x|l\rangle + \frac{i\hbar k}{im\omega_c}\langle k|-im\omega_c x|l\rangle$$

$$= (E_l - E_k + \hbar\omega_c)\langle k|p|l\rangle, \tag{8.188}$$

or

$$(E_l - E_k + \hbar\omega_c)\langle k | p + im\omega_c x | l \rangle = 0. \tag{8.189}$$

Thus, $\langle k | p + im\omega_c x | l \rangle \neq 0$ only if $(E_l - E_k + \hbar\omega_c) = 0$. That is, the operator $p + im\omega_c x$ operating on the ket $|l\rangle$ is a multiple of the ket $|k\rangle$ with a higher energy $E_k = E_l + \hbar\omega_c$. We define

$$a^+ = p + im\omega_c x, \tag{8.190}$$

and

$$a^- = p - im\omega_c x, \tag{8.191}$$

to be the raising and lowering operators, respectively. a^+ raises the energy by $\hbar\omega_c$, $a-$ lowers the energy by $\hbar\omega_c$. We also require that

$$(p - i\hbar\omega_c x)|0\rangle = 0. \tag{8.192}$$

To find the energy eigenvalues, consider

$$a^+ a^- = (p + i\hbar\omega_c x)(p - i\hbar\omega_c x), \tag{8.193}$$

and

$$a^+ a^- |0\rangle = \left(p^2 + m^2\omega_c^2 x^2 - m\hbar\omega_c\right)|0\rangle = 2m\left(H - \frac{1}{2}\hbar\omega_c\right)|0\rangle = 0. \tag{8.194}$$

$|0\rangle$ is an eigenstate of H with energy $\frac{1}{2}\hbar\omega_c$. Similarly,

$$(p + i\hbar\omega_c x)|0\rangle = (E_0 + \hbar\omega_c)|1\rangle = E_1|1\rangle. \tag{8.195}$$

Therefore, $E_1 = \frac{3}{2}\hbar\omega_c$ and $E_n = \left(n + \frac{1}{2}\right)\hbar\omega_c$. We have

$$a^- a^+ = \frac{H}{\hbar\omega_c} + \frac{1}{2}, \tag{8.196}$$

and

$$a^+ a^- = \frac{H}{\hbar\omega_c} - \frac{1}{2}, \tag{8.197}$$

therefore

$$a^- a^+ - a^+ a^- = 1, \tag{8.198}$$

and

$$H = \left(a^+ a^- + \frac{1}{2} \right) \hbar\omega . \tag{8.199}$$

The eigenvalues of $a^+ a^-$ are the integers, and $a^+ a^-$ is called the *number operator*. In terms of a^+, a^-

$$x = \left(\frac{\hbar}{2m\omega_c} \right)^{1/2} \left(a^+ + a^- \right), \tag{8.200}$$

and

$$p = i \left(\frac{m\hbar\omega_c}{2} \right)^{1/2} \left(a^+ - a^- \right). \tag{8.201}$$

8.4 Applications: Three-Dimensional Systems

8.4.1 Spherical Potentials

It is generally impossible to obtain analytic solutions of the three-dimensional wave equation unless it can be separated into total differential equations in each of the three space coordinates. One system where this can be done is the spherical polar system, where $x = r\sin\theta\cos\phi$, $y = r\sin\theta\sin\phi$, and $z = r\cos\theta$. By definition,

$$\frac{\partial}{\partial x} = \frac{\partial r}{\partial x}\frac{\partial}{\partial r} + \frac{\partial\theta}{\partial x}\frac{\partial}{\partial\theta} + \frac{\partial\phi}{\partial x}\frac{\partial}{\partial\phi}, \tag{8.202}$$

where

$$r = \left(x^2 + y^2 + z^2\right)^{1/2}, \tag{8.203}$$

$$\theta = \cos^{-1}\left[\frac{z}{\left(x^2 + y^2 + z^2\right)^{1/2}}\right], \tag{8.204}$$

and

$$\phi = \tan^{-1}\left[\frac{y}{x}\right]. \tag{8.205}$$

Thus,

$$\frac{\partial}{\partial x} = \frac{1}{2}\left(x^2 + y^2 + z^2\right)^{-1/2} 2x \frac{\partial}{\partial r} + \frac{xz}{r^2 \sin\theta} \frac{\partial}{\partial \theta} - \frac{y}{x^2} \cos^2\phi \frac{\partial}{\partial \phi} \tag{8.206}$$

or

$$\frac{\partial}{\partial x} = \sin\theta \cos\phi \frac{\partial}{\partial r} + \frac{\cos\theta \cos\phi}{r} \frac{\partial}{\partial \theta} - \frac{\sin\phi}{r \sin\theta} \frac{\partial}{\partial \phi}. \tag{8.207}$$

In like manner, it can be shown that

$$\nabla^2 = \frac{1}{r^2} \frac{\partial}{\partial r}\left(r^2 \frac{\partial}{\partial r}\right) + \frac{1}{r^2 \sin\theta} \frac{\partial}{\partial \theta}\left(\sin\theta \frac{\partial}{\partial \theta}\right) + \frac{1}{r^2 \sin\theta} \frac{\partial^2}{\partial \phi^2}. \tag{8.208}$$

If we let

$$\psi(r,\theta,\phi) = R(r)Y(\theta,\phi), \tag{8.209}$$

the Schrödinger equation (eq. 8.46) becomes

$$\frac{1}{R} \frac{\partial}{\partial r}\left(r^2 \frac{\partial R}{\partial r}\right) + \frac{2mr^2}{\hbar^2}(E - V) = -\frac{1}{Y}\left[\frac{1}{\sin\theta} \frac{\partial}{\partial \theta}\left(\sin\theta \frac{\partial Y}{\partial \theta}\right) + \frac{1}{\sin\theta} \frac{\partial^2 Y}{\partial \phi^2}\right]. \tag{8.210}$$

This can only be true if both sides equal a constant. Therefore,

$$\frac{1}{r^2}\frac{\partial}{\partial r}\left(r^2\frac{\partial R}{\partial r}\right)+\left\{\frac{2m}{\hbar^2}\left(E-V\right)-\frac{\lambda}{r^2}\right\}R=0\,, \qquad (8.211)$$

and

$$\frac{1}{\sin\theta}\frac{\partial}{\partial\theta}\left(\sin\theta\frac{\partial Y}{\partial\theta}\right)+\frac{1}{\sin\theta}\frac{\partial^2 Y}{\partial\phi^2}+\lambda Y=0\,. \qquad (8.212)$$

If the angular equation can in turn be separated as

$$Y(\theta,\phi)=\Theta(\theta)\Phi(\phi)\,, \qquad (8.213)$$

then equation 8.197 reduces to

$$\left[\frac{1}{\sin\theta}\frac{\partial}{\partial\theta}\left(\sin\theta\frac{\partial\Theta}{\partial\theta}\right)+\lambda\Theta\right]\Phi\sin^2\theta=-\frac{\partial^2\Phi}{\partial\phi^2}\,. \qquad (8.214)$$

Again, this can only be true if both sides of the equation equal a constant and we have

$$\frac{1}{\sin\theta}\frac{\partial}{\partial\theta}\left(\sin\theta\frac{\partial\Theta}{\partial\theta}\right)+\left(\lambda-\frac{\nu}{\sin^2\theta}\right)\Theta=0\,, \qquad (8.215)$$

and

$$\frac{\partial^2\Phi}{\partial\phi^2}+\nu\Phi=0\,. \qquad (8.216)$$

This last equation has solutions

$$\Phi(\phi)=Ae^{i\sqrt{\nu}\phi}+Be^{-i\sqrt{\nu}\phi}\,, \qquad (8.217)$$

for $\nu\neq 0$, and

$$\Phi(\phi)=A+B\phi\,, \qquad (8.218)$$

for $\nu=0$. Both Φ and $d\Phi/d\phi$ must be continuous over the range $0-2\pi$, therefore $\sqrt{\nu}$ must be an integer. The general solution is

$$\Phi(\phi) = \frac{1}{\sqrt{2\pi}} e^{im\phi}. \tag{8.219}$$

We can now solve equation 8.200, which reduces to

$$\frac{1}{\sin\theta} \frac{\partial}{\partial\theta}\left(\sin\theta \frac{\partial\Theta}{\partial\theta}\right) + \left(\lambda - \frac{m^2}{\sin^2\theta}\right)\Theta = 0. \tag{8.220}$$

If we let $\omega = \cos\theta$, the equation further simplifies to

$$\frac{d}{d\omega}\left((1-\omega^2)\frac{dP}{d\omega}\right) + \left(\lambda - \frac{m^2}{1-\omega^2}\right)P = 0. \tag{8.221}$$

If $m = 0$, the physically acceptable solutions to this equation are the Legendre polynomials $P_l(\omega)$. Their generating function is

$$T(\omega,s) = \left(1 - 2s\omega + s^2\right)^{-1/2} = \sum_{l=0}^{\infty} P_l(\omega)s^l \tag{8.222}$$

for $s < 1$. In the manner done for Hermite polynomials,

$$\frac{\partial T}{\partial\omega} = -\frac{1}{2}\left(1 - 2s\omega + s^2\right)^{-3/2}(-2s) = \frac{s}{\left(1 - 2s\omega + s^2\right)^{3/2}}, \tag{8.223}$$

and

$$\frac{\partial T}{\partial s} = -\frac{1}{2}\left(1 - 2s\omega + s^2\right)^{-3/2}(-2\omega + 2s) = \frac{\omega - s}{\left(1 - 2s\omega + s^2\right)^{3/2}}, \tag{8.224}$$

or equivalently,

$$\frac{\partial T}{\partial\omega} = \sum_{l=0}^{\infty} P_l^3(\omega)s^{3l+1} = \sum_{l=0}^{\infty} P_l'(\omega)s^l, \tag{8.225}$$

and

$$\frac{\partial T}{\partial s} = \sum_{l=0}^{\infty}\left(\omega P_l^3(\omega)s^{3l} - P_l^3(\omega)s^{3l+1}\right) = \sum_{l=0}^{\infty} lP_l(\omega)s^{l-1}. \quad (8.226)$$

It can be shown that

$$\left(1-\omega^2\right)P_l' = -l\omega P_l + lP_{l-1}, \quad (8.227)$$

and

$$(l+1)P_l = (2l+1)\omega P_l - lP_{l-1}. \quad (8.228)$$

If $\lambda = l(l+1)$, $m = 0$, this satisfies the differential equation that gives the Legendre polynomials. If $m \neq 0$, there are physically admissible solutions for $|m| \leq l$. These are the associated Legendre functions,

$$P_l^m(\omega) = \left(1-\omega^2\right)^{\frac{|m|}{2}} \frac{d^{|m|}}{d\omega^{|m|}} P_l(\omega). \quad (8.229)$$

The total solution for Y is called a *spherical harmonic,*

$$Y_{lm}(\theta,\phi) = N_{lm}P_l^m(\cos\theta)\Phi_m(\phi). \quad (8.230)$$

We now look at the radial wave equation (eq. 8.210) which is

$$\frac{1}{r^2}\frac{\partial}{\partial r}\left(r^2\frac{\partial R}{\partial r}\right) + \left\{\frac{2m}{\hbar^2}(E-V) - \frac{l(l+1)}{r^2}\right\}R = 0. \quad (8.231)$$

If we make the substitution $R = \chi/r$, we obtain

$$-\frac{\hbar^2}{2m}\frac{d^2\chi}{dr^2} + \left[V(r) + \frac{l(l+1)\hbar^2}{2mr^2}\right]\chi = E\chi. \quad (8.232)$$

If a classical particle has angular momentum L about an axis through the origin, $\omega = L/mr^2$, it experiences an inward force,

$$m\omega^2 r = \frac{L^2}{mr^3}. \quad (8.233)$$

As a result, the term

$$V(r) + \frac{l(l+1)\hbar^2}{2mr^2}, \tag{8.234}$$

has the form of $V(r)$ plus an angular momentum term, if $L^2 = l(l+1)\hbar^2$. To see this rigorously, consider

$$L_x = yp_z - zp_y$$

$$= -i\hbar\left(y\frac{\partial}{\partial z} - z\frac{\partial}{\partial y}\right)$$

$$= i\hbar\left(\sin\phi\frac{\partial}{\partial\theta} + \cot\theta\cos\phi\frac{\partial}{\partial\theta}\right), \tag{8.235}$$

$$L_y = zp_x - xp_z$$

$$= -i\hbar\left(z\frac{\partial}{\partial x} - x\frac{\partial}{\partial z}\right)$$

$$= i\hbar\left(-\cos\phi\frac{\partial}{\partial\theta} + \cot\theta\sin\phi\frac{\partial}{\partial\theta}\right), \tag{8.236}$$

and

$$L_z = xp_y - yp_x.$$

$$= -i\hbar\left(x\frac{\partial}{\partial y} - y\frac{\partial}{\partial x}\right)$$

$$= -i\hbar\frac{\partial}{\partial\phi} \tag{8.237}$$

Consequently,

$$L^2 = L_x^2 + L_y^2 + L_z^2 = -\hbar^2\left[\frac{1}{\sin\theta}\frac{\partial}{\partial\theta}\left(\sin\theta\frac{\partial}{\partial\theta}\right) + \frac{1}{\sin^2\theta}\frac{\partial^2}{\partial\phi^2}\right]. \tag{8.238}$$

This is similar to the equation for Y, that is,

$$-\frac{L^2}{\hbar^2}Y + \lambda Y = 0. \qquad (8.239)$$

Thus,

$$L^2 Y = l(l+1)\hbar^2 Y, \qquad (8.240)$$

where l is the orbital angular momentum quantum number. Similarly,

$$L_z Y = m\hbar Y, \qquad (8.241)$$

where m is the magnetic quantum number.

8.4.2 Free Particle in Spherical Coordinates

For a free particle, $V(r) = 0$, the radial equation is

$$\frac{1}{r^2}\frac{\partial}{\partial r}\left(r^2\frac{\partial R}{\partial r}\right) + \left\{\frac{2mE}{\hbar^2} - \frac{l(l+1)}{r^2}\right\}R = 0, \qquad (8.242)$$

or simply

$$\frac{d^2\chi}{dr^2} + \left[k^2 - \frac{l(l+1)}{r^2}\right]\chi = 0, \qquad (8.243)$$

with $k^2 = \dfrac{2mE}{\hbar^2}$. If $l = 0$, then the solution $\chi = A\sin kr$ preserves the constraint that $\chi(r = 0) = 0$. That is,

$$R = \frac{A\sin r}{r}. \qquad (8.244)$$

If $l \neq 0$, then $R(r)$ is equivalent to the spherical Bessel functions,

$$j_l(kr) = (-1)^l\left(\frac{r}{k}\right)^l\left(\frac{1}{r}\frac{d}{dr}\right)^l\frac{\sin kr}{kr}. \qquad (8.245)$$

For example,

$$j_0(kr) = \frac{\sin kr}{kr}, \qquad (8.246)$$

$$j_1(kr) = \frac{\sin kr}{(kr)^2} - \frac{\cos kr}{kr}, \qquad (8.247)$$

and so on.

8.4.3 Radial Spherical Potential

Consider the potential shown in figure 8.4. The radial equation is

$$-\frac{\hbar^2}{2m}\frac{d^2\chi}{dr^2} - V_o\chi = E\chi, \qquad (8.248)$$

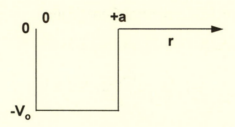

Figure 8.4 A radially spherical potential.

for $r < a$, and

$$-\frac{\hbar^2}{2m}\frac{d^2\chi}{dr^2} = E\chi \qquad (8.249)$$

for $r > a$. The solutions are

$$\chi(r) = A\sin\alpha r + B\cos\alpha r \qquad (8.250)$$

for $r < a$, and

$$\chi(r) = Ce^{-\beta r} \qquad (8.251)$$

for $r > a$, with

$$\alpha = \left[\frac{2m(V_o - E)}{\hbar^2}\right]^{1/2}, \tag{8.252}$$

and

$$\beta = \left[\frac{2mE}{\hbar^2}\right]^{1/2}. \tag{8.253}$$

Because $R(r = 0) \neq \infty$, we must set $B = 0$. Continuity of R and $\frac{1}{R}\frac{dR}{dr}$ at $r = a$ requires

$$\alpha \cot \alpha a = -\beta. \tag{8.254}$$

If we let $\rho = \alpha r$, the radial solution for $r < a$ is

$$\frac{d^2 R}{d\rho^2} + \frac{2}{\rho}\frac{dR}{d\rho} + \left[1 - \frac{l(l+1)}{\rho^2}\right]R = 0. \tag{8.255}$$

Comparing this to Bessel's equation,

$$\rho \frac{d}{d\rho}\left(\rho \frac{dP}{d\rho}\right) + \left(\gamma^2 p^2 - m^2\right)P = 0, \tag{8.256}$$

we see that the solutions to $R(r)$ are given in terms of Bessel's functions.

8.5 Approximation Methods

8.5.1 Stationary Perturbation Theory

Suppose that we have a simple Hamiltonian H_0 subjected to a small perturbation H', so that the total Hamiltonian can be written as

$$H = H_0 + \lambda H'. \tag{8.257}$$

where λ is simply a convenient parameter introduced to track the measure of the perturbation. That is, the unperturbed case can be recovered by letting λ tend to zero. We define the unperturbed eigenfunctions and eigenvalues by

$$H_0 \phi_n = E_n \phi_n, \tag{8.258}$$

while the perturbed wave functions and energy levels are given by

$$H \psi_n = W_n \psi_n. \tag{8.259}$$

We further make the assumption that

$$\lim_{\lambda \to 0} W_n = E_n, \tag{8.260}$$

and

$$\lim_{\lambda \to 0} \psi_n = \phi_n. \tag{8.261}$$

If we express the perturbed wave function as a superposition of unperturbed states, we have

$$\psi_n = \sum_i a_{in} \phi_i . \tag{8.262}$$

From the definition of H, W_n, and ψ_n, we have

$$H \psi_n = (H_o + \lambda H') \sum_i a_{in} \phi_i = \sum_i a_{in} E_i \phi_i + \lambda \sum_i a_{in} H' \phi_i, \tag{8.263}$$

and

$$W_n \psi_n = W_n \sum_i a_{in} \phi_i , \tag{8.264}$$

so that equation 8.259 reduces to

$$W_n \sum_i a_{in} \phi_i = \sum_i a_{in} E_i \phi_i + \lambda \sum_i a_{in} H' \phi_i . \tag{8.265}$$

If we multiply through by ϕ_j^* and integrate, we obtain

$$\langle j | W_n | a_{in} \phi_i \rangle = \langle j | E_i | a_{in} \phi_i \rangle + \lambda \langle j | H' | a_{in} \phi_i \rangle, \tag{8.266}$$

or simply

$$a_{jn}W_n = a_{jn}E_j + \lambda \sum_i H'_{ji}a_{in}, \qquad (8.267)$$

with

$$H'_{ji} = \langle \phi_j | H' | \phi_i \rangle. \qquad (8.268)$$

Equation 8.267 is equivalent to

$$a_{nn}W_n = a_{nn}E_n + \lambda a_{nn}H'_{nn} + \lambda \sum_{i \neq n} H'_{ni}a_{in}, \qquad (8.269)$$

for $j = n$, and

$$a_{jn}W_n = a_{jn}E_j + \lambda a_{nn}H'_{jn} + \lambda \sum_{i \neq n} H'_{ji}a_{in}, \qquad (8.270)$$

for $j \neq n$. Dividing equation 8.269 by a_{nn} gives

$$W_n = E_n + \lambda H'_{nn} + \lambda \sum_{i \neq n} H'_{ni} \frac{a_{in}}{a_{nn}}, \qquad (8.271)$$

for $j = n$. Performing a similar operation on equation 8.270 and rearranging terms gives

$$\frac{a_{jn}}{a_{nn}} = \frac{\lambda}{W_n - E_j} H'_{jn} + \frac{\lambda}{W_n - E_j} \sum_{i \neq n} H'_{ji} \frac{a_{in}}{a_{nn}}. \qquad (8.272)$$

Equation 8.272 indicates that the ratio a_{jn}/a_{nn} is of order λ, so that the summation term in equation 8.272 contributes terms of order λ^2 and greater. As a result, the energy eigenvalues W_n and the state functions ψ_n can be expressed as a power series in λ. That is, for each nth state,

$$W_n = W_{n,0} + \lambda W_{n,1} + \lambda^2 W_{n,2} + \ldots, \qquad (8.273)$$

and

$$\psi_n = \psi_{n,0} + \lambda \psi_{n,1} + \lambda^2 \psi_{n,2} + \ldots \ldots \tag{8.274}$$

where, from equations 8.246 and 8.247,

$$W_{n,0} = E_n, \tag{8.275}$$

and

$$\psi_{n,0} = \phi_n. \tag{8.276}$$

Comparison of equations 8.271 and 8.272 indicates that the first-order correction to the energy (the λ term) is simply

$$W_{n,1} = H'_{nn} = \left\langle \phi_n | H' | \phi_n \right\rangle. \tag{8.277}$$

Similarly, the second-order correction, (the λ^2 term), is obtained by selecting the term of order one from equation 8.274 and plugging it into equation 8.271 to obtain

$$W_{n,2} = \sum_{i \neq n} \frac{H'_{ni} H'_{in}}{E_n - E_i}. \tag{8.278}$$

Equation 8.271 is therefore equivalent to

$$W_n = E_n + \lambda H'_{nn} + \lambda^2 \sum_{i \neq n} \frac{H'_{ni} H'_{in}}{E_n - E_i} + \ldots \ldots \tag{8.279}$$

In like manner, it can be shown that the state function is given by

$$\psi_n = \phi_n + \lambda \sum_{j \neq n} \frac{H'_{jn}}{E_n - E_j} \phi_j + \ldots \ldots \tag{8.280}$$

Example 8.2
A plane rigid rotator with a moment of inertia I, and electric dipole moment d is placed in a homogeneous electric field E. Determine the lowest order nonvanishing correction to the energy levels due to the electric field. The unperturbed Hamiltonian is

$$H_0 = -\frac{\hbar^2}{2I}\frac{d^2}{d\phi^2}.$$ (8.281)

From the definition of the problem, the perturbation is given by

$$H_1 = -\vec{d} \cdot \vec{E} = -dE\cos\theta.$$ (8.282)

We first solve the unperturbed equation (eq. 8.259). Using the form of H_0 given in the problem, we have

$$-\frac{\hbar^2}{2I}\psi_0'' - W_0\psi_0 = 0,$$ (8.283)

which has the general solution

$$\psi_0 = \frac{1}{\sqrt{2\pi}}e^{ik\phi},$$ (8.284)

with

$$n = \sqrt{\frac{2IE_n}{\hbar^2}}.$$ (8.285)

From equation 8.277, we have

$$W_{n,1} = \frac{1}{2\pi}\int_0^{2\pi} e^{2in\phi}(-dE\cos\theta)d\phi$$

$$= \frac{-dE}{4\pi}\int_0^{2\pi}\left[e^{i(2n+1)\phi} + e^{i(2n-1)\phi}\right]d\phi$$

$$= 0.$$ (8.286)

Look at the second-order term, from equation 8.278,

$$W_{n,2} = \sum_{i\neq n}\frac{H'_{ni}H'_{in}}{E_n - E_i}$$

$$= \left(\frac{dE}{\hbar\pi}\right)^2 \frac{I}{2} \sum_{m\neq k} \frac{\left|\frac{1}{2}\int_0^{2\pi}\left[e^{i(m-n+1)\phi} + e^{i(m-n-1)\phi}\right]d\phi\right|}{n^2 - m^2}$$

$$= \left(\frac{dE}{\hbar\pi}\right)^2 \frac{I}{8}\left[\frac{2\pi}{n^2-(n-1)^2} + \frac{2\pi}{n^2-(n+1)^2}\right]^2$$

$$= 8I\left(\frac{dEn}{\hbar}\right)^2\left[\frac{1}{4n^2-1}\right]^2. \tag{8.287}$$

8.5.2 The Variational (Rayleigh-Ritz) Method

The variation method can be used for the approximate determination of the lowest or ground-state energy of a system, when there is no closely related problem that is capable of exact solution, so that the perturbation method is inapplicable. We have (eq. 8.5)

$$\psi = \sum_E A_E u_E, \tag{8.288}$$

where $Hu_E = Eu_E$, and

$$\langle H\rangle = \langle\psi|H|\psi\rangle = \sum_E E|A_E|^2. \tag{8.289}$$

If we replace each eigenvalue E with the lowest eigenvalue E_0, then obviously

$$\langle H\rangle \geq \sum_E E_0|A_E|^2 = E_0\sum_E|A_E|^2 = E_0, \tag{8.290}$$

or, more generally,

$$E_o \leq \frac{\langle\psi|H|\psi\rangle}{\langle\psi|\psi\rangle}. \tag{8.291}$$

Evaluate E_0 with a trial function ψ that depends on a number of parameters. Vary these parameters until the expectation value of the energy is a minimum.

8.5.3 WKB Approximation

A solution to the Schrödinger equation can be written in the form

$$\psi(\vec{r},t) = A(r)e^{iS(r)/\hbar}. \tag{8.292}$$

In one dimension, $\dfrac{d^2\psi}{dx^2}$ is

$$\frac{d^2\psi}{dx^2} = \left[\frac{d^2A}{dx^2} + 2i\frac{dA}{dx}\frac{dS}{dx} + \frac{i}{\hbar}A\frac{d^2S}{dx^2} - \frac{A}{\hbar^2}\left(\frac{dS}{dx}\right)^2 \right] e^{-iS/\hbar}. \tag{8.293}$$

Plugging into the Schrödinger equation gives

$$-\frac{\hbar^2}{2\mu}\left[A'' + 2iA'S' + \frac{i}{\hbar}AS'' - \frac{A}{\hbar^2}(S')^2 \right] e^{-iS/\hbar}$$

$$+ VAe^{-iS/\hbar} = EAe^{-iS/\hbar}, \tag{8.294}$$

or

$$A\left[\frac{1}{2\mu}(S')^2 + V - E \right] - \frac{i\hbar}{2\mu}[2A'S' + AS''] - \frac{\hbar^2}{2m}A'' = 0. \tag{8.295}$$

To solve for A, S in the limit as $\hbar \to 0$, set the first two terms separately to zero and neglect the third term.

$$\frac{dS}{dx} = \pm[2\mu(E - V)] = \pm p(x), \tag{8.296}$$

$$S(x) = \pm\int p(x)dx, \tag{8.297}$$

and

$$2\frac{dA}{dx}p(x) + A\frac{dP}{dx} = 0, \tag{8.298}$$

with $\frac{d}{dx}\left(A^2 p\right) = 0$ and $A = \frac{c}{\sqrt{p}}$. The solution is

$$\psi_E^{\pm} = \frac{c_{\pm}}{\sqrt{p}} e^{\pm i \int p \, dx / \hbar}, \qquad (8.299)$$

with

$$p = \left(2\mu(E - V)\right)^{1/2}. \qquad (8.300)$$

In terms of a definite integral,

$$\psi_E^{\pm}(x) = \psi_E^{\pm}(x_o) \sqrt{\frac{P(x_o)}{P(x)}} \exp\left[\pm i \int_{x_o}^{x} p \frac{dx}{\hbar}\right]. \qquad (8.301)$$

The WKB solutions are useful when the potential energy changes so slowly that the momentum of the particle is sensibly constant over many wavelengths.

8.5.4 Time-Dependent Perturbation Theory

If we have a Hamiltonian of the form

$$H = H_0 + H'(t), \qquad (8.302)$$

we wish to examine solutions to

$$H\psi = \left(H_0 + H'(t)\right)\psi = -\frac{\hbar}{i} \frac{\partial \psi}{\partial t}. \qquad (8.303)$$

We can assume

$$\psi(x,t) = \sum_n a_n(t) \phi_n(x). \qquad (8.304)$$

If $H' = 0$, then $a_n \propto e^{-iE_n t / \hbar}$, so let

$$\psi(x,t) = \sum_n C_n(t)e^{-iE_nt/\hbar}\phi_n(x). \qquad (8.305)$$

Solving equation 8.303 now gives

$$H\psi = \sum_n C_n(t)E_n\phi_n e^{-iE_nt/\hbar} + \sum_n C_n(t)e^{-iE_nt/\hbar}H'\phi_n, \qquad (8.306)$$

which is equivalent to

$$-\frac{\hbar}{i}\frac{\partial\psi}{\partial t} = -\frac{\hbar}{i}\sum_n \frac{dc_n}{dt}e^{-iE_nt/\hbar}\phi_n + \sum_n C_n E_n e^{-iE_nt/\hbar}\phi_n. \qquad (8.307)$$

The Schrödinger equation is

$$-\frac{\hbar}{i}\sum_n \frac{dc_n}{dt}e^{-iE_nt/\hbar}\phi_n = \sum_n C_n(t)e^{-iE_nt/\hbar}H'\phi_n. \qquad (8.308)$$

Multiplying by ϕ_m^* and integrating gives

$$\frac{dc_m}{dt} = -\frac{i}{\hbar}\sum_n H'_{mn(t)}e^{-i(E_n-E_m)t/\hbar}c_n(t). \qquad (8.309)$$

If we assume a weak perturbation or look only at small values of t, then all c_m except c_k will be small. Where k is the initial unperturbed state, we have

$$\frac{dc_k}{dt} = -\frac{i}{\hbar}H'_{kk}(t)c_k(t), \qquad (8.310)$$

for $m = k$, and

$$\frac{dc_m}{dt} = -\frac{i}{\hbar}H'_{mk}(t)e^{-i(E_k-E_m)t/\hbar}c_k(t) \qquad (8.311)$$

for $m \neq k$. This gives

$$c_k = \exp^{-\frac{i}{\hbar}\int_0^t H'_{kk}(t)dt} \qquad (8.312)$$

for $m = k$, and

$$c_m \approx -\frac{i}{\hbar} \int_0^t H'_{mk}(t) e^{-i(E_k - E_m)t/\hbar} dt \qquad (8.313)$$

for $m \neq k$, where we have set $c_k \approx 1$.

If $H'(t)$ is independent of time except for being switched on at $t = 0$ and off at $t = t$, then

$$c_k = \exp^{-\frac{i}{\hbar} H'_{kk} t} \qquad (8.314)$$

for $m = k$, and

$$c_m \approx \frac{H'_{mk}}{E_k - E_m} \left[e^{-i(E_k - E_m)t/\hbar} - 1 \right] \qquad (8.315)$$

for $m \neq k$. Equivalently,

$$|c_m(t)|^2 = \frac{4|H'_{mk}|^2}{(E_k - E_m)^2} \sin^2 \left(\frac{(E_k - E_m)t}{2\hbar} \right) \qquad (8.316)$$

if we use degenerate, or nearly degenerate, states. For $E_k \sim E_m$ we can expand the $\sin^2$ term to get

$$|c_m(t)|^2 \approx \frac{|H'_{mk}|^2 t^2}{\hbar^2}. \qquad (8.317)$$

Consider the probability for transitions to a dense group of final states, whose energy E'_k lies within a given interval $(E_k + \Delta E, E_k - \Delta E)$. The transition probability is

$$P = \sum_{E_m \approx E_k} |c_m(t)|^2 = \sum_{E_m \approx E_k} 4|H'_{mk}|^2 \frac{\sin^2\left[(E_k - E)t/2\hbar\right]}{(E_k - E)^2}. \qquad (8.318)$$

If we neglect changes in $\rho(E)$ and H'_{mk} over this interval, letting $x = (E_k - E)t/2\hbar$ gives

$$P \approx \frac{2\pi}{\hbar} \rho(E) |H'_{mk}|^2 \int\limits_{-t\Delta E/2\hbar}^{+t\Delta E/2\hbar} \frac{\sin^2 x}{x^2} dx, \qquad (8.319)$$

or

$$P \approx \frac{2\pi}{\hbar} \rho(E) |H'_{mk}|^2 t. \qquad (8.320)$$

The transition rate is

$$W = \frac{dP}{dt} = \frac{2\pi}{\hbar} \rho(E) |H'_{mk}|^2. \qquad (8.321)$$

8.6 Scattering

8.6.1 Scattering Cross Section

Suppose that we bombard a group of n particles or scattering centers with a parallel flux of N particles per unit area per unit time and count the number that emerge per unit time in a small solid angle ΔW_0 centered about a direction that has polar angles θ_0, ϕ_0 with respect to the bombarding direction as polar axis. The number of incident particles that emerge per unit time in ΔW_0 is

$$nN\sigma_0(\theta_0, \phi_0)\Delta W_0. \qquad (8.322)$$

This is the differential scattering cross section for the laboratory coordinate system. In the center-of-mass coordinate system, the differential cross section in the direction θ, ϕ is $\sigma(\pi - \theta, \phi + \pi)$. The total scattering cross section is

$$\sigma_0 = \int \sigma_0(\theta_0, \phi_0) dw_0. \qquad (8.323)$$

The relation between the cross sections in the different coordinate systems is seen to be

$$\sigma_0(\theta_0, \phi_0)\sin\theta_0 d\theta_0 d\phi_0 = \sigma(\theta, \phi)\sin\theta d\theta d\phi, \qquad (8.324)$$

where the left-hand side of the equation is the laboratory frame and the right-hand side is the center of mass frame. From the relationship between the two systems, we can find

$$\sigma_0(\theta_0,\phi_0) = \frac{\left(1+\gamma^2 + 2\gamma\cos\theta\right)^{3/2}}{\left|1+\gamma\cos\theta\right|} \; \sigma(\theta,\phi), \qquad (8.325)$$

where

$$\gamma = \left(\frac{m_1 m_3}{m_2 m_4}\frac{E}{E+Q}\right)^{1/2}. \qquad (8.326)$$

The differential scattering cross section $\sigma(\theta,\phi)$ can be found, in center-of-mass coordinates, from

$$-\frac{\hbar^2}{2\mu}\nabla^2 u + Vu = Eu, \qquad (8.327)$$

the wave equation for relative motion, with

$$\mu = \frac{m_1 m_2}{m_1 + m_2} \qquad (8.328)$$

and

$$E = \frac{m_2}{m_1 + m_2} E_o. \qquad (8.329)$$

We can think of this equation as representing the elastic collision of a particle of mass μ, initial speed v, and kinetic energy $E = \frac{1}{2}\mu v^2$, with a fixed scattering center described by $V(\vec{r})$.

When the colliding particles are far apart, we want u to contain a part representing the initial particle and a radially outgoing particle. That is,

$$u(r,\theta,\phi) \underset{r\to\infty}{\to} A\left[e^{ikz} + \frac{f(\theta,\phi)e^{ikr}}{r}\right], \qquad (8.330)$$

with $k = \dfrac{\mu v}{\hbar}$ and $f(\theta, \phi)$ the scattering amplitude.

The particle flux is related to the probability current density

$$S(\vec{r},t) = \frac{\hbar}{2im}\left[\psi^* \nabla \psi - \left(\nabla \psi^*\right)\psi\right]. \tag{8.331}$$

To find the scattered flux, use

$$\psi = \frac{Af(\theta, \phi)e^{ikr}}{r}. \tag{8.332}$$

The leading term goes as

$$\nabla\psi = Af(\theta, \phi)\left\{ik\frac{e^{ikr}}{r} - \frac{e^{ikr}}{r^2}\right\}, \tag{8.333}$$

therefore,

$$\psi^* \nabla \psi - \left(\nabla \psi^*\right)\psi \approx |A|^2|f(\theta, \phi)|^2\left(\frac{ik}{r} - \frac{1}{r^2}\right)\left(-\frac{ik}{r} - \frac{1}{r^2}\right), \tag{8.334}$$

which reduces to

$$\psi^* \nabla \psi - \left(\nabla \psi^*\right)\psi \approx |A|^2|f(\theta, \phi)|^2\frac{k^2}{r^2}. \tag{8.335}$$

The probability current density becomes

$$S(\vec{r},t) \approx \frac{v^2|A|^2|f(\theta, \phi)|^2}{r^2}, \tag{8.336}$$

and

$$\sigma(\theta, \phi) = |f(\theta, \phi)|^2. \tag{8.337}$$

Example 8.3
A spinless particle of mass M is scattered by the potential shown below. At low energies the wave function for large r is well described by

$$\psi(\vec{r}) \to e^{i\vec{k}\cdot\vec{r}} + \frac{e^{i(kr+\delta)}}{kr}\sin\delta,$$

where δ(k) is a phase shift. Find (1) the total low-energy scattering cross section in terms of the quantities defined, and (2) the form of the Schrödinger equation for large r.

(1) By comparison with equation 8.330, it is obvious that

$$f(\theta) = \frac{e^{i\delta}}{k}\sin\delta. \qquad (8.338)$$

Consequently

$$\frac{d\sigma}{d\Omega} = |f(\theta)|^2 = \frac{\sin^2\delta}{k^2}, \qquad (8.339)$$

and

$$\sigma = \int |f(\theta)|^2 d\Omega = 4\pi\frac{\sin^2\delta}{k^2}. \qquad (8.340)$$

(2) The Schrödinger equation is

$$-\frac{\hbar^2}{2m}\nabla^2\psi + V(r)\psi = E\psi. \qquad (8.341)$$

Knowing the form of ψ, the first term in the Schrödinger equation reduces to

$$-\frac{\hbar^2}{2m}\nabla^2\psi = -\frac{\hbar^2}{2m}\nabla^2\left(e^{i\vec{k}\cdot\vec{r}} + \frac{e^{i(kr+\delta)}}{kr}\sin\delta\right)$$

$$= -\frac{\hbar^2}{2m}\frac{1}{r^2}\frac{\partial}{\partial r}\left[r^2\frac{\partial}{\partial r}\left(e^{ikr\cos\theta} + \frac{e^{i(kr+\delta)}}{kr}\sin\delta\right)\right]$$

$$-\frac{\hbar^2}{2m}\frac{1}{r^2\sin\theta}\frac{\partial}{\partial\theta}\left[\sin\theta\frac{\partial}{\partial\theta}\left(e^{ikr\cos\theta}\right)\right]$$

$$+ \text{ other terms that} \to 0. \qquad (8.342)$$

After simplification, the expression reduces to

$$-\frac{\hbar^2}{2m}\left[\left(\frac{2ik\cos\theta e^{ikr\cos\theta}}{r}-k^2\cos^2\theta e^{ikr\cos\theta}-\frac{ke^{ikr}e^{i\delta}\sin\delta}{r}\right)\right]$$

$$-\frac{\hbar^2}{2m}\left[\left(\frac{-2ik\cos\theta e^{ikr\cos\theta}}{r}-k^2\sin^2\theta e^{ikr\cos\theta}\right)\right], \qquad (8.343)$$

or simply

$$-\frac{\hbar^2}{2m}\left[\left(-k^2 e^{ikr\cos\theta}-\frac{ke^{ikr}e^{i\delta}\sin\delta}{r}\right)\right]. \qquad (8.344)$$

Consequently, the Schrödinger equation simplifies to

$$\left(\frac{\hbar^2 k^2}{2m}+V(r)-E\right)\left(e^{ikr\cos\theta}+\frac{e^{ikr}e^{i\delta}\sin\delta}{kr}\right)=0. \qquad (8.345)$$

8.6.2 Method of Partial Waves

If V is a function only of r, the method of partial waves allows us to find the connection between the solutions separated in spherical polar coordinates and the asymptotic form of equation 8.330. Obviously, the solution is independent of ϕ. Therefore we may write, for scattering off of a fixed center,

$$u(r,\theta)=\sum_{l=0}^{\infty}(2l+1)i^l R_l(r)P_l(\cos\theta), \qquad (8.346)$$

or

$$u(r,\theta)=\sum_{l=0}^{\infty}(2l+1)i^l \frac{\chi_l(r)}{r}P_l(\cos\theta), \qquad (8.347)$$

where χ_l satisfies

$$\frac{d^2\chi_l}{dr^2} + \left[k^2 - U(r) - \frac{l(l+1)}{r^2} \right]\chi_l = 0, \tag{8.348}$$

with

$$U(r) = \frac{2\mu V(r)}{\hbar}. \tag{8.349}$$

As $r \to \infty$, we may ignore $U(r)$ and $1/r^2$. We can try a solution of the form

$$\chi_l(r) = A \exp\left[\int_a^r f(r')dr' \right] e^{\pm ikr}, \tag{8.350}$$

where $\phi(r)$ falls off more rapidly than $1/r$ as $r \to \infty$. Using this gives

$$f' + f^2 \pm 2ikf = U(r) + \frac{l(l+1)}{r^2} = W(r). \tag{8.351}$$

If $W(r)$ varies as r^{-s}, then χ_l varies as $e^{\pm ikr}$. The general asymptotic solution is

$$\chi_l(r) \xrightarrow[r \to \infty]{} A_l' \sin(kr + \delta_l'), \tag{8.352}$$

where δ_l = phase shift of the lth partial wave. It is the difference in phase between the actual radial function and the function that would have been if $U = 0$. The most general form for $R_l(r)$ is, for $r < a$,

$$R_l(r) = A_l\left[\cos\delta_l j_l(kr) - \sin\delta_l n_l(kr) \right]. \tag{8.353}$$

We match the two solutions at $r = a$, subject to the constraint that $\dfrac{1}{R_l}\dfrac{dR_l}{dr}$ is continuous. We find that if ϕ_l is the ratio of slope to value of the interior wave function,

$$\gamma_l = \frac{k\left[j_l'(ka)\cos\delta_l - n_l'(ka)\sin\delta_l \right]}{j_l(ka)\cos\delta_l - n_l(ka)\sin\delta_l}, \tag{8.354}$$

or

$$\tan\delta_l = \frac{kj_l'(ka) - \gamma_l j_l(ka)}{kn_l'(ka) - \gamma_l n_l(ka)}. \tag{8.355}$$

8.7 Approximation Methods in Collision Theory

8.7.1 Scattering Matrix

We have previously seen, Chapter 1, that ϕ must satisfy

$$\psi(r',t') = i \int G(r',t':r,t)\psi(r,t)d^3r, \qquad (8.356)$$

where $G(r',t':r,t)$ is the Green's function. We define the retarded Green's function as

$$G^+(r',t':r,t) = G(r',t':r,t) \qquad (8.357)$$

for $t' < t$, and

$$G^+(r',t':r,t) = 0 \qquad (8.358)$$

for $t' > t$. We define the advanced Green's function as

$$G^-(r',t';r,t) = -G(r',t';r,t) \qquad (8.359)$$

for $t' < t$, and

$$G^-(r',t';r,t) = 0 \qquad (8.360)$$

for $t' > t$. We can also see that

$$G_0(r',t';r,t) = -i\left[\frac{\mu}{2\pi i\hbar(t'-t)}\right]^{3/2} \exp\left[\frac{i\mu|\vec{r}'-\vec{r}|^2}{2\hbar(t'-t)}\right]. \qquad (8.361)$$

Thus

$$G_0^+(r',t';r,t) = \Theta(t'-t)G_0(r',t';r,t), \qquad (8.362)$$

and

$$G_0^-(r',t';r,t) = -\Theta(t'-t)G_0(r',t';r,t). \qquad (8.363)$$

Imagine that V is turned off except for a number of very short intervals of time between t and t': from $t_1 \rightarrow t_1 + \Delta t_1$, $t_2 \rightarrow t_2 + \Delta t_2$, etc. G_0^+ may be used to propagate from each time interval. We may obtain a relation of the form

$$G^\pm = G_0^\pm + \hbar^{-1} G_0^\pm V G_0^\pm + \hbar^{-2} G_0^\pm V G_0^\pm V G_0^\pm + \; \ldots \qquad (8.364)$$

If V is not effective in the remote past or for the future, we may replace H with H_0 for $t < -T_1$ and $t > T_2$, where T_1, T_2 are large but finite. Associated with each free particle wave function ϕ_α is a wave function $\psi_\alpha^+(\vec{r}', t')$ that grows out of starting when $t < -T_1$. Thus,

$$\psi_\alpha^+(\vec{r}', t') = i \int G^+(\vec{r}', t'; \vec{r}, t) \, \phi_\alpha(\vec{r}, t) d^3 r, \qquad (8.365)$$

for $t < -T_1$. When $t > T_2$, V is ineffective and $\psi_\alpha^+(\vec{r}', t')$ must satisfy the free particle Schrödinger equation, and we have (eq. 8.5)

$$\psi_\alpha = \sum_n c_n \phi_\beta^n . \qquad (8.366)$$

Therefore,

$$\left(\phi_\beta, \psi_\alpha^+ \right) = \langle \beta | S | \alpha \rangle . \qquad (8.367)$$

This is the amplitude of the state β, after scattering, that grew out of what was the state a before scattering. S is the scattering matrix, T is the transition matrix with

$$\langle \beta | S - 1 | \alpha \rangle = -\frac{i}{\hbar} \langle \beta | T | \alpha \rangle \int_{-\infty}^{+\infty} g(t) e^{\pm i \omega \beta_\alpha t} dt , \qquad (8.368)$$

and

$$\langle \beta | T | \alpha \rangle = \int u_\beta^* V \chi_\alpha^+ d^3 r , \qquad (8.369)$$

where $V(r,t) = V(r)g(t)$.

8.7.2 The Born Approximation

The time-dependent Schrödinger equation is of the form

$$\left(\nabla^2 + k^2\right)\psi(r) = \frac{2m}{\hbar^2}V(r)\psi(r), \tag{8.370}$$

where $k^2 = 2mE / \hbar^2$ and $\psi(r) = e^{ik\cdot r} + f(\hat{n})\frac{e^{ikr}}{r}$. The Green's function satisfies

$$\left(\nabla^2 + k^2\right)G(r,r') = -\delta(r-r'). \tag{8.371}$$

We then have

$$\psi(\vec{r}) = e^{i\vec{k}\cdot\vec{r}} - \frac{2m}{\hbar^2}\int G(\vec{r},\vec{r}')V(\vec{r}')\psi(\vec{r}')d^3\vec{r}'. \tag{8.372}$$

As seen in chapter 1, we find that

$$G(\vec{r},\vec{r}') = \frac{e^{ik|r-r'|}}{4\pi|\vec{r}-\vec{r}'|}, \tag{8.373}$$

which gives

$$\psi(\vec{r}) = e^{i\vec{k}\cdot\vec{r}} - \frac{m}{2\pi\hbar^2}\int \frac{e^{ik|r-r'|}}{|\vec{r}-\vec{r}'|}V(\vec{r}')\psi(\vec{r}')d^3\vec{r}. \tag{8.374}$$

We expand

$$\int \frac{e^{ik|r-r'|}}{|\vec{r}-\vec{r}'|} = \frac{e^{ikr}}{r}e^{ik\hat{n}\cdot\vec{r}'} + 0\left(\frac{1}{r^2}\right), \tag{8.375}$$

to obtain

$$\psi(\vec{r}) = e^{i\vec{k}\cdot\vec{r}} - \frac{m}{2\pi\hbar^2}\int e^{ik\hat{n}\cdot\vec{r}}V(\vec{r}')\psi(\vec{r}')d^3\vec{r}'\cdot\frac{e^{ikr}}{r}. \tag{8.376}$$

Thus

$$f(\hat{n}) = -\frac{m}{2\pi\hbar^2}\int e^{-ik\hat{n}\cdot\vec{r}}V(r)\psi(r)d^3r,\qquad(8.377)$$

or for spherically symmetric potentials,

$$f_B(G) = -\frac{2\mu}{\hbar^2 q}\int_0^\infty r\sin qr V(r)dr,\qquad(8.378)$$

with $q = 2K\sin\frac{1}{2}\theta$.

8.8 Angular Momentum

8.8.1 Orbital Angular Momentum

The orbital angular momentum is defined by

$$\vec{L} = \vec{r}\times\vec{p},\qquad(8.379)$$

where we have the relations

$$L_x = yp_z - zp_y,\qquad(8.380)$$

$$L_y = zp_x - xp_z,\qquad(8.381)$$

and

$$L_z = xp_y - yp_x.\qquad(8.382)$$

We can easily see that

$$\left(L_x, L_y\right) = i\hbar L_z,\qquad(8.383)$$

and cyclically. We can also see that

$$\left(L_x, L^2\right) = \left(L_y, L^2\right) = \left(L_z, L^2\right) = 0.\qquad(8.384)$$

In our analysis of the spherical potential, we found that

$$L^2 Y_l^m = \hbar^2 l(l+1) Y_l^m, \qquad (8.385)$$

and

$$L_z Y_l^m = \hbar m Y_l^m, \qquad (8.386)$$

because $L^2 - L_z^2 = L_x^2 + L_y^2 > 0$, $l(l+1) - m^2 \geq 0$. We define

$$L_+ = L_x + iL_y, \qquad (8.387)$$

and

$$L_- = L_x - iL_y. \qquad (8.388)$$

This implies that

$$L^2 L_\pm Y_l^m = L_\pm L^2 Y_l^m = \hbar^2 l(l+1) L_\pm Y_l^m, \qquad (8.389)$$

where $L_\pm Y_l^m$ is an eigenfunction of Y_l^m. Consider

$$\left(L_\pm, L_z \right) = \left(L_x \pm iL_y, L_z \right), \qquad (8.390)$$

which is equivalent to

$$\left(L_\pm, L_z \right) = -i\hbar L_y \pm i\left(i\hbar L_x \right) = \mp \hbar L_x. \qquad (8.391)$$

Therefore,

$$L_z L_\pm Yl_\ell^m = L_\pm \left(L_z \pm \hbar \right) Y_l^m = \hbar\left(m \pm 1 \right) L_\pm Y_l^m, \qquad (8.392)$$

and $L_\pm Y_l^m$ is one eigenfunction of L_z.

$L_\pm Y_l^m$ is a simultaneous eigenfunction of L^2 and L_z with eigenvalues $\hbar^2 l(l+1)$, $\hbar^2 l(l+1)$, thus

$$L_\pm Y_l^m = Y_l^{m \pm 1}. \qquad (8.393)$$

Therefore, L_+ is the raising operator and L_- is the lowering operator. Similarly, the spin angular momentum S is defined by

$$S^2 \chi_s^m = \hbar^2 s(s+1)\chi_s^m, \tag{8.394}$$

and

$$S_z \chi_s^m = \hbar m \chi_s^m. \tag{8.395}$$

8.8.2 Total Angular Momentum

The total angular momentum is

$$\vec{J} = \vec{L} + \vec{S}, \tag{8.396}$$

with

$$J^2 Y_l^m = \hbar^2 j(j+1)Y_l^m, \tag{8.397}$$

$$J_z Y_l^m = \hbar m Y_l^m, \tag{8.398}$$

and

$$J_\pm = J_x \pm iJ_y. \tag{8.399}$$

All satisfy the relations

$$\vec{L} \times \vec{L} = i\hbar\vec{L}, \ \vec{S} \times \vec{S} = i\hbar\vec{S}, \ \vec{J} \times \vec{J} = i\hbar\vec{J}. \tag{8.400}$$

We can choose a representation for the matrices J^2, J_z as shown by

$$\left\langle jm \left| J^2 \right| j'm' \right\rangle = j(j+1)\hbar^2 \delta_{jj'}\delta_{mm'}, \tag{8.401}$$

and

$$\left\langle jm \left| J_z \right| j'm' \right\rangle = m\hbar \delta_{jj'}\delta_{mm'}. \tag{8.402}$$

As a check of the raising/lowering operators, we should have $(J_z, J_+) = \hbar J_+$.

$$\left\langle jm\left|J_z\right|j'm'\right\rangle\left\langle j'm'\left|J_+\right|j''m''\right\rangle - \left\langle jm\left|J_+\right|j'm'\right\rangle\left\langle j'm'\left|J_z\right|j''m''\right\rangle$$
$$= \hbar\left\langle jm\left|J_+\right|j''m''\right\rangle, \tag{8.403}$$

and

$$m\hbar\left\langle jm\left|J_+\right|j''m''\right\rangle - m''\hbar\left\langle jm\left|J_+\right|j''m''\right\rangle = \hbar\left\langle jm\left|J_+\right|j''m''\right\rangle, \tag{8.404}$$

so that

$$\left(m - m'' - 1\right)\left\langle jm\left|J_+\right|j''m''\right\rangle = 0. \tag{8.405}$$

The matrix is nonzero if $m = m'' + 1$. Consequently,

$$\left\langle j,m+1\left|J_+\right|j,m\right\rangle = \lambda_m\hbar, \tag{8.406}$$

and

$$\left\langle j,m\left|J_-\right|j,m+1\right\rangle = \lambda_m^*\hbar, \tag{8.407}$$

because

$$\left(J_+, J_-\right) = 2\hbar J_z. \tag{8.408}$$

This gives

$$\left|\lambda_{m-1}\right|^2 - \left|\lambda_m\right|^2 = 2m, \tag{8.409}$$

and

$$\left|\lambda_m\right|^2 = C - m(m+1). \tag{8.410}$$

It can be shown that

$$\left\langle j,m+1\left|J_+\right|jm\right\rangle = \left[j(j+1) - m(m+1)\right]^{1/2}\hbar, \tag{8.411}$$

$$\left\langle j,m-1\left|J_-\right|jm\right\rangle = \left[j(j+1) - m(m-1)\right]^{1/2}\hbar, \tag{8.412}$$

$$J_z|jm\rangle = m\hbar|jm\rangle, \tag{8.413}$$

and

$$J_\pm|jm\rangle = \left[j(j+1) - m(m\pm1)\right]^{1/2}\hbar|j,m\pm1\rangle. \tag{8.414}$$

We can compute the matrices for J^2, J_x, J_y, J_z using these guidelines. For example, if $j = 1/2$

$$J_x = \frac{1}{2}\hbar\begin{pmatrix} 0 & 1 \\ 1 & 0 \end{pmatrix}, \tag{8.415}$$

$$J_y = \frac{1}{2}\hbar\begin{pmatrix} 0 & -i \\ i & 0 \end{pmatrix}, \tag{8.416}$$

$$J_z = \frac{1}{2}\hbar\begin{pmatrix} 1 & 0 \\ 0 & -1 \end{pmatrix}, \tag{8.417}$$

and

$$J^2 = \frac{3}{4}\hbar^2\begin{pmatrix} 1 & 0 \\ 0 & 1 \end{pmatrix}. \tag{8.418}$$

Note the relation to the Pauli spin matrices (chapter 2),

$$\sigma_x = \begin{pmatrix} 0 & 1 \\ 1 & 0 \end{pmatrix}, \tag{8.419}$$

$$\sigma_y = \begin{pmatrix} 0 & -i \\ i & 0 \end{pmatrix}, \tag{8.420}$$

$$\sigma_z = \begin{pmatrix} 1 & 0 \\ 0 & 1 \end{pmatrix}. \tag{8.421}$$

The expressions for $\langle j,m+1|J_+|jm\rangle = \ldots$ are developed as follows. We have

$$\left|\lambda_m\right|^2 = C - m(m+1), \tag{8.422}$$

with $\left|\lambda_m\right|^2 \geq 0$. We need two differing values of m which satisfy $\lambda_m = 0$. Choose m_1 to be the larger value of m and m_2 will be the smaller value. From the preceding,

$$\left\langle j, m_1 + 1 \middle| J_+ \middle| j m_1 \right\rangle = 0, \tag{8.423}$$

and

$$\left\langle j, m_2 + 1 \middle| J_- \middle| j m_2 \right\rangle = 0. \tag{8.424}$$

We have a finite series of values of m ranging from m_1 down to $m_2 + 1$, where m_1, m_2 satisfy $C - m(m + 1) = 0$. That is,

$$m_1 = -\frac{1}{2} + \frac{1}{2}(1 + 4C)^{1/2}, \tag{8.425}$$

and

$$m_2 = -\frac{1}{2} - \frac{1}{2}(1 + 4C)^{1/2}. \tag{8.426}$$

Therefore $m_2 + 1 = m_1$, and the series ranges from m_1 to $-m_1$ in unit steps. $2m_1$ is an integer or zero, $m_1 = 0, 1/2, 1, 3/2, \ldots$.

The eigenvalues of J^2 are found from

$$J^2 = \frac{1}{2}\left(J_+ J_- + J_- J_+\right) + J_z^2, \tag{8.427}$$

which has eigenvalues

$$\frac{1}{2}\left|\lambda_{m-1}\right|^2 \hbar^2 + \frac{1}{2}\left|\lambda_m\right|^2 \hbar^2 + m^2 \hbar^2. \tag{8.428}$$

By inspection, $m_1(m_1 + 1)\hbar^2 = j(j+1)\hbar^2$ are the eigenvalues of J^2. If we rename m_1 to be j, then for each value of j there are $2j + 1$ values of m, from $-j$ to $+j$.

8.8.3 Clebsch-Gordon Coefficients

Consider two commuting angular momentum operators $\vec{J}_1$ and $\vec{J}_2$, where all components of $\vec{J}_1$ commute with all components of $\vec{J}_2$. The orthonormal eigenstates of $\vec{J}_1^2$, J_{1z} are $|j_1m_1\rangle$, while for $\vec{J}_2^2$, J_{2z} the eigenstates are $|j_2m_2\rangle$.

The total angular momentum is $\vec{J} = \vec{J}_1 + \vec{J}_2$ with eigenstates of $\vec{J}^2$, J_z, being $|jm\rangle$. The dimensionality of the subspace is $(2j_1 + 1)(2j_2 + 1)$, so we can denote $|j_1m_1j_2m_2\rangle$ by $|m_1m_2\rangle$. That is,

$$|jm\rangle = |m_1m_2\rangle\langle m_1m_2|jm\rangle, \tag{8.429}$$

because $J_z = J_{1z} + J_{2z}$, $\langle m_1m_2|jm\rangle = 0$ unless $m = m_1 + m_2$. The largest value of m is $j_1 + j_2$, when $m_1 = j_1$ and $m_2 = j_2$. The second largest is $j_1 + j_2 - 1$, when $m_1 = j_1$ and $m_2 = j_2 - 1$ or $m_1 = j_1 - 1$ and $m_2 = j_2$. Consequently, m ranges from $-(j_1 + j_2)$ to $+(j_1 + j_2)$. The elements of the unitary matrix $\langle m_1m_2|jm\rangle$ are the Clebsch-Gordon coefficients. The matrix $\langle m_1m_2|jm\rangle$ has $(2j_1 + 1)(2j_2 + 1)$ rows and columns.

8.9 Relativistic Wave Equation

From the theory of special relativity we have

$$E^2 = p^2c^2 + m^2c^4. \tag{8.430}$$

Since $E \rightarrow i\hbar\dfrac{\partial}{\partial t}$ and $\vec{p} \rightarrow -i\hbar\vec{\nabla}$, a relativistic wave equation is

$$-\hbar^2\frac{\partial^2\psi}{\partial t^2} = -\hbar^2c^2\nabla^2\psi + m^2c^4\psi. \tag{8.431}$$

Dirac sought to modify the equation

$$i\hbar\frac{\partial\psi}{\partial t} = H\psi, \tag{8.432}$$

more specifically, he modified the Hamiltonian to be linear in the space derivatives. The simplest H is

$$H = c\vec{\alpha} \cdot \vec{p} + \beta mc^2, \tag{8.433}$$

which gives

$$\left(i\hbar \frac{\partial}{\partial t} + i\hbar c\vec{\alpha} \cdot \vec{\nabla} - \beta mc^2 \right)\psi = 0. \tag{8.434}$$

We must have α_x, α_y, α_z, β independent of r, t, p, and E. If the wave equation that satisfies this also satisfies the relativistic Schrödinger equation, we have

$$\left(E + c\vec{\alpha} \cdot \vec{p} + \beta mc^2 \right)\left(E - c\vec{\alpha} \cdot \vec{p} - \beta mc^2 \right)\psi = 0. \tag{8.435}$$

These two expressions agree if the following is true:

$$\alpha_x^2 = \alpha_y^2 = \alpha_z^2 = \beta^2 = 1, \tag{8.436}$$

$$\alpha_x \alpha_y + \alpha_y \alpha_x = \alpha_y \alpha_z + \alpha_z \alpha_y = \alpha_z \alpha_x + \alpha_x \alpha_z = 0, \tag{8.437}$$

and

$$\alpha_x \beta + \beta \alpha_x = \alpha_y \beta + \beta \alpha_y = \alpha_z \beta + \beta \alpha_z = 0. \tag{8.438}$$

The simplest matrices that satisfy these equations are

$$\beta = \begin{pmatrix} 1 & 0 \\ 0 & -1 \end{pmatrix}, \tag{8.439}$$

and

$$\vec{\alpha} = \begin{pmatrix} 0 & \vec{\sigma} \\ \vec{\sigma} & 0 \end{pmatrix}, \tag{8.440}$$

where $\vec{\sigma}$ are the Pauli spin matrices. We must have

$$\psi(\vec{r},t) = \begin{pmatrix} \psi_1 \\ \psi_2 \\ \psi_3 \\ \psi_4 \end{pmatrix}, \tag{8.441}$$

with $\psi_i = u_i \exp^{i(\vec{k}\cdot\vec{r}-\omega t)}$. This gives

$$\left(E - c\vec{\alpha}\cdot\vec{p} - \beta mc^2\right)\psi = 0, \tag{8.442}$$

or

$$\left(E - mc^2\right)u_1 - cp_z u_3 - c\left(p_x - ip_y\right)u_4 = 0, \tag{8.443}$$

$$\left(E - mc^2\right)u_2 - c\left(p_x - ip_y\right)u_3 - cp_z u_4 = 0, \tag{8.444}$$

$$\left(E - mc^2\right)u_3 - cp_z u_1 - c\left(p_x - ip_y\right)u_2 = 0, \tag{8.445}$$

and

$$\left(E - mc^2\right)u_4 - c\left(p_x - ip_y\right)u_1 - cp_z u_2 = 0. \tag{8.446}$$

These equations are homogeneous in the u_j and have solutions only if the determinant of the coefficients, which $= \left(E^2 - m^2c^4 - c^2p^2\right)^2$, is zero. If $E_+ = +\left(m^2c^4 + c^2p^2\right)^{1/2}$, then solutions are

$$\begin{pmatrix} 1 \\ 0 \\ \dfrac{cp_z}{E_+ + mc^2} \\ \dfrac{c\left(p_x + ip_y\right)}{E_+ + mc^2} \end{pmatrix}, \tag{8.447}$$

or

$$\begin{pmatrix} 0 \\ 1 \\ \dfrac{c(p_x - ip_y)}{E_+ + mc^2} \\ \dfrac{-cp_z}{E_+ + mc^2} \end{pmatrix}. \tag{8.448}$$

If $E_- = -\left(m^2c^4 + c^2p^2\right)^{1/2}$, then solutions are

$$\begin{pmatrix} \dfrac{cp_z}{E_- - mc^2} \\ \dfrac{c(p_x + ip_y)}{E_- - mc^2} \\ 1 \\ 0 \end{pmatrix}, \tag{8.449}$$

or

$$\begin{pmatrix} \dfrac{c(p_x - ip_y)}{E_- - mc^2} \\ \dfrac{-cp_z}{E_- - mc^2} \\ 0 \\ 1 \end{pmatrix}. \tag{8.450}$$

Dirac hypothesized that the negative energy states are completely filled. The removal of a particle from the negative energy "sea" results in the production of antimatter.

8.10 Bibliography

Bohm, D. *Quantum Theory*. Englewood Cliffs, N.J.: Prentice Hall, 1951.

Dirac, P. A. M. *The Principles of Quantum Mechanics*. 3d ed. Oxford, England: Clarendon Press, 1947.

Saxon, D. S. *Elementary Quantum Mechanics*. San Francisco: Holden-Day, 1968.

Schiff, L. I. *Quantum Mechanics*. 3d ed. New York: McGraw-Hill, 1968.

9 Atomic Physics

9.1 Experimental Atomic Physics

9.1.1 Thomson's Measurement of e/m

J. J. Thomson hypothesized that cathode rays were composed of an a stream of particles of mass m and charge e. Thomson designed the experiment shown in figure 9.1 to calculate the ratio e/m. If a voltage V is applied to the plates, which are separated by a distance D, the rays feel an acceleration $eE/m = eV/Dm$. If the initial velocity is v, the time to cross x_1 is $t_1 = x_1/v$, the time to cross x_2 is $t_2 = x_2/v$. During t_1, the y-direction deflection is

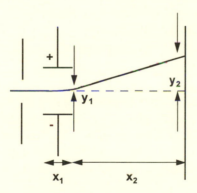

Figure 9.1 Deflection of charged particles by electric fields.

$$y_1 = \frac{1}{2}\left(\frac{eE}{m}\right)t_1^2 = \frac{1}{2}\left(\frac{eE}{m}\right)\left(\frac{x_1}{v}\right)^2. \qquad (9.1)$$

When the rays leave the area of the plates, they have a component of velocity in the y-direction given by

$$v_y = \left(\frac{eE}{m}\right)t_1 = \left(\frac{eE}{m}\right)\left(\frac{x_1}{v}\right). \tag{9.2}$$

The total y-deflection is

$$y_2 = \left(\frac{eE}{m}\right)t_1 + y_1 = \frac{e}{m}\left(\frac{Ex_1}{v_2}\right)\left(\frac{1}{2}x_1 + x_2\right). \tag{9.3}$$

To determine e/m we need to know v. If we place the apparatus within a Helmholtz coil, we can apply a magnetic field B in the z-direction. If both E and B fields are applied simultaneously and adjusted so that no deflection is noticed, then

$$ev\tilde{B} = e\tilde{E}. \tag{9.4}$$

We now know v and we find

$$y_2 = \frac{e}{m}\left(\frac{Ex_1\tilde{B}^2}{\tilde{E}^2}\right)\left(\frac{1}{2}x_1 + x_2\right). \tag{9.5}$$

The ratio e/m can now be determined by experiment.

9.1.2 Millikan's Oil Drop Experiment

In Millikan's experiment, very small oil droplets a few microns in diameter are sprayed mechanically from a nozzle. Some droplets are charged by friction as they are formed (alternatively they could be ionized by X rays) and passed through a small hole between two parallel plates of a capacitor. If the capacitor is uncharged, the droplets will fall under the influence of gravity until they reach a terminal velocity v_t, when the force of gravity is balanced by the viscous drag of the air. This occurs when

$$mg = 6\pi\eta r v_t, \tag{9.6}$$

where η is the viscosity of the air and r is the radius of the drop. By definition, $m = \frac{4}{3}\pi r^3 \rho_o$, where ρ_o is the density of the oil. If we allow for the buoyancy of the air, $m = \frac{4}{3}\pi r^3 (\rho_o - \rho_A)$, with ρ_a being the density of air. If we apply a potential V to the plate the drop will move upward until a new terminal velocity is reached. This occurs when

$$q\frac{V}{D} - mg = 6\pi\eta r v_2. \tag{9.7}$$

We find that the charge on the drop is given by

$$q = 6\pi\eta r\left(\frac{D}{V}\right)(v_1 + v_2). \tag{9.8}$$

9.1.3 The Compton Effect

In 1923 A. H. Compton irradiated a graphite target with a nearly monochromatic beam of X rays of wavelength λ_o. As shown in figure 9.2, part of the scattered radiation had the same wavelength as the incident beam, λ_o, but part had a different wavelength.

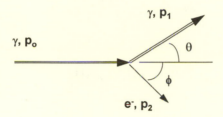

Figure 9.2 Compton scattering.

It was suggested that the incident EM radiation caused the atomic electrons to vibrate at the same frequency as the incident wave. Consequently, the scattered radiation would be produced from the vibrations of the electrons and have the same wavelength, λ_0. Compton suggested that the second component of the radiation, λ_1, was due to interactions with free electrons. From conservation of momentum

$$p_0 = p_1 \cos\theta + p_2 \cos\phi, \tag{9.9}$$

$$0 = p_1 \sin\theta - p_2 \sin\phi, \tag{9.10}$$

and from conservation of energy

$$E_0 + mc^2 = E_1 + \left(p_2^2 c^2 + m^2 c^4\right)^{1/2}. \tag{9.11}$$

Solving these equations for p_2 gives

$$p_2^2 = p_0^2 + p_1^2 - 2p_0 p_1 \cos\theta. \tag{9.12}$$

From conservation of energy

$$p_0 c + mc^2 = p_1 c + \left(p_2^2 c^2 + m^2 c^4\right)^{1/2}, \tag{9.13}$$

which gives

$$p_2 = \left[(p_0 - p_1)^2 + 2mc(p_0 - p_1)\right]^{1/2}. \tag{9.14}$$

Using this in the expression from conservation of momentum gives

$$\left[(p_0 - p_1)^2 + 2mc(p_0 - p_1)\right] = p_0^2 + p_1^2 - 2 p_0 p_1 \cos\theta, \tag{9.15}$$

or

$$mc(p_0 - p_1) = p_0 p_1 (1 - \cos\theta). \tag{9.16}$$

Because

$$(p_0 - p_1) = h\left(\frac{1}{\lambda_0} - \frac{1}{\lambda_1}\right), \tag{9.17}$$

the previous expression may be rewritten as

$$\frac{p_o - p_1}{p_o p_1} = \frac{h\left(\dfrac{\lambda_1 - \lambda_o}{\lambda_1 \lambda_o}\right)}{h^2 \dfrac{1}{\lambda_1 \lambda_o}} = h(\lambda_1 - \lambda_o) = \left(\frac{1}{mc}\right)(1 - \cos\theta) = \left(\frac{1}{mc}\right)\sin^2\left(\frac{\theta}{2}\right). \tag{9.18}$$

The wavelength shift, or Compton shift, is given by

$$\Delta\lambda = \lambda_1 - \lambda_2 = \lambda_c \sin^2\left(\frac{\theta}{2}\right), \tag{9.19}$$

where the Compton wavelength, λ_c, is defined by

$$\lambda_c = \frac{h}{mc}. \tag{9.20}$$

9.1.4 The Stern-Gerlach Experiment

An electron moving in a closed orbit forms a current loop that in turn creates a magnetic dipole moment, given by

$$\mu = IdA. \tag{9.21}$$

The current in this case is $I = \dfrac{ev}{2\pi r}$ and $dA = \pi r^2$, so

$$\mu = \frac{evr}{2} = \frac{eL}{2m}. \tag{9.22}$$

Because the direction of current is opposite the direction of rotation of the electron

$$\vec{\mu} = -\frac{e}{2m}\vec{L}, \tag{9.23}$$

or

$$\vec{\mu} = -\mu_B \vec{L}/\hbar, \tag{9.24}$$

where $\mu_B = \dfrac{e\hbar}{2m}$ is defined as the *Bohr magneton*. In general, if a system of electrons has a total angular momentum $\vec{J}$, then

$$\vec{\mu} = -g\mu_B \vec{J}/\hbar, \tag{9.25}$$

where g is the gyromagnetic ratio.

If an atom with a magnetic moment $\vec{\mu}$ is placed in a magnetic field $\vec{B}$, the energy of interaction is

$$W = -\vec{\mu} \cdot \vec{B}. \tag{9.26}$$

Therefore,

$$\vec{F} = -\nabla W, \tag{9.27}$$

or

$$F_x = \mu \frac{\partial B}{\partial x}, \; F_y = \mu \frac{\partial B}{\partial y}, \; F_z = \mu \frac{\partial B}{\partial z}. \tag{9.28}$$

If the B field is constant, there will be no force on the dipole, but it will precess because

$$\vec{\tau} = \vec{\mu} \times \vec{B} = \frac{d\vec{L}}{dt} = \frac{\mu_B}{\hbar} BL .$$ (9.29)

Therefore, by definition,

$$W_L = \frac{\mu_B}{\hbar} B .$$ (9.30)

In 1912, Stern suggested that magnetic moments of atoms could be measured by detecting the deflection of an atomic beam by an inhomogeneous B field that satisfied

$$\frac{\partial B_z}{\partial y} = \frac{\partial B_z}{\partial x} = 0 ,$$ (9.31)

so that

$$F_z = \mu_z \frac{\partial B_z}{\partial z} .$$ (9.32)

The experimental results indicated that the angular momentum was quantized, that is,

$$L_z = m\hbar ,$$ (9.33)

where m is the magnetic quantum number.

9.2 One Electron Atoms

9.2.1 Bohr Model of Hydrogen

In 1913 Bohr assumed that the electrons in an atom moved in a circular orbit about the nucleus under the influence of the electrostatic attraction of the nucleus. He postulated that only a certain number of these orbits were allowed and that they satisfied

$$h\nu = E_b - E_a .$$ (9.34)

Bohr also assumed that the angular momentum was quantized

$$L = n\hbar. \tag{9.35}$$

The force on the electron is given by

$$\frac{1}{4\pi\varepsilon_o} \frac{Ze^2}{r^2} = \frac{mv^2}{r}. \tag{9.36}$$

From the quantization of the angular momentum,

$$L^2 = m^2 v^2 r^2 = n^2\hbar^2 = \frac{1}{4\pi\varepsilon_o} \frac{Ze^2}{r^2} mr, \tag{9.37}$$

so that

$$r = \frac{(4\pi\varepsilon_o)\hbar^2 n^2}{Ze^2 m}. \tag{9.38}$$

Similarly, the velocity must satisfy

$$v = \frac{n\hbar}{mr} = \frac{1}{4\pi\varepsilon_o} \frac{Ze^2}{\hbar n}. \tag{9.39}$$

The energy is given by

$$T = \tfrac{1}{2}mv^2 = \frac{m}{2} \frac{1}{\hbar^2} \left(\frac{Ze^2}{4\pi\varepsilon_o}\right)^2 \frac{1}{n^2}, \tag{9.40}$$

$$V = -\frac{1}{4\pi\varepsilon_o} \frac{Ze^2}{r} = -\frac{m}{\hbar^2} \left(\frac{Ze^2}{4\pi\varepsilon_o}\right)^2 \frac{1}{n^2}, \tag{9.41}$$

and

$$E_n = T + V = -\frac{m}{2\hbar^2} \left(\frac{Ze^2}{4\pi\varepsilon_o}\right)^2 \frac{1}{n^2}, \tag{9.42}$$

with n being defined as the principle quantum number. For hydrogen,

$$E_n = -(13.6\text{eV})\frac{1}{n^2}. \tag{9.43}$$

9.2.2 Solutions to the Schrödinger Equation

A hydrogen-like atom moves in a potential given by (now in cgs units)

$$V = \frac{-Ze^2}{r}. \tag{9.44}$$

The radial wave equation reduces to

$$\frac{-\hbar^2}{2\mu} \frac{1}{r^2} \frac{d}{dr}\left(r^2 \frac{dR}{dr}\right) - \frac{Ze^2}{r} R + \frac{l(l+1)\hbar^2}{2\mu r^2} R = ER, \tag{9.45}$$

where $E < 0$ for bound states. We substitute $\rho = \alpha r$ so that

$$\frac{-\hbar^2}{2\mu} \frac{\alpha^2}{\rho^2} \alpha \frac{d}{d\rho}\left(\frac{\rho^2}{\alpha^2} \alpha \frac{dR}{d\rho}\right) - \frac{Ze^2}{\rho} \alpha R + \frac{l(l+1)\hbar^2}{2\mu\rho^2} \alpha^2 R = ER. \tag{9.46}$$

This reduces to

$$\frac{\alpha^2}{\rho^2} \frac{d}{d\rho}\left(\rho^2 \frac{dR}{d\rho}\right) + \left[\frac{2\mu}{\hbar^2}\right]\left(E + \frac{Ze^2\alpha}{\rho} - \frac{l(l+1)}{\rho^2} \frac{\hbar^2}{2\mu} \alpha^2\right) R = 0, \tag{9.47}$$

or

$$\frac{1}{\rho^2} \frac{d}{d\rho}\left(\rho^2 \frac{dR}{d\rho}\right) + \frac{1}{\alpha^2}\left[\frac{2\mu E}{\hbar^2} + \frac{1}{\rho} \frac{2\mu Ze^2\alpha}{\hbar^2} - \frac{l(l+1)}{\rho^2} \alpha^2\right] R = 0. \tag{9.48}$$

If we define $\alpha^2 = \frac{8\mu|E|}{\hbar}$ and $\lambda = \frac{2\mu Ze^2}{\hbar^2\alpha}$ the expression simplifies to

$$\frac{1}{\rho^2} \frac{d}{d\rho}\left(\rho^2 \frac{dR}{dp}\right) + \left(\frac{\lambda}{\rho} - \frac{1}{4} - \frac{l(l+1)}{\rho^2}\right) R = 0. \tag{9.49}$$

Note that the term $-\frac{1}{4}$ appears because $E < 0$. By definition,

$$\frac{1}{\rho^2} \frac{d}{d\rho}\left(\rho^2 R'\right) = \frac{1}{\rho^2}\left[2\rho R' + \rho^2 R''\right], \tag{9.50}$$

so that the equation is

$$R'' + \frac{2}{\rho} R' + \left(\frac{\lambda}{\rho} - \frac{1}{4} - \frac{l(l+1)}{\rho^2} \right) R = 0. \tag{9.51}$$

As $\rho \to \infty$, R must vary as $e^{-\frac{1}{2}\rho}$. Try a solution of the form

$$R(\rho) = F(\rho) e^{-\frac{1}{2}\rho}. \tag{9.52}$$

This gives

$$R' = F' e^{-\frac{1}{2}\rho} - \frac{1}{2} F e^{-\frac{1}{2}\rho}, \tag{9.53}$$

and

$$R'' = F'' e^{-\frac{1}{2}\rho} - F' e^{-\frac{1}{2}\rho} + \frac{1}{4} F e^{-\frac{1}{2}\rho}, \tag{9.54}$$

so that

$$\left(F'' e^{-\frac{1}{2}\rho} - F' e^{-\frac{1}{2}\rho} + \frac{1}{4} F e^{-\frac{1}{2}\rho} \right) + \frac{2}{\rho} \left(F' e^{-\frac{1}{2}\rho} - \frac{1}{2} F e^{-\frac{1}{2}\rho} \right)$$
$$+ \left(\frac{\lambda}{\rho} - \frac{1}{4} - \frac{l(l+1)}{\rho^2} \right) F e^{-\frac{1}{2}\rho} = 0. \tag{9.55}$$

This expression simplifies to

$$F'' + \left(\frac{2}{\rho} - 1 \right) F' + \left[\frac{\lambda - 1}{\rho} - \frac{l(l+1)}{\rho^2} \right] F = 0. \tag{9.56}$$

Try a solution of the form

$$F(\rho) = \rho^s \left(a_0 + a_1 \rho + a_2 \rho^2 + \ldots \right) = \rho^s L(\rho). \tag{9.57}$$

We have

$$F' = s\rho^{s-1}L + \rho^s L', \tag{9.58}$$

and

$$F'' = s(s-1)\rho^{s-2}L + 2s\rho^{s-1}L' + \rho^s L'', \tag{9.59}$$

so that the equation of interest reduces to

$$L''\rho^s + L'\left(2s\rho^{s-1} + 2\rho^{s-1} - \rho^s\right)\rho^s L'' + \rho^{s-1}2sL' + \rho^{s-2}s(s-1)L$$

$$+\left(\frac{2}{\rho}-1\right)\left(s\rho^{s-1}L + \rho^s L'\right) - \left[\frac{\lambda-1}{\rho} - \frac{l(l+1)}{\rho^2}\right]\rho^s L = 0, \tag{9.60}$$

or

$$L''\rho^s + L'\left(2s\rho^{s-1} + 2\rho^{s-1} - \rho^s\right)$$

$$+ L\left(\rho^{s-2}s(s-1) + 2s\rho^{s-2} - s\rho^{s-1} - (\lambda-1)\rho^{s-1} + l(l+1)\rho^{s-2}\right) = 0. \tag{9.61}$$

This must be true for equal powers of s, so

$$L''\rho^2 + L'\left(2s(s+1) - \rho\right)\rho + L\left(s(s-1) - (s+\lambda-1)\rho + l(l+1)\rho\right) = 0. \tag{9.62}$$

This can only be true if $s(s+1) = l(l+1)$. The two solutions are $s = l$ and $s = -(l+1)$, but the boundary conditions are only satisfied for $s = l$. If we use

$$L = \sum_n a_n \rho^n, \tag{9.63}$$

then

$$a_{n+1} = \frac{n+l+1-\lambda}{(n+1)(n+2l+2)} a_n. \tag{9.64}$$

As $n \to \infty$, $\dfrac{a_{n+1}}{a_n} \to \dfrac{1}{n}$, so the series must terminate. If we choose $\lambda = n + \ell - 1 = n$, the energy eigenvalues are, for $E < 0$,

$$n = \lambda = \frac{2\mu Z e^2}{\hbar^2 \alpha} = \frac{2\mu Z e^2}{\hbar^2}\sqrt{\frac{\hbar^2}{8\mu - E}}, \tag{9.65}$$

so that

$$E = -\frac{\mu Z^2 e^4}{2\hbar^2 n^2}. \tag{9.66}$$

The physically admissible solutions for L are the Laguerre polynomials with generating function

$$U(\rho, s) = \frac{e^{-\rho s/(1-s)}}{1-s} = \sum_{q=0}^{\infty} \frac{L_q(\rho)}{q!} s^\rho, \tag{9.67}$$

for $s < 1$.

9.3 Interaction of One-Electron Atoms with Electromagnetic Radiation

9.3.1 Principle of Detailed Balancing

The Hamiltonian of a spinless particle of mass m and charge q is

$$H = \frac{1}{2m}\left(\vec{p} - q\vec{A}\right)^2 + q\phi, \tag{9.68}$$

where

$$\left(\vec{p} - qA\right)^2 = \left(-i\hbar\vec{\nabla} + e\vec{A}\right)^2 = -\hbar^2\nabla^2 - 2ie\hbar\vec{\nabla}\cdot\vec{A} + e^2\vec{A}^2, \tag{9.69}$$

and

$$\vec{\nabla}\cdot\left(\vec{A}\psi\right) = \left(\vec{\nabla}\cdot\vec{A}\right)\psi + \vec{A}\cdot\left(\vec{\nabla}\psi\right) = \vec{A}\cdot\vec{\nabla}\psi. \tag{9.70}$$

The equivalent Hamiltonian is given by

$$H_{equiv.} = \frac{1}{2m}\left(-\hbar^2\nabla^2 - 2ie\hbar\vec{A}\cdot\vec{\nabla} + e^2 A^2\right), \tag{9.71}$$

and the Schrödinger equation is

$$i\hbar \frac{\partial \psi}{\partial t} = \left[\frac{-\hbar^2}{2m} \nabla^2 - \frac{Ze^2}{r} - \frac{i\hbar e}{m} \vec{A} \cdot \vec{\nabla} + \frac{e^2}{2m} A^2 \right] \psi . \qquad (9.72)$$

If we neglect the A^2 term, this is equivalent to

$$i\hbar \frac{\partial}{\partial t} \psi = (H_o + H') \psi , \qquad (9.73)$$

where

$$H' = \frac{-i\hbar e}{m} \vec{A} \cdot \vec{\nabla} . \qquad (9.74)$$

Treating this as a time-dependent perturbation theory problem gives

$$H_o \psi_k = E_k \psi_k , \qquad (9.75)$$

and

$$\psi = \sum_k C_k(t) \psi_k(r) e^{-iE_k t/\hbar} . \qquad (9.76)$$

Using this in the Schrödinger equation gives

$$i\hbar \left[\sum_k \left(\dot{C}_k \psi_k(r) e^{-iE_k t/\hbar} - i \frac{E_k}{\hbar} C_k \psi_k(r) e^{-iE_k t/\hbar} \right) \right]$$

$$= \sum_k (E_k + H') C_k \psi(r) e^{-iE_k t/\hbar} , \qquad (9.77)$$

or

$$\sum_k \dot{C}_k(t) \psi_k(r) e^{-iE_k t/\hbar} = \frac{1}{i\hbar} \sum_k H' C_k(t) \psi_k(r) e^{-iE_k t/\hbar} . \qquad (9.78)$$

Explicitly,

$$\sum_k \int \psi_b(r)\dot{C}_k(t)\psi_k(r)e^{-iE_k t/\hbar} = \sum_k \dot{C}_k(t)e^{-iE_k t/\hbar}\delta_{bk}, \quad (9.79)$$

and

$$\frac{1}{i\hbar}\sum_k \int \psi_b(r)H'\psi_k(r)C_k(t)e^{-iE_k t/\hbar} = \frac{1}{i\hbar}\sum_k H'_{bk}C_k(t)e^{-iE_k t/\hbar}. \quad (9.80)$$

This gives

$$\dot{C}_b(t) = \frac{1}{i\hbar}\sum_k H'_{bk}C_k(t)e^{-i\omega_{bk}t}, \quad (9.81)$$

with $E_b - E_k = \hbar\omega_{bk}$. We can expand $C_b(t)$ in a power series

$$C_b(t) = C_b^0 + \lambda C_b^1 + \lambda^2 C_b^2 + \ldots, \quad (9.82)$$

and

$$\dot{C}_b = \dot{C}_b^0 + \lambda \dot{C}_b^1 + \lambda^2 \dot{C}_b^2 + \ldots + \lambda C_b^1 + \ldots. \quad (9.83)$$

If we treat λ as a parameter of smallness, then $H' \to \lambda H'$. Grouping terms of like order in λ gives

$$\dot{C}_b^0 = 0, \quad (9.84)$$

$$\dot{C}_b^1 = \frac{1}{i\hbar}\sum_k H'_{bk}C_k^0 e^{i\omega_{bk}t}, \quad (9.85)$$

and so on. Because C_b^0 is time independent, if we assume the system is initially in a well-defined state, $C_b^0 = \delta_{sa}$. Thus

$$\frac{dC_b^1}{dt} = \frac{1}{i\hbar}\sum_k H'_{bk}\delta_{ka}e^{i\omega_{bk}t}, \quad (9.86)$$

or

$$C_b^1(t) = \frac{1}{i\hbar} \int_o^t H_{ba}^t e^{i\omega_{ba}t} dt, \qquad (9.87)$$

which reduces to

$$C_b^1(t) = \frac{1}{i\hbar} \int_o^t \left\langle \psi_b \left| \frac{-i\hbar e}{m} \vec{A} \cdot \vec{\nabla} \right| \psi_a \right\rangle e^{i\omega_{ba}t} dt$$

$$= \frac{-e}{m} \int_o^t \left\langle \psi_b \left| \vec{A} \cdot \vec{\nabla} \right| \psi_a \right\rangle e^{i\omega_{ba}t} dt. \qquad (9.88)$$

We have, in general, $\vec{A} = \int A_o \hat{e} e^{i(\vec{k} \cdot \vec{r} - \omega t + \delta)} d\omega$ + complex conjugate ($c.c.$), so that

$$C_b^1(t) = \frac{-e}{m} \int d\omega A_o(\omega) e^{i\delta} \left\langle \psi_b \left| e^{i\vec{k} \cdot \vec{r}} \hat{\varepsilon} \cdot \vec{\nabla} \right| \psi_a \right\rangle \int_o^t dt' e^{i(\omega_{ba} - \omega)t'} + c.c., \qquad (9.89)$$

and

$$\psi = \sum_k C_k(t) \psi_k(r) e^{iE_k t/\hbar}. \qquad (9.90)$$

We can solve the problem to find

$$\dot{C}_b(t) = \frac{1}{i\hbar} \sum_k H_{bk}^t C_k(t) e^{iw_{bk}t}. \qquad (9.91)$$

Expanding $C_b(t)$ in a power series about λ, we find that we can, to first approximation, set $C_k(t) = \delta_{ka}$ so that

$$C_b(t) = \frac{1}{i\hbar} \int_o^t H_{ba}^t(t') e^{i\omega_{ba}t'} dt'. \qquad (9.92)$$

We have

$$C_b(t) = \frac{-e}{m} \int_o^t \left\langle \psi_b \left| \vec{A} \cdot \vec{\nabla} \right| \psi_a \right\rangle e^{i\omega_{ba}t'} dt', \qquad (9.93)$$

where $\vec{A}(\vec{r}, t)$ is a superposition of plane waves, i.e.,

$$\vec{A}(\vec{r},t) = \int_{\Delta\omega} A_0(\omega)\hat{E}\left[e^{i\left(\vec{k}\cdot\vec{r}-\omega t+\delta\right)} + c.c.\right]. \tag{9.94}$$

Thus,

$$C_b^1(t) = \frac{-e}{m}\int_{\Delta\omega} d\omega A_0(\omega)\left[e^{i\delta}\left\langle\psi_b\left|e^{i\vec{k}\cdot\vec{r}}\hat{\varepsilon}\cdot\vec{\nabla}\right|\psi_a\right\rangle\int_0^t dt' e^{i(\omega_b-\omega)t'} + c.c.\right], \tag{9.95}$$

and

$$\left|C_b^1(t)\right|^2 = 2\int_{\Delta\omega} d\omega\left[\frac{eA_0(\omega)}{m}\right]^2\left|M_{ba}(\omega)\right|^2 F(t,\omega-\omega_{ba}). \tag{9.96}$$

The first term of $C_b^1(t)$ is an absorption term and the *c.c.* is the emission term. We find that for both

$$W_{ba} = \frac{4\pi^2}{m^2 c}e^2\frac{I(\omega_{ba})}{\omega_{ba}^2}\left|M_{ba}(\omega_{ba})\right|^2. \tag{9.97}$$

This is the principle of detailed balancing, where

$$M_{ba} = \left\langle\psi_b\left|e^{i\vec{k}\cdot\vec{r}}\hat{E}\cdot\vec{\nabla}\right|\psi_{aA}\right\rangle. \tag{9.98}$$

In cases of practical interest, we may expand $e^{i\vec{k}\cdot\vec{r}}$,

$$e^{i\vec{k}\cdot\vec{r}} = 1 + \left(i\vec{k}\cdot\vec{r}\right) + \frac{1}{2!}\left(i\vec{k}\cdot\vec{r}\right)^2 + \ldots. \tag{9.99}$$

Usually $\lambda \sim 10^3$ Å, so that $k = 2\pi/\lambda$ can be neglected in comparison to 1 if $r < 1$ Å. This is the electric dipole approximation. We have

$$M_{ba} = \hat{\varepsilon}\cdot\left\langle\psi_b|\nabla|\psi_a\right\rangle = \hat{\varepsilon}\cdot\frac{im}{\hbar}\left\langle\psi_b|\vec{r}|\psi_a\right\rangle, \tag{9.100}$$

with

$$\dot{r} = \frac{1}{i\hbar}[r, H_0], \tag{9.101}$$

and

$$\langle \psi_b | \dot{r} | \psi_a \rangle = \frac{1}{\hbar i}(E_a - E_b)\langle \psi_b | r | \psi_a \rangle.$$ (9.102)

The result is

$$W_{ba} = \frac{4\pi^2}{c\hbar^2}\left(\frac{e^2}{4\pi\varepsilon_o}\right)I(\omega_{ba})|\hat{\varepsilon} \cdot r_{ba}|^2.$$ (9.103)

If $|\hat{\varepsilon} \cdot r_{ba}|$ vanishes the transition is forbidden.

9.3.2 Einstein Coefficients

Consider an enclosure containing a single kind of atom and radiation in equilibrium at temperature T, with two energy states E_a and E_b, where $E_b > E_a$. The number of atoms absorbing radiated energy is

$$\dot{N}_{ba} = B_{ba}N_a\rho(\omega_{ba}),$$ (9.104)

where $\dot{N}_{ba}$ is the number of transitions from a to b, and B_{ba} is the Einstein coefficient for absorption. Similarly, for emission,

$$\dot{N}_{ab} = A_{ab}N_b + B_{ab}N_b\rho(\omega_{ba}),$$ (9.105)

where A_{ab} is the coefficient for spontaneous emission and B_{ab} is the coefficient for stimulated emission. At equilibrium, $\dot{N}_{ba} = \dot{N}_{ab}$, and

$$\frac{N_a}{N_b} = \frac{A_{ab} + B_{ab}\rho(\omega_{ba})}{B_{ba}\rho(\omega_{ba})}.$$ (9.106)

In thermal equilibrium

$$\frac{N_a}{N_b} = e^{-(E_a-E_b)/kT} = e^{\hbar\omega_{ba}/kT},$$ (9.107)

and

$$\rho\left(\omega_{ba}\right) = \frac{A_{ab}}{B_{ba}e^{\hbar\omega_{ba}/kT} - B_{ab}}. \tag{9.108}$$

This must equal $\dfrac{\hbar\omega^3_{ba}}{\pi^2 c^3}\dfrac{1}{e^{\hbar\omega_{ba}/kT} - 1}$, so

$$B_{ba} = B_{ab}, \tag{9.109}$$

and

$$A_{ab} = \frac{\hbar\omega^3_{ba}}{\pi^2 c^3} B_{ab}. \tag{9.110}$$

9.3.3 Selection Rules

Selection rules for the various atomic transitions are summarized in table 9.1.

Table 9.1
Summary of Transition Selection Rules

Transition	Selection Rule	Constraints
Electric dipole	$\Delta m = 0 \pm 1$ $\Delta\ell = \pm 1$	—
Magnetic dipole	$\Delta\ell = 0$ $\Delta j = 0, \pm 1$ $\Delta m_j = 0, \pm 1$	No $j = 0 \rightarrow j' = 0$
Electric quadrupole	$\Delta\ell = 0, \pm 2$ $\Delta m = 0, \pm 1, \pm 2$	No $l = 0 \rightarrow l' = 0$

9.3.4 Oscillator Strengths

The oscillator strength is defined by

$$f_{ka} = \frac{2m\omega_{ka}}{3\hbar}\left|r_{ka}\right|^2. \tag{9.111}$$

We have

$$f^x_{ka} = \frac{2m\omega_{ka}}{3\hbar}\left|x_{ka}\right|^2 = \frac{2m\omega_{ka}}{3\hbar}\left\langle a|x|k\right\rangle\left\langle k|x|a\right\rangle, \tag{9.112}$$

where

$$x_{ka} = \langle k|x|a \rangle = \frac{-i}{m\omega_{ka}} \langle k|p_x|a \rangle, \qquad (9.113)$$

and

$$x_{ka} = \langle a|x|k \rangle = \frac{i}{m\omega_{ka}} \langle a|p_x|k \rangle. \qquad (9.114)$$

Thus,

$$f_{ka}^x = \frac{2i}{3\hbar} \langle a|p_x|k \rangle \langle k|x|a \rangle = \frac{i}{3\hbar} [\langle a|p_x|k \rangle \langle k|x|a \rangle - \langle a|x|k \rangle \langle k|p_x|a \rangle], \quad (9.115)$$

and

$$\sum_k f_{ka}^x = \frac{i}{3\hbar} \langle a|p_x x - x p_x|a \rangle = \frac{1}{3}. \qquad (9.116)$$

The oscillator strength sum rule is

$$\sum_k f_{ka} = 1. \qquad (9.117)$$

9.3.5 Line Shapes

All atomic levels, except the ground state, decay with a finite lifetime τ. By the uncertainty principle, the energy of such a level cannot be precisely determined; it is uncertain by an amount of order $\hbar/\tau$. Therefore, there is a finite probability that photons will be emitted within an energy band of width $\left(\frac{\hbar}{\tau_a} + \frac{\hbar}{\tau_b} \right)$ about $(E_b - E_a)$, where τ_a and τ_b are the lifetimes of the states a and b. We use $C_b(t) = 1$, for $t < 0$, and $C_b(t) = e^{-t/2\tau}$, for $t \geq 0$. We find that the intensity of the emitted radiation reaches a maximum when $\omega = \omega_{ba}$ and decreases to one-half of maximum at $\omega = \omega_{ba} \pm \frac{1}{2\tau}$, where $E = E_b - E_a \pm \frac{\Gamma}{2}$ and $\Gamma = \hbar/\tau$. If a is the ground state, in general $\Gamma = \left(\frac{\hbar}{\tau_a} + \frac{\hbar}{\tau_b} \right)$.

When the atoms in a gas undergo collisions, there is a probability that an atom in an excited state will decay to a lower state, so that the lifetime of the excited state is decreased. If the number of collisions per second that result in radiationless transitions is ω_c, the number of transitions per second is $\omega_c + 1/\tau_b$. The breadth of the line is $\Gamma = \hbar(\omega_c + 1/\tau)$. This is *pressure broadening*.

If the atoms are moving the wavelength of the emitted light is shifted, $\lambda = \lambda_o \left(1 \pm \dfrac{v_x}{c}\right)$. As a result, the line will be broadened. This is called *Doppler broadening*.

9.3.6 Fine Structure

The fine structure of the energy levels of hydrogenic atoms is due to relativistic effects. When we solve the Dirac equation for an electron in a central field we can use perturbation theory to find energy corrections. We can show that

$$H = H_0 + H', \qquad (9.118)$$

where

$$H_0 = \frac{p^2}{2m} - \frac{Ze^2}{r}, \qquad (9.119)$$

and

$$H' = H_1' + H_2' + H_3', \qquad (9.120)$$

with

$$H_1' = -\frac{p^4}{8m^3 c^2}, \qquad (9.121)$$

$$H_2' = \frac{1}{2m^2 c^2} \frac{1}{r} \frac{dV}{dr} \vec{L} \cdot \vec{S}, \qquad (9.122)$$

and

$$H_3' = \frac{\pi \hbar^2}{2m^2 c^2} \left(Ze^2\right) \delta(r). \qquad (9.123)$$

$H_1{}'$ is a relativistic correction to the kinetic energy, $H_2{}'$ is the spin orbit term, and $H_3{}'$ is the Darwin term.

The relativistic correction is seen to be

$$\Delta E_1 = \left\langle \psi_{nlm_l m_s} \left| -\frac{p^4}{8m^3c^2} \right| \psi_{nlm_l m_s} \right\rangle$$

$$= \frac{-1}{2mc^2} \left\langle \psi_{nlm_l} \left| T^2 \right| \psi_{nlm_l} \right\rangle, \tag{9.124}$$

with $T = \dfrac{p^2}{2m}$. Because $T = H_0 + \dfrac{Ze^2}{r}$, we have

$$\Delta E_1 = -E_n \frac{(Z\alpha)^2}{n^2} \left[\frac{3}{4} - \frac{n}{l+\frac{1}{2}} \right]. \tag{9.125}$$

The spin orbit term gives

$$\Delta E_2 = \left\langle \psi_{nlm_l m_s} \left| \frac{1}{2m^2c^2} \frac{1}{r} \frac{dV}{dr} \vec{L} \cdot \vec{S} \right| \psi_{nlm_l m_s} \right\rangle. \tag{9.126}$$

We have $v = \dfrac{-Ze^2}{r}$ and $\vec{J} = \vec{L} + \vec{S}$, so $\vec{L} \cdot \vec{S} = \dfrac{1}{2}\left(\vec{J}^2 - \vec{L}^2 - \vec{S}^2 \right)$. The energy correction becomes

$$\Delta E_2 = \hbar^2 \left[j(j+1) - l(l+1) - \frac{3}{4} \right] \left\langle \frac{1}{4m^2c^2} \frac{1}{r} \frac{Ze^2}{r^2} \right\rangle. \tag{9.127}$$

The effect of the Darwin term is

$$\Delta E_3 = \left\langle \psi_{nlm_l m_s} \left| \delta(r) \right| \psi_{nlm_l m_s} \right\rangle = -E_n \frac{(Z\alpha)^2}{n}, \tag{9.128}$$

for $l = 0$.

As the result, a nonrelativistic energy level E_n splits into n different levels, one for each value $j = \dfrac{1}{2}, \dfrac{3}{2}, \ldots, n-\dfrac{1}{2}$ of the total angular momentum quantum number j. The splitting is controlled by the constant $\alpha = \dfrac{1}{137}$, hence it is called the *fine structure constant*. The splitting for levels $n = 1, 2, 3$ is seen in figure 9.3.

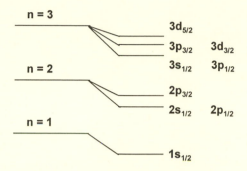

Figure 9.3 Fine structure splitting.

Transitions $nlj \rightarrow n'l'j'$ obey the rules: $\Delta l = \pm 1$ and $\Delta j = 0, \pm 1$ with no $j = 0 \rightarrow j' = 0$. Under most circumstances the initial states are excited in proportion to their statistical weights.

9.3.7 Zeeman Effect

In 1896, P. Zeeman observed that the spectral lines of atoms were split in the presence of an external magnetic field. We have

$$\left[\frac{-\hbar^2}{2m} \nabla^2 - \frac{Ze^2}{r} - \frac{i\hbar e}{m} \vec{A} \cdot \vec{\nabla} + \frac{e^2}{2m} A^2 \right] \psi = E\psi, \qquad (9.129)$$

where

$$\frac{-i\hbar e}{m} \vec{A} \cdot \vec{\nabla} = \frac{-i\hbar e}{2m} \left(\vec{B} \times \vec{r} \right) \cdot \vec{\nabla} = \frac{-i\hbar e}{2m} \vec{B} \cdot \left(\vec{r} \times \vec{\nabla} \right) = \frac{e}{2m} \vec{B} \cdot \vec{L}. \quad (9.130)$$

The interaction term takes the form $H' = -\vec{\mu} \cdot \vec{B}$. The complete Schrödinger equation, neglecting some terms, is

$$\left[-\frac{\hbar^2}{2m} \nabla^2 - \frac{Ze^2}{r} + \xi(r) \vec{L} \cdot \vec{S} + \frac{\mu_B}{\hbar} \left(\vec{L} + 2\vec{S} \right) \cdot \vec{B} \right] \psi = E\psi. \quad (9.131)$$

9.3.7.1 Strong Fields

If $B > Z^4$ we find

$$E = E_n + \mu_B B_z (m_l + 2m_s), \tag{9.132}$$

with $m_s = \pm\dfrac{1}{2}$. For electric dipole transitions we require $\Delta m_s = 0$ and $\Delta m_l = 0, \pm 1$. $\Delta m_l = 0$ is the π line and $\Delta m_l = \pm 1$ is the σ line; they correspond to frequencies $\nu_{nn'}$, and $\nu_{nn'} \pm \nu_L$, respectively, where

$$\nu_L = \frac{\mu_B B_z}{h} \tag{9.133}$$

is the Larmour frequency.

9.3.7.2 Weak Fields (Anomalous Zeeman Effect)

When the interaction caused by the magnetic field is small compared to the spin-orbit term we have the anomalous Zeeman effect. The energy shift is

$$\Delta E m_j = g\mu_B B_z m_j, \tag{9.134}$$

where

$$g = 1 + \frac{j(j+1) + s(s+1) - l(l+1)}{2j(j+1)}. \tag{9.135}$$

The total energy of the level with quantum numbers n, j, m_j of a hydrogenic atom in a constant magnetic field is

$$E_{njm_j} = E_n + \Delta E_{nj} + \Delta E_{m_j}. \tag{9.136}$$

9.3.8 Stark Effect

The splitting of spectral lines by a static electric field is known as the Stark effect. We have

$$H' = eEz, \tag{9.137}$$

and

$$E'_{nlm} = eE\langle \psi_{nlm}|z|\psi_{nlm}\rangle. \tag{9.138}$$

The matrix elements $\langle nlm|z|n'l'm'\rangle$ vanishes unless $m = m'$, and $l = l'\pm1$. Consequently, the ground state exhibits no Stark effect. The quadratic Stark effect is

$$E^2_{100} = e^2 E^2 \sum_{n+l} \frac{\left|\langle \psi_{nlm}|z|\psi_{100}\rangle\right|^2}{E_1 - E_n}. \tag{9.139}$$

9.3.9 Lamb Shift

A quantized radiation field in its lowest energy state is not one with zero electromagnetic field. Even in a vacuum there are fluctuations in the zero point radiation field which can act on the electron. This will cause the electron to oscillate rapidly and the charge is averaged out over a small radius. If the electron is bound by a nonuniform field, as is the case in multi-electronic atoms, its potential will be slightly different than that for a point charge. In particular, the electron is not so strongly attracted at short distances. As a result s-states are raised in energy with respect to other states, by about 4.37×10^{-6} eV. This energy difference is called a *Lamb shift*. The need to explain the Lamb shift led to the development of quantum electrodynamics (QED).

9.3.10 Hyperfine Structure

Hyperfine effects arise because the nucleus of the atom is not truly a point charge, but has a finite size. Hyperfine structure effects split the energy levels, isotope shifts shift the energy of the levels. In 1924 Pauli suggested that the nucleus had a total angular momentum I. For bosons, I is an integral, for fermions I is half-integral. I^2 has eigenvalues $I(I+1)\hbar^2$, and I_z has eigenvalues $M_I\hbar$, $M_I = -I, \ldots, 0, \ldots, I$. The total angular momentum of the atom, nucleus + electrons, is $\vec{F} = \vec{I} + \vec{S}$. That is, $F = |I - j|, \ldots, |I + j|$. The energy shift is found from

$$\vec{\mu}_N = \frac{g_I \mu_N}{\hbar} \vec{I}, \tag{9.140}$$

where μ_N is the nuclear magneton and $\mu_N = \frac{m}{m_p}\mu_B$. The perturbation due to μ_N is

$$H'_{MD} = H'_1 + H'_2, \qquad (9.141)$$

with

$$H'_1 = -\frac{i\hbar e}{m} \vec{A} \cdot \vec{\nabla}, \qquad (9.142)$$

and

$$H'_2 = -\vec{\mu}_s \cdot \vec{B} = \frac{2\mu_s}{\hbar} \vec{S} \cdot \vec{B}. \qquad (9.143)$$

The energy shift for s-states is

$$\Delta E = \frac{c}{2} \left[F(F+1) - I(I+1) + S(S+1) \right]. \qquad (9.144)$$

The hyperfine separation is

$$\Delta E(F) - \Delta E(F-1) = cF. \qquad (9.145)$$

and is thus proportional to F. For electric dipole transitions, $\Delta l = \pm 1$, $\Delta j = 0, \pm 1$, and $\Delta F = 0, \pm 1$, with no $F = 0 \rightarrow F' = 0$.

Example 9.1

A spectral line arises from a transition, $^3D_1 \rightarrow {}^3P_0$, for an atom which has nuclear spin 3/2. Under high resolution the "line" reveals hyperfine structure. Draw a diagram showing the hyperfine splitting of the levels and indicate the allowed transitions. How many lines would be contained in the hyperfine structure of the original unresolved "line"? Determine the relative spacings and intensities.

In hyperfine splitting the energy shift is given by equation 9.144, where C is a structure constant and F takes the values $F = |I - J|, \ldots, |I + J|$. For $^3D_1$, $J = 1$ and $F = 1/2, 3/2, 5/2$. For $^3P_0$, $J = 0$ and $F = 3/2$. The splitting of the terms is shown in table 9.2.

There are three transitions which obey the rule $\Delta F = 0, \pm 1$. As a result, three hyperfine lines are contained in the original unresolved line. The spacings of the lines are proportional to the energy differences. The sum rule tells us that the intensities are proportional to the statistical weight, $2F + 1$. Consequently, the intensities are $a:b:c = 11:7:3$.

Table 9.2

Hyperfine Structure of Spectral Lines

$^3D_1$				$^3P_0$			
I	*J*	*F*	*ΔE*	*I*	*J*	*F*	*ΔE*
3/2	1	1/2	−5C/2	3/2	0	3/2	0
3/2	1	3/2	−C				
3/2	1	5/2	+3C/2				

9.4 Many-Electron Atoms

9.4.1 The Schrödinger Equation

The Schrödinger equation for n identical particles is

$$i\hbar\frac{\partial}{\partial t}\psi(1,2,\ldots,n,t) = H(1,2,\ldots,n,t)\psi(1,2,\ldots,t). \quad (9.146)$$

A wave function is symmetric if the interchange of any pair of particles leaves the wave functions unchanged. A wave function is antisymmetric if the interchange of any pair of particles changes the sign of ψ. Exchange degeneracy occurs whenever two different wave functions, describing a system of particles, have the same energy. For a system without interactions, a stationary state solution of the A-particle system is simply the product of A single particle state functions. This is given by the Slater determinant

$$\psi_{AS}(1,2,\ldots,n) = \begin{vmatrix} v_\alpha(1) & v_\alpha(2) & \ldots & v_\alpha(n) \\ v_\beta(1) & v_\beta(2) & \ldots & v_\beta(n) \\ \vdots & \vdots & \ddots & \vdots \\ v_\nu(1) & v_\nu(2) & \ldots & v_\nu(n) \end{vmatrix}. \quad (9.147)$$

The Slater determinant vanishes if two or more of the v's are the same. This is known as the *Pauli exclusion principle*. Particles that are described by antisymmetric wave functions are *fermions*, particles with symmetric wave function are *bosons*.

9.4.2 Two-Electron Atoms

The states described by space symmetric wave functions are called *para* states, those described by space antisymmetric wave functions are called *ortho* states. The total wave function is

$$\Psi(q_1, q_2) = \psi(r_1, r_2)\chi(1,2), \tag{9.148}$$

where ϕ is the spatial eigenfunction and χ is the spin wave function. For two particles, the antisymmetric solution is

$$\chi_{o,o}(1,2) = \frac{1}{\sqrt{2}}(\alpha(1)\beta(2) - \beta(1)\alpha(2)), \tag{9.149}$$

and the symmetric solutions are

$$\chi_{1,1}(1,2) = \alpha(1)\alpha(2), \tag{9.150}$$

$$\chi_{1,o}(1,2) = \frac{1}{\sqrt{2}}(\alpha(1)\beta(2) - \beta(1)\alpha(2)), \tag{9.151}$$

$$\chi_{1,-1}(1,2) = \beta(1)\beta(2). \tag{9.152}$$

Each solution χ_{sm_s} is denoted by quantum numbers s, m_s. Terms are denoted by $^{2s+1}_{L}X$, with $L = s, p, d, f, \ldots$. The multiplicity of states is $2s + 1$. There can be no transitions from ortho to para states.

9.4.3 Central Field Approximation

For a many-electron atom, the presence of the additional electrons effectively screens the nucleus. Thus, instead of

$$V(r) = \frac{-Ze^2}{r}, \tag{9.153}$$

we have

$$V(r) = \frac{-Ze^2}{r} + S(r). \tag{9.154}$$

However, as $r \to 0$, the electron is close to the nucleus, and $V(r) = \dfrac{-Ze^2}{r}$.
As $r \to \infty$, the electron is far away from the nucleus and
$V(r) = \dfrac{-Z+N-1}{r}e^2$. Perturbation theory may be used to solve for the
energy.

9.4.4 Thomas-Fermi Model

Treat the electrons surrounding the nucleus as a Fermi gas. The energy of
the most energetic electrons in the system is

$$E_{max} = E_F + V(r), \tag{9.155}$$

which gives

$$k\overset{2}{F}(r) = \frac{2m}{\hbar^2}\left[E_{max} - V(r)\right]. \tag{9.156}$$

Because

$$\rho(r) = \frac{1}{3\pi^2}\left(\frac{2m}{\hbar^2}\right)^{3/2}\left[E_{max} - V(r)\right]^{3/2}, \tag{9.157}$$

we have

$$\phi(r) = \frac{-1}{e}V(r), \tag{9.158}$$

and

$$\phi(r) = \phi(r) - \phi_o. \tag{9.159}$$

Thus

$$\rho(r) = \frac{1}{3\pi^2}\left(\frac{2m}{\hbar^2}\right)^{3/2}\left[e\Phi(r)\right]^{3/2}, \tag{9.160}$$

for $\Phi \geq 0$ and

$$\rho(r) = 0, \tag{9.161}$$

for $\Phi < 0$. This gives

$$\nabla^2 \Phi(r) = \frac{e}{\varepsilon}\rho(r) = \frac{e}{3\pi^2\varepsilon}\left(\frac{2m}{\hbar^2}\right)^{3/2}\left[e\,\Phi(r)\right]^{3/2}, \tag{9.162}$$

subject to the constraint

$$4\pi \int_0^{r_o} \rho(r)\alpha^2 dr = N, \tag{9.163}$$

and

$$\lim_{r\to\infty} r\Phi(r) = Ze. \tag{9.164}$$

Let $r = bx$, $r\Phi(r) = Ze\chi(x)$ and it can be seen that

$$\rho = \frac{Z}{b^3}\left(\frac{\chi}{x}\right)^{3/2}, \tag{9.165}$$

for $\chi \ge 0$, and

$$\rho = 0 \tag{9.166}$$

for $\chi < 0$. This is the *Thomas-Fermi equation*.

The Thomas-Fermi model is based on a treatment of a Fermi electron gas. A free particle in a cube of length L satisfies

$$\frac{-\hbar^2}{2m}\left(\nabla^2\right)\psi = E\psi, \tag{9.167}$$

so that

$$\psi = C\sin\left(\frac{n_x\pi}{L}x\right)\sin\left(\frac{n_y\pi}{L}y\right)\sin\left(\frac{n_z\pi}{L}z\right), \tag{9.168}$$

and

$$E = \frac{\pi^2 \hbar^2}{2mL^2}\left(n_x^2 + n_y^2 + n_z^2\right) = \frac{\pi^2 \hbar^2}{2mL^2} n^2. \tag{9.169}$$

The number of electrons with energies of up to E is

$$N_s = 2\left(\frac{1}{8}\right)\frac{4}{3}\pi n^3 = \frac{1}{3}\pi n^3 = \frac{1}{3\pi^2}\left(\frac{2m}{\hbar^2}\right)^{3/2} VE^{3/2}. \tag{9.170}$$

The factor of two is present because of the possibility of two spin states. The factor 1/8 is present because we are only in one octant of a sphere with a radius n in n-space. The density of states is

$$D(E) = \frac{dN}{dE} = \frac{1}{2\pi^2}\left(\frac{2m}{\hbar^2}\right)^{3/2} VE^{3/2}, \tag{9.171}$$

with

$$N = \int_0^{E_F} D(E)dE = \frac{1}{3\pi^2} = \left(\frac{2m}{\hbar^2}\right)^{3/2} VE_F^{3/2}, \tag{9.172}$$

and

$$E_{tot} = \int_0^{E_F} ED(E)dE = \frac{3}{5} NE_F. \tag{9.173}$$

The maximum kinetic energy of an electron in a Fermi gas at $T = 0$ is E_F. The total energy of the most energetic electrons in a atom would be modified by the potential, i.e.,

$$E_{max} = E_F + V(r), \tag{9.174}$$

or

$$k_F^2(r) = \frac{2m}{\hbar^2}\left[E_{max} - V(r)\right]. \tag{9.175}$$

9.4.5 Hund Rule

1. The term with the largest possible value of S for a given configuration has the lowest energy, for other terms the energy increases with decreasing S.

2. For a given value of S, the term having the maximum possible value of L has the lowest value.

Example 9.2

1. *Write down the electronic configuration for carbon.*
2. *Determine the L-S coupling states that can result from this configuration and give their spectroscopic designations.*
3. *Sketch the approximate level scheme for the five levels found above. The Lande interval rule applies to some levels.*
4. *Electromagnetic transitions between these levels will be by magnetic dipole radiation. Discuss why this is true and indicate, by means of arrows, the expected transitions.*

(1.) The electronic configuration for carbon is $1s^2 2s^2 2p^2$.

(2.) The $1s^2$ and $2s^2$ shell is full, so we need to calculate the transitions for two electrons in the $2p^2$ shell. Because the electrons are in the same subshell, the Pauli exclusion principle must be observed. We have $l_1 = 1$, $l_2 = 1$, $s_1 = 1/2$, $s_2 = 1/2$, so $m_{l_1} = -1, 0, +1$, $m_{l_2} = -1, 0, +1$, $m_{s_1} = -1/2, +1/2$, and $m_{s_2} = -1/2, +1/2$. The possible configurations are shown in table 9.3.

For $L = 2$, $S = 0$ we have the ^1D term. For $L = 1$, $S = 1$ we have the ^3P term. For $L = 0$, $S = 0$ we have the ^1S term. Each fine structure term $^{2S+1}L_J$ is $(2J + 1)$ fold degenerate with respect to $M_j = -J, \ldots, +J$. The possible values of J are $|L - S|, \ldots, |L + S|$. The ^1D term satisfies $J = 2$, so it is $^1D_2$. The ^3P term satisfies $J = 0, 1, 2$ so we have three terms: $^3P_1$, $^3P_2$, $^3P_3$. Similarly, the ^1S term satisfies $J = 0$, so it is $^1S_0$. As a result, there are five possible levels.

(3.) The level scheme may be drawn with the help of Hund's rules and the Lande interval rule and are shown in figure 9.4.

(4.) Magnetic dipole transitions will obey the rules $\Delta l = 0$, $\Delta J = 0, \pm 1$ (but no $J = 0$ to $J = 0$), and $\Delta m_j = \pm 1$. The transition is a magnetic dipole transition because the interaction between the nucleus and the moving electrons is magnetic in nature.

Table 9.3
Possible Electronic Configurations for Carbon

m_{l_1}	m_{s_1}	m_{l_2}	m_{s_2}	$m_L = m_{l_1} + m_{l_2}$	$m_S = m_{s_1} + m_{s_2}$
−1	−1/2	−1	+1/2	−2	0
−1	−1/2	0	+1/2	−1	0
−1	−1/2	0	−1/2	−1	−1
−1	−1/2	+1	+1/2	0	0
−1	−1/2	+1	−1/2	0	−1
−1	+1/2	0	+1/2	−1	+1
−1	+1/2	0	−1/2	−1	0
−1	+1/2	+1	+1/2	0	+1
−1	+1/2	+1	−1/2	0	0
0	−1/2	0	+1/2	0	0
0	−1/2	+1	+1/2	+1	0
0	−1/2	+1	−1/2	+1	−1
0	+1/2	+1	−1/2	+1	0
+1	+1/2	0	+1/2	+1	+1
+1	+1/2	+1	−1/2	+2	0

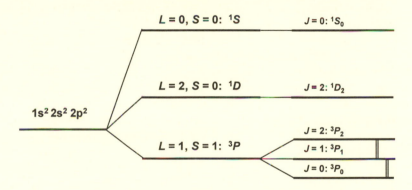

Figure 9.4 Magnetic dipole transitions for carbon.

9.5 Bibliography

Bransden, B. H., and Joachain, C. J. *Physics of Atoms and Molecules.* New York: Longman, 1983.

10   Nuclear Physics

10.1   Properties of the Nucleus

10.1.1   Nomenclature

We define:

Z = number of protons in the nucleus,
N = number of neutrons in the nucleus,
$A = N + Z$ = number of nucleons.

Nuclides with identical Z are called *isotopes*, nuclides with identical N are called *isotones*, nuclides with identical A are called *isobars*.

Protons, neutrons, and the particles that hold them together — π mesons — are the only particles believed to exist in the atomic nucleus. When other particles are observed to come from the nucleus (such as alpha, beta, or gamma rays), they are assumed to have been formed in the instant of emission.

The total binding energy of the nuclei, the energy that is released when the nucleus is assembled from Z protons and N neutrons, is given by

$$BE = \left[Zm_p + (A - Z)m_n - m_A \right]c^2 , \qquad (10.1)$$

where m_p, m_n, and m_A are the rest masses of the proton, neutron, and nucleus, respectively.

A semi-empirical formula for the mass of an atomic nucleus is given by

$$m_A = Zm_H + (A - Z)m_n - c_v A + c_a A^{2/3} + c_c \frac{Z^2}{A^{1/3}} + c_s \frac{(A - 2Z)^2}{A} \pm \delta, \quad (10.2)$$

where m_H is the mass of hydrogen, c_v is a coefficient referring to the volume of the nucleus, ($\sim 1.5 \times 10^{-2}$ amu), c_a is a coefficient relating to the area of the nucleus, ($\sim 1.4 \times 10^{-2}$ amu), c_c is a coefficient relating to the Coulomb energy, ($\sim 6.4 \times 10^{-4}$ amu), c_s is a coefficient relating to the symmetry of the nucleus, and δ is a constant equaling 0 for odd A nuclei, $-\delta$ for even Z even N nuclei, and $+\delta$ for odd Z odd N nuclei, ($\delta \sim 1.3 \times 10^{-3}$ amu).

Nuclei with certain values of N or Z (2, 8, 20, 28, 50, 82, and 126) have an exceptional number of stable nuclei in comparison to their neighbors on the periodic table. These numbers are referred to as *magic numbers*. Nuclei with magic numbers of neutrons are more stable against beta decay and have smaller nuclear reaction cross sections than usual. Some nuclei, such as He^4, O^{16}, and Pb^{128}, have magic numbers of both protons and neutrons. These elements are called *doubly magic* and are unusually stable.

10.1.2 Meson Theory of Nuclear Forces

Nuclear forces are produced by a meson field which is similar in origin to the electromagnetic field, but is of much shorter range. The scalar potential for the electric potential U in free space is

$$\left(\nabla^2 - \frac{1}{c^2} \frac{\partial^2}{\partial t^2} \right) U = 0. \tag{10.3}$$

If we let $p = -i\hbar \frac{\partial}{\partial x}$ and $E = -i\hbar \frac{\partial}{\partial t}$, for a photon this is

$$-p^2 + \frac{1}{c^2} E^2 = 0, \tag{10.4}$$

while for a particle of nonzero mass it reduces to

$$-p^2 - m^2 c^2 + \frac{1}{c^2} E^2 = 0. \tag{10.5}$$

This suggests that, in general,

$$\left(\nabla^2 - \frac{m^2 c^2}{\hbar^2} - \frac{1}{c^2} \frac{\partial^2}{\partial t^2} \right) \Phi = 0. \tag{10.6}$$

This is the equation for a spinless meson field Φ. Separating out time, $\Phi(r,t) = \phi(r)T(t)$ gives

$$\phi = g \frac{e^{-\mu r}}{r}, \tag{10.7}$$

with $\mu = \dfrac{mc}{\hbar}$ and g being a constant that is the equivalent of the source term if compared to the electrostatic potential, $U = q/r$. Similarly, the mechanical potential resulting from the interaction of two charges goes as q^2/r so the mechanical potential for the interaction of two nucleons goes as

$$V = g^2 \frac{e^{-\mu r}}{r}. \tag{10.8}$$

From experiment, μ = mass of pions. The interpretation of this is that the force field between two nucleons can be carried only by pions. For interactions between two protons or two neutrons, the field is carried by a π^0 meson. For interactions between a proton and a neutron the field may be carried by a π^+ or π^- meson.

10.1.2 Nuclear Magnetic Resonance

In 1936 Garter pointed out that it should be possible to measure the Larmour precession frequency of nuclei in a magnetic field by detecting the induced electromagnetic fields produced when the magnetic moments of a large number of nuclei simultaneously are forced to change direction. An RF signal produced by a driver coil and having its magnetic field perpendicular to the direction of the external field is timed to resonance with the Larmour precession frequency of the nucleus. The interaction energy between the field B and the magnetic moment μ in substate m_I is

$$E = \mu B \left(\frac{m_I}{I} \right). \tag{10.9}$$

We have $\Delta E = h\nu = \mu B/I$, corresponding to $\Delta m_I = \pm 1$. This gives an angular frequency $\omega = \mu B / \hbar I$, which is identical to the Larmor precession frequency. A driver coil is used to detect a very weak RF signal induced when millions of nuclei simultaneously change the direction of their magnetic moment.

10.2 Stopping Nuclear Radiation

10.2.1 Stopping Charged Particles

Consider a collision, with an impact parameter a, between an electron and a nucleus as shown in figure 10.1. The particle moves by the atom with a velocity v, which we assume is so great that the electron will not appreciably move before the particle has passed by.

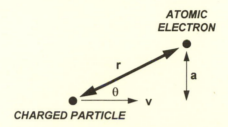

Figure 10.1 Nuclear interaction geometry.

The net impulse is

$$p = \int_{-\infty}^{+\infty} F \sin\theta \, dt, \qquad (10.10)$$

where $F = \dfrac{Ze^2}{r}$, $r = \dfrac{a}{\sin\theta}$ and $dt = \dfrac{dx}{v} = \dfrac{a d\theta}{v \sin^2\theta}$. We have

$$p = \int_{0}^{\pi} \frac{Ze^2}{av} \sin\theta \, d\theta = \frac{Ze^2}{2av}. \qquad (10.11)$$

Thus,

$$T = \frac{p^2}{2m_e} = \frac{Z^2 e^4}{8 m_e a^2 v^2}. \qquad (10.12)$$

We define the stopping cross section as

$$\sigma_{stop} = \int \Delta T \, dA, \qquad (10.13)$$

where ΔT is the loss of kinetic energy sustained when moving through an area dA. We have

$$\sigma_{stop} \approx \int_{a_1}^{a_2} T_e \, 2\pi \, a \, da = \frac{Z^2 e^4 \pi}{4 m_e v^2} \ln \frac{a_2}{a_1}. \qquad (10.14)$$

The upper and lower limits of integration are

$$a_2 = \frac{v}{\nu}, \qquad (10.15)$$

and

$$a_1 = \frac{h}{2 m_e v}, \qquad (10.16)$$

so that the stopping cross section reduces to

$$\sigma_{stop} = \frac{Z^2 e^4 \pi}{4 m_e v^2} \ln \frac{2 m_e v^2}{h\nu}. \qquad (10.17)$$

With relativistic corrections, the expression is

$$\sigma_{stop} = \frac{Z^2 e^4 \pi}{4 m_e v^2} \left[\ln \frac{2 m_e v^2}{I} - \ln\left(1 - \frac{v^2}{c^2}\right) - \frac{v^2}{c^2} \right]. \qquad (10.18)$$

10.2.1.1 Stopping Power

Stopping power is defined as the amount of energy lost by a particle per unit length of path traversed through the stopping medium. It is defined by

$$-\frac{dT}{dx} = n\sigma_{stop} = \frac{n Z^2 e^4 \pi}{4 m_e v^2} \ln \frac{2 m_e v^2}{h\nu}. \qquad (10.19)$$

10.2.1.2 Range

The range of a particle is the line-of-sight, straight-line distance it travels from birth to rest. It differs from the path length, which is the length of the actual route traversed by the particle during its numerous nuclear/atomic collisions, because the path is not a straight line. For heavy charged particles, such as protons or alpha particles, the path length is ~10% greater than the range. Conversely, for electrons the path length can be twice as long as the range. High-energy electrons deviate less from straight-line paths than do low-energy electrons. By analogy with equation 10.19, the range can be defined as

$$R = \int dx = -\int_{T}^{0} \frac{dT}{n\sigma_{stop}}. \tag{10.20}$$

Note that removing the density n from equation 10.20 allows the range to be specified in the units g/cm^2. Empirically, it is seen that the range can be approximated by

$$R = \delta E^{\gamma}, \tag{10.21}$$

where E is the energy in MeV and δ and γ are constants, the specific value of which are a function of particle type.

10.2.1.3 Brehmsstrahlung Radiation

Brehmsstrahlung (braking) radiation may occur when a charged particle undergoes rapid deceleration. In order to conserve both energy and momentum, a charged particle passing near a nucleus with relativistic velocities will often lose energy in the form of energetic photons. This brehmsstrahlung radiation is significant only for particles whose e/m ratio is high, such as electrons and positrons. For thick targets, which can stop the charged particle, the fraction of initial particle energy which is converted into brehmsstrahlung is

$$\sim 7 \times 10^{-4} \left(\frac{m_e}{M} \right)^2 ZE, \tag{10.22}$$

where m_e is the mass of the electron, M is the mass of the charged particle, E is the kinetic energy of the charged particle, and Z is the atomic number of the target material.

10.2.2 Stopping Photons

Because there is no electrostatic force between a photon and a nucleus, the photon will move in a straight line unless intercepted directly by the nucleus. The probability for an interaction is given by

$$-\frac{dN}{N} = \frac{(nAdx)\sigma_{tot}}{A} = n\sigma_{tot}dx, \tag{10.23}$$

where n is the material density and σ_{tot} is the total cross section for absorption. The photon density a distance x into the material is given by

$$N(x) = N\exp(-n\sigma_{tot}x). \tag{10.24}$$

Photons may be absorbed by three processes. The *photoelectric effect* is when a gamma ray transfers energy to an atomic electron, which is then expelled. This process is usually significant for energies $< \sim 0.1$ MeV. The *Compton effect* is when a gamma ray is scattered off an electron. This process is usually significant between 0.01 and 10.0 MeV. *Pair production* is when a gamma ray decays into an electron-positron pair. This process is usually significant above 10.0 MeV.

10.2.2.1 Compton Scattering

When a gamma ray scatters off an electron, the gamma ray will exhibit a change in frequency. If the initial direction of travel is defined to be horizontal, conservation of momentum gives

$$\frac{E_\gamma}{c} = \frac{E_\gamma{}'}{c}\cos\phi + p_m{}'\cos\theta, \tag{10.25}$$

and

$$0 = \frac{E_\gamma{}'}{c}\sin\phi - p_m{}'\sin\theta. \tag{10.26}$$

Conservation of energy gives

$$E_\gamma + mc^2 = E_\gamma{}' + E_m{}', \tag{10.27}$$

or

$$E_m' = E_\gamma + mc^2 - E_\gamma' = \left(m^2 c^4 + p_m'^2 c^2\right)^{1/2}, \qquad (10.28)$$

with the assumption that the initial kinetic energy of the electron is negligible. Equations 10.25 and 10.26 can be rearranged to give

$$p_m'^2 = \left(\frac{E_\gamma}{c}\right)^2 - 2\left(\frac{E_\gamma E_\gamma'}{c^2}\right)\cos\phi + \left(\frac{E_\gamma'}{c}\right)^2. \qquad (10.29)$$

Using this expression in equation 10.28 gives

$$E_\gamma' = \frac{E_\gamma}{\left[1 + \left(\dfrac{E_\gamma}{mc^2}\right)(1 - \cos\phi)\right]}. \qquad (10.30)$$

As a result, photons that scatter off atomic electrons will exhibit a change in direction and a change in energy (frequency).

10.2.2.2 Pair Production

Pair production, where a photon disintegrates into an electron-positron pair, requires the photon to have a minimum kinetic energy given by

$$E_\gamma = m_- c^2 + m_+ c^2 + KE_- + KE_+. \qquad (10.31)$$

Consequently the minimum energy needed for pair production is twice the rest energy of an electron, or 1.02 MeV. To conserve momentum, pair production can take place only in the vicinity of the nucleus.

10.3 Nuclear Disintegration Processes—Radioactivity

Many heavy nuclei are intrinsically unstable and decay to lighter, more stable elements through three kinds of disintegration processes called alpha, beta, or gamma emission. Regardless of the specific decay mode, the number of decays occurring between time t and time $t + dt$ is seen to follow the relation

$$-dN = \lambda N dt, \qquad (10.32)$$

which has solution

$$N(t) = N(0)\exp^{-\lambda t}. \tag{10.33}$$

There are four radioactive decay series. The $4n$ thorium series, the $4n + 1$ neptunium series, the $4n + 2$ uranium series, and the $4n + 3$ actinium series. The neptunium series is not observed in nature because the longest-lived member, Np ^{237}, has a half-life of only 2.2×10^6 years, far shorter than the 4.5×10^9 year age of the Earth. For the same reason, the actinium series is almost extinct, making it a difficult process to obtain enriched U^{235}.

Each series decays by a sequence of alpha and beta decay modes to a final stable isotope. For a general decay scheme,

$$A \to B \to C \to \ldots. \tag{10.34}$$

A is called the *parent nuclei*, B is the *daughter*, C the *granddaughter*, and so on. The relative equilibrium between the various generations is a function of the relative half-lives. The number of nuclei present at a given time, times the half-life of the nuclei, is called the *activity*.

Example 10.1
At $t = 0$ there are N_0 radioactive nuclei A which decay with a mean life t_A into nuclei B. Nuclei B decay with a mean life t_B into stable nuclei C. Find the number of nuclei N_B as a function of time. For the special case where $t_A = t_B$, what is the maximum value of N_B?

From equation 10.33 we have

$$\frac{dN_A}{dt} = -\lambda_A N_A, \tag{10.35}$$

and by comparison

$$\frac{dN_B}{dt} = \lambda_A N_A - \lambda_B N_B, \tag{10.36}$$

where $\lambda_A = 1/t_A$ and $\lambda_B = 1/t_B$. From equation 10.35,

$$N_A(t) = N_A(0)e^{-\lambda_A t}. \tag{10.37}$$

Using this expression in equation 10.36 gives

BARNES & NOBLE
STORE 2579 PITTSBURGH, PA 412-521-3600

REG#05   BOOKSELLER#022
RECEIPT#   31203 10/18/97   1:34 PM

CUSTOMER COPY
S 0385319460   FERMATS LAST THEOREM
                    1 @ 9.95   9.95

S 0691026629   PRINCETON GT ADVD PHYSIC
                    1 @ 19.95   19.95

S 0201409569   6 EASY PIECES BK CASS
LIST PRICE: 49.95   1 @ 44.95   44.95

SUBTOTAL   74.85
SALES TAX - 7%   5.24
TOTAL   80.09
DISCOVER CARD PAYMENT   80.09
ACCOUNT# 6011008670039061   EXP 0998
AUTHORIZATION# 018319   CLERK 22

BOOKSELLERS SINCE 1873

$$\frac{dN_B}{dt} = \lambda_A N_A(0) e^{-\lambda_A t} - \lambda_B N_B. \tag{10.38}$$

Multiplying both sides by $e^{\lambda_B t}$ and integrating gives

$$N_B(t) e^{\lambda_B t} = \frac{\lambda_A}{\lambda_B - \lambda_A} N_A(0) e^{(\lambda_B - \lambda_A)t} + C, \tag{10.39}$$

where C is a constant. Applying the initial condition, $N_B = 0$ at $t = 0$ and we find that

$$C = -\frac{\lambda_A}{\lambda_B - \lambda_A} N_A(0), \tag{10.40}$$

and

$$N_B(t) = \frac{\lambda_A}{\lambda_B - \lambda_A} N_A(0) \left[e^{-\lambda_A t} - e^{-\lambda_B t} \right]. \tag{10.41}$$

In general, the maximum/minimum of equation 10.40 are given when $dN_B/dt = 0$, or when

$$\lambda_A e^{-\lambda_A t} = \lambda_B e^{-t}. \tag{10.42}$$

It follows that the time of maximum activity is

$$t_{max} = \frac{\ln\left(\frac{\lambda_B}{\lambda_A}\right)}{(\lambda_B - \lambda_A)} = \frac{\ln\left(\frac{t_A}{t_B}\right)}{\left(\frac{1}{t_A} - \frac{1}{t_B}\right)} = \frac{t_B t_A}{t_B - t_A} \ln\left(\frac{t_A}{t_B}\right). \tag{10.43}$$

To solve this equation when $t_A = t_B$, we approximate

$$t_A = t_B(1 + \delta). \tag{10.44}$$

Substituting this in equation 10.43 gives

$$t_{max} = t_B \left(\frac{1+\delta}{\delta} \right) \ln \left(1+\delta \right)$$

$$= t_B \left(\frac{1+\delta}{\delta} \right) \left(\delta - \frac{\delta^2}{2!} + \cdots \right)$$

$$= t_B \left(1+\delta \right) \left(1 - \frac{\delta}{2} + \cdots \right)$$

$$\approx t_B \left(1 + \frac{\delta}{2} + \cdots \right)$$

$$\approx t_B \left(\frac{t_A}{t_B} \right)^{1/2}$$

$$\approx \left(t_A t_B \right)^{1/2}. \tag{10.45}$$

It follows that the maximum value of N_B occurs for $t = t_A = t_B$. Using a Taylor series expansion, it can be shown that equation 10.40 reduces to

$$N_B(t) = N_A(0) \frac{\lambda_A}{\lambda_B + \lambda_A} \left[1 - e^{-(\lambda_A + \lambda_B)t} \right], \tag{10.46}$$

so that for $t = t_A = t_B$

$$N_{B\,max} = \frac{N_A(0)}{2} \left[1 - e^{-2} \right] = (0.432) N_A(0). \tag{10.47}$$

10.3.1 Alpha Disintegration

An example of alpha disintegration is

$$U^{232} \rightarrow Th^{228} + He^4. \tag{10.48}$$

An alpha particle is a He^4 nucleus, $2p + 2n = \alpha$, and carries a binding energy of 28.3 MeV. The theory of alpha disintegration, which occurs only for heavy nuclei ($A > 200$), assumes that the alpha particle exists as a separate entity inside the nucleon itself. Because of its close confinement inside a volume of radius $\sim 1.5 \times 10^{-15}$ m, it has a velocity of $\sim 1.5 \times 10^7$ m/s due to the Heisenberg uncertainty principle, and an energy of ~ 5 MeV. Consequently, it hits the inside of the "Coulomb barrier" approximately 10^{21} times per second. At each impact, it has an escape probability given by the quantum mechanical tunneling probability. This probability is very low, $\sim 10^{-38}$ for U^{235} and $\sim 10^{-14}$ for Pb^{212}.

10.3.2 Beta Disintegration

When energetically possible, some heavy nuclides decay through the emission of an electron or a positron. This is referred to as *beta disintegration*. An example of β^- disintegration is

$$n \rightarrow p + e^- + \nu_e, \tag{10.49}$$

while an example of β^+ disintegration is

$$p \rightarrow n + e^+ + \bar{\nu}_e. \tag{10.50}$$

The disintegration energy is supplied through a change in the rest energy and the binding energy of the converted nucleus. Beta particles (electrons and positrons) have continuous energy spectra since they share the quantized disintegration energy with the neutrinos.

The half-lives of the nuclei undergoing beta decay depend on the degree of forbiddenness of the transition, which depends on the degree of overlap between the nuclear wave functions before and after the transitions. For allowed transitions, the half-life is proportional to

$$\tau_{1/2} \sim \left(E_{max} + m_o c^2 \right)^{-1/2}. \tag{10.51}$$

Inverse beta decay can occur when an electron or positron strikes a nucleus. The cross section for such interactions is quite small and decreases with beta particle energy.

10.3.3 Gamma Emission

As predicted by the shell model, nuclei have excited states as well as ground states. Each excited state has a spin I and a parity which are usually different from those of the ground state. Even-even nuclei usually have 0^+, 2^+, 4^+, . . . for the spins and parities of their ground state and their increasingly excited states. The energy of the excited states is usually several MeV above the energy of the ground state, and de-excitation of excited nuclear states can occur with the emission of a gamma ray. Gamma ray transitions are electric quadrupole transitions.

For example, N^{14} could break up into $C^{13} + H^1$ if it is excited to at least 7.54 MeV. Whether an excited N^{14} nucleus with this much energy will break up in this way depends on the other competing reactions available, such as gamma ray decay to a lower N^{14} state. In general, gamma ray decay is slow and occurs only because competing decay modes are not energetically possible.

10.3.3.1 Internal Conversion

When a nucleus emits a gamma ray, it may interact with a planetary electron in the same atom. The result is that the photon disappears and the electron leaves the atom with the photon's energy, less the electron's initial binding energy. This is called *internal conversion*, and is usually observed in high Z atoms involving K shell electrons. As a result of the vacancy produced in the K shell, one electron from each of the other shells falls inward to fill the vacancy. This produces fluorescent radiation.

10.4 Nuclear Reactions

10.4.1 Nuclear Scattering (Nonrelativistic)

Consider a collision in which a particle of mass m_1, initially moving with velocity v_1, interacts with a body of mass m_2 (assumed initially at rest) and produces two different bodies, m_3 and m_4. Conservation of energy gives

$$\frac{1}{2}m_1 v_1^2 = \frac{1}{2}m_3 v_3^2 + \frac{1}{2}m_4 v_4^2 - Q, \qquad (10.52)$$

where Q, called the *nuclear disintegration energy*, is equal to the final kinetic energy minus the initial kinetic energy. Conservation of momentum, in laboratory coordinates, gives

$$m_1 v_1 = m_3 v_3 \cos\theta + m_4 v_4 \cos\phi, \qquad (10.53)$$

and

$$0 = m_3 v_3 \sin\theta - m_4 v_4 \sin\phi. \qquad (10.54)$$

From the definition of momentum ($p^2 = 2mE$) equation 10.52 may be reduced to

$$Q = E_3\left(1 + \frac{m_3}{m_4}\right) - E_1\left(1 - \frac{m_1}{m_4}\right) - 2\frac{\sqrt{m_1 m_3 E_1 E_e}}{m_4}\cos\theta, \quad (10.55)$$

usually called the *Q-value equation*.

10.4.2 Fission

Some nuclei are observed to break up spontaneously into two or three large fragments, plus some neutrons. This process is called *fission*. The fission reaction may also be initiated by slow neutrons. For example,

$$U^{235} + n_{slow} \rightarrow U^{236} \rightarrow X + Y + \nu n_{fast}, \tag{10.56}$$

where X and Y are two intermediate mass nuclides and ν is the average number of neutrons released in the process ($\nu = 2.47$ for the reaction shown). These reactions occur only when an odd A nucleon acquires a neutron, making it an even A nucleon. The slow neutron cross section of U^{235} is so large, relatively speaking, in comparison to other nuclei that it was said to be "as big as a barn." This terminology stuck, and nuclear cross sections are typically measured in barns (10^{-24} cm^2). The U^{235} reaction can also produce additional neutrons, which can in turn produce more reactions. The result is a "chain reaction" provided the surface to volume ratio of the U^{235} is small enough.

The same compound U^{236} nucleus can be produced by

$$_{90}\text{Th}^{232} + _2\text{He}^4 \rightarrow _{92}\text{U}^{236} \rightarrow X + Y + \nu n, \tag{10.57}$$

but since the alpha particle must surmount a 28 MeV barrier, this reaction produces much more excitation energy than equation 10.36. When the U^{236} nucleus disintegrates, it gives off ~ 200 MeV in excess energy. This energy is distributed between the fission fragments (165 MeV); prompt neutrino (5 MeV); prompt gamma rays (7 MeV); neutrinos (10 MeV); delayed beta particles (7 MeV); and delayed gamma rays (6 MeV).

10.4.3 Fusion

If two smaller mass particles have enough kinetic energy to overcome the Coulomb repulsion forces, they can be combined into a single larger nucleus. If the final rest mass of the large particle is greater than the sum of its parts, kinetic energy must be added to make the reaction possible. In some cases, the final rest mass of the large particle is less than the sum of its parts and energy is released. For example, the reaction responsible for the majority of the energy in stars is

$$H^1 + H^1 \rightarrow H^2 + e^+ + \nu. \tag{10.58}$$

Because of the differences in the rest mass of each side of the equation, it is seen that this process releases 0.42 MeV of energy. When the positron annihilates with an electron, an additional 1.02 MeV is released. The prospect of controlling this reaction is the driving force behind most fusion research programs.

10.5 Elementary Particle Physics

At the most fundamental level, the particles of physics are subject to four basic forces: strong, electromagnetic, weak, and gravitational. All particles of interest interact through the weak force, a subset interacts through the electromagnetic force, and a smaller subset interacts through the strong force.

- Weak interactions are characterized by lifetimes from 10^{-10} s to 15 min.

- Electromagnetic interactions have lifetimes from 10^{-15} s to 10^{-18} s.

- Strong interactions have lifetimes from 10^{-20} s to 10^{-23} s.

Examples of particles and their methods of interaction are shown in table 10.1.

Table 10.1

Elementary Particles and Corresponding Interaction Modes

Weak	Electromagnetic	Strong
e	e	π
μ	μ	K
ν	π	p
π	K	n
K	p	Λ
p	n	
n	Λ	
Λ		

The lightest group of particles, *leptons*, can interact through the weak or electromagnetic force only. All leptons have spin 1/2, so they are fermions. The leptons, or their antiparticles, are fundamental particles.

Particles of intermediate mass, *mesons*, undergo all three types of interactions. Mesons have spin 0 or 1. All mesons are unstable and have no distinct antiparticles.

The heaviest group of particles, *baryons*, undergo all three types of interactions and have half-integer spin.

All nuclear reactions and decays are consistent with conservation of electric charge and nuclear number. In addition, the leptons conserve *Lepton number*. A property called *strangeness* was originally introduced to describe the properties of the K meson decays. Only the weak interaction does not conserve strangeness. However, weak interactions must still satisfy $\Delta S = \pm 1$. Similar to lepton number, *baryon number* is also conserved.

If we know the speed of a particle, we can find its lifetime by observing the length of its track. Some particles have lifetimes so short, 10^{-23} s, that we can't even witness a track in the laboratory. Such particles are called *resonance particles*. We detect them by looking at their decay products. The energy available for a decay, Q, is given by equation 10.55.

10.5.1 The Quark Model

We can understand the physics of mesons and baryons by treating these particles as though they were composed of fundamental particles called *quarks*. Mesons are a quark-antiquark pair, while baryons are bound states of three quarks. The three quarks are known as up (u), down (d), and strange (s). Properties of the u, d, and s quark are summarized in tables 10.2 and 10.3.

Table 10.2

Properties of the Up, Down, and Strange Quarks

quark	charge	spin	baryon #	strangeness
u	2/3	1/2	1/3	0
d	−1/3	1/2	1/3	0
s	−1/3	1/2	1/3	1

Table 10.3

Additional Properties of the Up, Down, and Strange Quarks

quark	spin	B	Q	I_3	S	Y
u	1/2	1/3	2/3	1/2	0	1/3
d	1/2	1/3	−1/3	−1/2	0	1/3
s	1/2	1/3	−1/3	0	1	−1/3

When we attempt to construct a model for the baryons, we find that we must introduce a separate property called *color* in order to avoid problems with the Pauli exclusion principle. The three colors are red, green, and blue. Particles must be constructed so that the bound states are colorless. The three possible combinations are (1) equal mixtures of RGB, (2) equal mixtures of anticolors, $\overline{R}\,\overline{G}\,\overline{B}$, or (3) equal mixtures of $R\overline{R}$, $G\overline{G}$, $B\overline{B}$. The eight possible quark combinations forming the baryons are illustrated in figure 10.2.

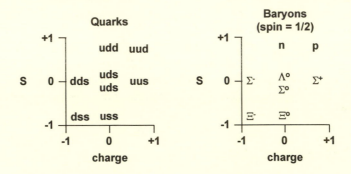

Figure 10.2 The eight quark combinations forming the baryons.

Finally, the mesons are described as a quark-antiquark bound state. The quark and antiquark are illustrated in figure 10.3. There are nine possible mesons, as shown in figure 10.4.

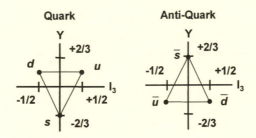

Figure 10.3 The quark and antiquark.

The quark-antiquark pair can be labeled an octet plus a singlet, as shown in figure 10.5. The values A, B, and C are defined by

$$A = \sqrt{\frac{1}{3}}\left(u\bar{u} - d\bar{d}\right), \tag{10.59}$$

$$B = \sqrt{\frac{1}{3}}\left(u\bar{u} + d\bar{d} + 2s\bar{s}\right), \qquad (10.60)$$

and

$$C = \sqrt{\frac{1}{3}}\left(u\bar{u} + d\bar{d} + s\bar{s}\right). \qquad (10.61)$$

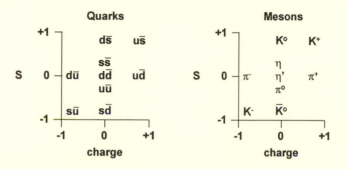

Figure 10.4 The nine quark-antiquark pairs forming the mesons.

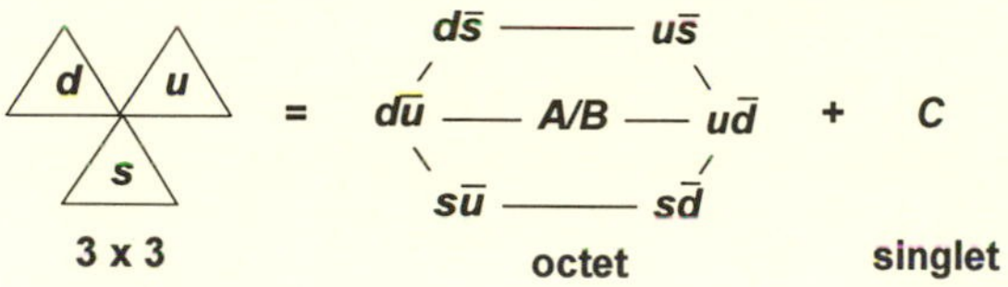

Figure 10.5 The quark-antiquark octet and singlet.

10.4 Bibliography

Bransden, B. H., and Joachain, C. J. *Physics of Atoms and Molecules.* New York: Longman, 1980.

Enge, H. *Introduction to Nuclear Physics.* Reading, Mass.: Addison-Wesley, 1966.

Particles and Fields. Readings from Scientific American. San Francisco: W. H. Freeman, 1980.

Perkins, D. H. *Introduction to High Energy Physics.* Reading, Mass.: Addison-Wesley, 1972.

Tipler, P. A. *Modern Physics.* New York: Worth, 1969.

11 Statistical Physics

11.1 Thermodynamics

11.1.1 The Basic Laws of Thermodynamics

Thermodynamics is the study of the motion of heat. The laws of thermodynamics are based on empirical observations of the conservation, and degradation, of energy.

Zeroth Law: If two systems are in thermodynamic equilibrium with a third system they must be in thermodynamic equilibrium with each other.

First Law: Heat is a form of energy.

Second Law: If a closed system is not in the equilibrium configuration, the most probable consequence is that the entropy of the system will increase monotonically in successive instants of time.

Third Law: The entropy of a system approaches a constant value as the temperature approaches zero.

11.1.2 The Carnot Cycle

Empirically, it is seen that there is a relationship between pressure, volume, temperature, and number density of a gas. The simplest relationship between these variables, which assumes the gas is composed of noninteracting particles, is called the *ideal gas law*. The simplest model that takes into account the interactions, and is capable of predicting a phase transition, is the van der Waals model. Several equations of state are possible, as shown in table 11.1.

Table 11.1

Equations of State for Various Gas Models

Model	Equation of State
$PV = NkT$	Ideal Gas
$\left(P + \dfrac{N^2 a}{V^2}\right)(V - Nb) = NkT$	van der Waals Gas
$\left(P + \dfrac{N^2 a}{TV^2}\right)(V - Nb) = NkT$	Berthelot Gas
$P(V - Nb) = NkT \exp\left(-\dfrac{C}{N^2 kTV}\right)$	Dieterici Gas
$P = \dfrac{NkT}{V^2}(1 - \varepsilon)(V + NB) - \dfrac{N^2 A}{V^2}$	Beattie-Bridgeman Gas
$PV = A + \dfrac{B}{V} + \dfrac{C}{V^2} + \ldots$	Virial Gas

a and b are constants; A, B, C are functions of temperature

The first law of thermodynamics is a statement of the fact that heat can be used to generate work. The simplest process for generating work from heat is known as the *Carnot cycle*. In the Carnot cycle, an ideal gas is expanded and compressed in four stages, as shown in figure 11.1. The laws of thermodynamics can be applied to this cycle to determine the amount of heat converted to work.

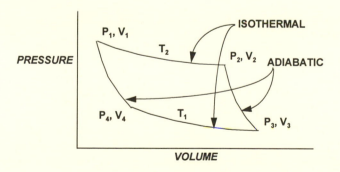

Figure 11.1 The Carnot gas cycle.

The first law of thermodynamics can be written in equation form as,

$$dE = dq - dW, \tag{11.1}$$

where dE is the change in the internal energy of the system, dq is the heat absorbed by the system, and dW is the work done by the system.

I. The first leg of the curve is an isothermal expansion at the higher temperature T_2. In a reversible isothermal expansion $dq = dW$, so there is no change in internal energy E. By definition,

$$dW = \vec{F} \cdot d\vec{x} = PAdx = PdV, \tag{11.2}$$

which gives

$$q_1 = W_1 = \int_{V_1}^{V_2} PdV = \int_{V_1}^{V_2} \frac{RT_2}{V} dV = RT_2 \ln\frac{V_2}{V_1}. \tag{11.3}$$

II. The second leg is an adiabatic expansion from T_2 to T_1. No heat is added, so $q_2 = 0$. As a result, the work done is equal to the negative of the change in potential energy. In equation form, this is

$$W_2 = \int_{V_2}^{V_3} PdV = -\int_{T_2}^{T_3} dE = -\int_{T_2}^{T_1} C_v dT = -C_v(T_1 - T_2). \tag{11.4}$$

III. The third leg is an isothermal compression at T_1. By comparison with the first leg,

$$q_3 = W_3 = \int_{V_3}^{V_4} PdV = \int_{V_3}^{V_4} \frac{RT_1}{V} dV = RT_1 \ln\frac{V_4}{V_3}. \tag{11.5}$$

IV. The fourth leg is an adiabatic compression to complete the cycle. By comparison with the second leg,

$$W_4 = \int_{V_4}^{V_1} P dV = -\int_{T_1}^{T_2} dE = -\int_{T_1}^{T_2} C_v dT = -C_v(T_2 - T_1). \qquad (11.6)$$

In the complete cycle, the engine does work

$$W = W_1 + W_2 + W_3 + W_4 = W_1 + W_3. \qquad (11.7)$$

In doing this work, the engine absorbs heat q_1 from the reservoir at T_2 and gives off the heat q_3 to the reservoir at T_1. It is readily seen that

$$\Delta E = q - W = q_1 + q_3 + W_1 + W_3 = 0. \qquad (11.8)$$

The efficiency of the engine is the ratio of the total work to the heat absorbed,

$$\eta = \frac{W_1 + W_3}{q_1} = \frac{q_1 + q_3}{q_1}. \qquad (11.9)$$

In the adiabatic process,

$$dq = dE + dW = C_v dT + P dV = 0. \qquad (11.10)$$

When this is combined with the ideal gas law, it reduces to

$$-C_v \int_{T_2}^{T_1} \frac{dT}{T} = \int_{V_2}^{V_3} \frac{R}{V} dV, \qquad (11.11)$$

or

$$-C_v \ln\frac{T_2}{T_1} = R \ln\frac{V_3}{V_2}. \qquad (11.12)$$

Similarly, we find that

$$-C_v \ln\frac{T_2}{T_1} = R \ln\frac{V_1}{V_4}, \qquad (11.13)$$

or

$$\frac{V_1}{V_4} = \frac{V_2}{V_3}.$$ (11.14)

Using this and the expressions for q_1, q_3 we see that

$$\eta = \frac{T_2 - T_1}{T_2}.$$ (11.15)

Combining this with the previous expression for η gives

$$\frac{q_3}{T_1} + \frac{q_1}{T_2} = 0.$$ (11.16)

11.1.3 Thermodynamic Functions

11.1.3.1 Entropy

For the Carnot cycle $q_2 = q_4 = 0$, and equation 11.16 may be generalized to

$$\oint \frac{dq}{T} = 0.$$ (11.17)

Thus, dq/T is a state function; that is, it depends only on the initial and final states. Therefore, we define entropy, S, as

$$dS = \frac{dq}{T}.$$ (11.18)

If $dS = 0$, the process is reversible, if $dS > 0$ the process is irreversible. The third law is

$$\lim_{T \to 0} \Delta S = 0.$$ (11.19)

11.1.3.2 Internal Pressure

The first and second laws of thermodynamics are, in essence,

$$dE = TdS - PdV.$$ (11.20)

However, by definition,

$$dE = \left(\frac{\partial E}{\partial S}\right)_V dS + \left(\frac{\partial E}{\partial V}\right)_S dV, \qquad (11.21)$$

therefore we conclude that

$$T = \left(\frac{\partial E}{\partial S}\right)_V, \qquad (11.22)$$

and

$$P = -\left(\frac{\partial E}{\partial V}\right)_S. \qquad (11.23)$$

An equally acceptable solution is

$$dE = \left(\frac{\partial E}{\partial T}\right)_V dT + \left(\frac{\partial E}{\partial V}\right)_T dV. \qquad (11.24)$$

We define

$$C_v = \left(\frac{\partial E}{\partial T}\right)_V, \qquad (11.25)$$

and

$$P_i = \left(\frac{\partial E}{\partial V}\right)_T, \qquad (11.26)$$

where C_v is the heat capacity and P_i is the internal pressure. Similarly,

$$dS = \left(\frac{\partial S}{\partial T}\right)_V dT + \left(\frac{\partial S}{\partial V}\right)_T dV, \qquad (11.27)$$

which gives

$$dE = T\left(\frac{\partial S}{\partial T}\right)_V dT + \left[T\left(\frac{\partial S}{\partial V}\right)_T - P\right]dV. \qquad (11.28)$$

Consequently, heat capacity and internal pressure are also given by

$$C_V = T\left(\frac{\partial S}{\partial T}\right)_V,$$ (11.29)

and

$$P_i = T\left(\frac{\partial S}{\partial V}\right)_T - P.$$ (11.30)

11.1.3.3 Enthalpy

We define enthalpy, H, by the relation

$$H = E + PV,$$ (11.31)

which gives

$$dH = dE + VdP + PdV = TdS + VdP.$$ (11.32)

For a system at constant pressure, the change in enthalpy is equal to the heat absorbed by the system.

11.1.3.4 Helmholtz Free Energy

We define the Helmholtz free energy, A, by

$$A = E - TS,$$ (11.33)

which gives

$$dA = dE - SdT - SdT = -PdV - SdT.$$ (11.34)

For isothermal processes, the change in the Helmholtz free energy is the negative of the work done.

11.1.3.5 Gibbs Free Energy

We define the Gibbs free energy, G, by

$$G = H - TS = E - TS + PV,$$ (11.35)

which gives

$$dG = VdP - SdT - (dW - PdV). \qquad (11.36)$$

The Gibbs free energy is often called the *thermodynamic potential*. The Gibbs free energy per particle is termed the chemical potential, μ. For a multicomponent gas,

$$G = \sum_i N_i \mu_i. \qquad (11.37)$$

The change in the Gibbs free energy is the negative of the work done in excess of the pressure-volume work at constant T, P. The Gibbs free energy is a minimum for a system in equilibrium at a constant pressure when in thermal contact with a reservoir.

11.1.3.6 VAT-VUS (NAT-NUS) Diagram

All of the preceding relationships may be reconstructed with the help of the VAT-VUS or NAT-NUS diagram shown in figure 11.2. Move either horizontally or vertically across the diagram. The first variable encountered is held constant, and the partial differential of the second variable with respect to the third variable is equal to the variable that is diagonal from the third variable. If the movement is against the arrow, a minus sign is incurred.

Figure 11.2 The VAT-VUS and NAT-NUS learning aids.

We see this with the help of the following examples:

$$\left(\frac{\partial U}{\partial S}\right)_V = \left(\frac{\partial E}{\partial S}\right)_V = T, \qquad (11.38)$$

$$\left(\frac{\partial U}{\partial V}\right)_S = \left(\frac{\partial E}{\partial V}\right)_S = -P, \qquad (11.39)$$

or

$$\left(\frac{\partial U}{\partial S}\right)_N = \left(\frac{\partial E}{\partial S}\right)_N = T, \tag{11.40}$$

$$\left(\frac{\partial U}{\partial N}\right)_S = \left(\frac{\partial E}{\partial N}\right)_S = \mu. \tag{11.41}$$

11.1.3.7 Heat Capacity of Gases

From the definition of heat capacity, we have

$$C_v = \left(\frac{\partial E}{\partial T}\right)_V = T\left(\frac{\partial S}{\partial T}\right)_V, \tag{11.42}$$

and

$$C_p = \left(\frac{\partial H}{\partial T}\right)_P = \left(\frac{\partial E}{\partial T}\right)_P + P\left(\frac{\partial V}{\partial T}\right)_P. \tag{11.43}$$

Similarly, from the definition

$$dE = \left(\frac{\partial E}{\partial T}\right)_V dT + \left(\frac{\partial E}{\partial V}\right)_T dV, \tag{11.44}$$

we obtain

$$\left(\frac{\partial E}{\partial T}\right)_P = C_v + \left(\frac{\partial E}{\partial V}\right)_T\left(\frac{\partial V}{\partial T}\right)_P, \tag{11.45}$$

and

$$C_p = C_v + \left[\left(\frac{\partial E}{\partial V}\right)_T + P\right]\left(\frac{\partial V}{\partial T}\right)_P. \tag{11.46}$$

Using

$$P_i = \left(\frac{\partial E}{\partial V}\right)_T = T\left(\frac{\partial P}{\partial T}\right)_V - P, \tag{11.47}$$

reduces the expression for C_p to

$$C_p = C_v + T\left(\frac{\partial P}{\partial T}\right)_V \left(\frac{\partial V}{\partial T}\right)_P. \tag{11.48}$$

If we define

$$\alpha = \frac{1}{V}\left(\frac{\partial V}{\partial T}\right)_P, \tag{11.49}$$

and

$$\beta = -\frac{1}{V}\left(\frac{\partial V}{\partial P}\right)_T, \tag{11.50}$$

then

$$\frac{\alpha}{\beta} = -\frac{\left(\frac{\partial V}{\partial T}\right)_P}{\left(\frac{\partial V}{\partial P}\right)_T} = \left(\frac{\partial P}{\partial T}\right)_V. \tag{11.51}$$

For an ideal gas, the expression for heat capacity reduces to

$$C_p = C_v + \frac{\alpha^2 VT}{\beta}. \tag{11.52}$$

11.2 Ideal Gas

11.2.1 Kinetic Theory

If a single molecule of velocity v impacts a surface at an angle θ and is scattered elastically, the magnitude of the change in momentum of the molecule, Δp, is given by

$$\Delta p = mv\cos\theta - (-mv\cos\theta) = 2mv\cos\theta. \tag{11.53}$$

Envision a cylinder of length vdt inclined at an angle θ to the surface element dA, as shown in figure 11.3.

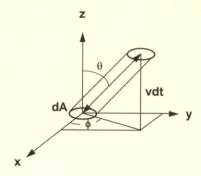

Figure 11.3 A cylindrical volume element of gas.

Consider those molecules contained by the volume element θ and $\theta + d\theta$, ϕ and $\phi + d\phi$, with velocity v to $v + dv$. All molecules satisfying the constraints on θ, f, and v will strike the surface area dA in time dt. The total number of molecules in the cylinder is $dN = ndV = nv\cos\theta \, dt \, dA$. If we let dn/dv represent the number of molecules with speed between v and $v + dv$, the number of molecules with the required speed is given by

$$\frac{dn}{dv} v \cos\theta \, dt \, dA. \tag{11.54}$$

The number of molecules having both the required speed and direction is

$$\frac{dn}{dv} v \cos\theta \, dt \, dA \frac{d\omega}{4\pi}. \tag{11.55}$$

Therefore, the change in momentum due to all collisions, from all directions, per unit area per unit time is given by

$$\int_0^{2\pi} \int_0^{\pi/2} (2mv \cos\theta) \left(\frac{dn}{dv} v \cos\theta \, \frac{1}{4\pi} \right) \sin\theta \, d\theta \, d\phi, \tag{11.56}$$

which reduces to

$$\frac{1}{3} mv^2 \frac{dn}{dv}. \tag{11.57}$$

The total pressure exerted on the walls of the vessel is found by integrating over all molecules, or equivalently, over the entire distribution of molecular velocities, and is given by

$$P = \frac{dF}{dA} = \frac{1}{3}m\left(\int_0^n v^2 dn\right) = \frac{1}{3}m\left(\int_0^n v^2 \frac{dn}{dv} dv\right). \tag{11.58}$$

We define the average value of the square of the velocity by

$$\bar{v}^2 = \frac{1}{n}\int_0^n v^2 \frac{dn}{dv} dv, \tag{11.59}$$

and the expression for pressure reduces to

$$P = \frac{1}{3}mn\bar{v}^2. \tag{11.60}$$

The empirically determined equation of state for an ideal gas is

$$PV = NkT. \tag{11.61}$$

Combining equations 11.60 and 11.61 with the definition $nV = N$, we see that the translational kinetic energy of the molecules is related the temperature of the gas by

$$\frac{1}{2}m\bar{v}^2 = \frac{3}{2}kT. \tag{11.62}$$

11.2.2 Maxwell-Boltzmann Velocity Distribution

We define a one-dimensional normalized distribution function of a single velocity component, v_i, by

$$f(v_i) = \frac{1}{n}\frac{dn}{dv_i}. \tag{11.63}$$

For any values of v_x and v_y, it is possible to define a rotation to a new set of coordinates, α and β, where $v_\alpha^2 = v_x^2 + v_y^2$ and $v_\beta^2 = 0$. Consequently, it follows that

$$f(v_\alpha)f(0) = f(v_x)f(v_y). \tag{11.64}$$

Differentiating the expression with respect to v_x gives

$$f'(v_x)f(v_y) = f'(v_\alpha)\frac{\partial v_\alpha}{\partial v_x}f(0). \tag{11.65}$$

Because of the relationship between v_α and v_x

$$\frac{\partial v_\alpha}{\partial v_x} = \frac{v_x}{v_\alpha}. \tag{11.66}$$

Plugging this into equation 11.65 and developing similar relationships for v_y gives

$$f'(v_\alpha)f(0) = \frac{v_\alpha}{v_x}f'(v_x)f(v_y) = \frac{v_\alpha}{v_y}f(v_x)f'(v_y). \tag{11.67}$$

Dividing this expression by $f(v_x)\,f(v_y)$ gives

$$\frac{1}{v_x}\frac{f'(v_x)}{f(v_x)} = \frac{1}{v_y}\frac{f'(v_y)}{f(v_y)}. \tag{11.68}$$

Because each side of equation 11.68 is independent of the other, the only way the relation can be satisfied is if both sides are equal to the same constant, C. As a result we see that

$$f(v_{i,j}) = A\,\exp\left(\frac{-C}{2}v_{i,j}^2\right). \tag{11.69}$$

If we require $f(v_{i,j})$ to be normalized, it follows that

$$\int_{-\infty}^{\infty} f(v_i)dv_i = A\sqrt{\frac{2}{C}}\int_{-\infty}^{\infty}\exp\left(-t^2\right)dt = A\sqrt{\frac{2\pi}{C}} = 1, \tag{11.70}$$

so that

$$f(0) = A = \left(\frac{C}{2\pi}\right)^{1/2}. \tag{11.71}$$

In three dimensions, the density of states with velocity between v and $v + dv$ goes as $4\pi v^2 dv$. By analogy, in three dimensions,

$$\frac{dn}{dv} = 4\pi n f(v) f^2(0) dv. \tag{11.72}$$

By definition, the average of the square of the velocities is given by

$$\overline{v^2} = 4\pi f^2(0) \int_0^\infty v^4 f(v) dv = 4\pi \left(\frac{C}{2\pi}\right)^{3/2} \int_0^\infty v^4 \exp\left(-\frac{Cv^2}{2}\right) dv. \tag{11.73}$$

Comparing this expression to equation 11.63 and solving for the constant C, from equation 11.62, gives

$$f(v) = 4\pi v^2 \left(\frac{m}{2\pi kT}\right)^{3/2} \exp\left(-\frac{mv^2}{2kT}\right), \tag{11.74}$$

which is the Maxwell-Boltzmann velocity distribution.

11.3 Density of States

11.3.1 General Case

Consider the three-dimensional Schrödinger equation,

$$\frac{p^2}{2m}\psi = -\frac{\hbar^2}{2m}\nabla^2\psi = E\psi, \tag{11.75}$$

subject to the periodic boundary condition

$$\psi(x + L, y, z) = \psi(x, y + L, z) = \psi(x, y, z + L) = \psi(x, y, z). \tag{11.76}$$

(L is visualized as being the interatomic spacing distance.) It is easily see that a solution to equation 11.75 is of the form

$$\psi \sim \sin\left(\frac{\pi}{L}n_x x\right)\sin\left(\frac{\pi}{L}n_y y\right)\sin\left(\frac{\pi}{L}n_z z\right), \tag{11.77}$$

where

$$k^2 = \frac{2mE}{\hbar^2} = \frac{\pi^2}{L^2}\left(n_x^2 + n_y^2 + n_z^2\right). \qquad (11.78)$$

In k-space, we have

$$\bar{k} = \frac{\pi}{L}\left(n_x, n_y, n_z\right). \qquad (11.79)$$

This is a cubic lattice with the spacing between points having the value π/L. Consequently, in k-space the volume per point is $(\pi/L)^3$, and the density of points, per unit volume, is $(L/\pi)^3$. If we envision a spherical surface of radius k in k-space, in the octant where n_x, n_y, and n_z are all positive, we see that the volume of the region between k and $k + dk$ is $(1/8)(4\pi k^2 dk)$. The density of states is therefore given by

$$f(k)dk = \frac{1}{8}\frac{\left(4\pi k^2\right)}{\left(\frac{\pi}{L}\right)^3}dk = \frac{L^3 k^2}{2\pi^2}dk = \frac{V k^2}{2\pi^2}dk. \qquad (11.80)$$

Knowing that $v = \omega/k$, we have, in frequency space,

$$f(\omega)d\omega = \frac{V\omega^2}{2\pi^2 v^3}d\omega. \qquad (11.81)$$

11.3.2 The Boltzmann Factor

Consider two vessels of equal volume with equal temperatures and pressures, each containing one mole of two different perfect gases. Let us further suppose that the two gases are separated by a partition. When we remove the partition we also introduce an element of randomness into the system. Because of this we may postulate a relationship between the entropy of the system and the randomness of the system. That is,

$$S = f(\Omega), \qquad (11.82)$$

where w is the number of a priori equally probable states accessible to the system. If the gases are composed of identical molecules, then

$$S_{tot} = S_1 + S_2. \tag{11.83}$$

However, because the systems are independent, the number of a priori equally probable states is $\Omega_1 \Omega_2$ so that

$$S_1 + S_2 = f(\Omega_1 + \Omega_2). \tag{11.84}$$

Consequently, S must have the form

$$S = k' \ln \Omega + C, \tag{11.85}$$

where k' and C are constants. It can be seen that C must be zero. That is,

$$S = k' \ln \Omega. \tag{11.86}$$

Consider a reservoir at temperature T in contact with a gas having n_2 particles in state E_2 and n_1 particles in state E_1. Since $N = n_1 + n_2$, the number of ways to have n_1 particles in state E_1 and $N - n_1$ particles in state E_2, without regard to order, is

$$n_1!(N - n_1)!. \tag{11.87}$$

Thus, we have initially

$$S_i = k' \ln\left[n_1!(N - n_1)!\right]. \tag{11.88}$$

After one particle makes a transition from state 2 to state 1, we have

$$S_f = k' \ln\left[(n_1 + 1)!(N - n_1 - 1)!\right]. \tag{11.89}$$

We make use of the mathematical identities

$$\ln AB \approx \ln A + \ln B, \tag{11.90}$$

and

$$\ln N! \approx N \ln N - N, \tag{11.91}$$

to see that

$$\frac{S_i}{k'} = n_1 \ln n_1 - n_1 + (N - n_1) \ln(N - n_1) - (N - n_1), \quad (11.92)$$

and

$$\frac{S_f}{k'} = (n_1 + 1) \ln(n_1 + 1) - (n_1 + 1)$$

$$+ (N - n_1 - 1) \ln(N - n_1 - 1) - (N - n_1 - 1). \quad (11.93)$$

Consequently, the change in the entropy of the gas is given by

$$\frac{\Delta S}{k'} = \frac{S_f - S_i}{k'} = (n_1 + 1) \ln(n_1 + 1) - n_1 \ln n_1$$

$$- (N - n_1) \ln(N - n_1) + (N - n_1 - 1) \ln(N - n_1 - 1). \quad (11.94)$$

If we assume

$$\ln(n_1 + 1) \approx \ln n_1, \quad (11.95)$$

then equation 11.94 reduces to

$$\frac{\Delta S}{k'} \approx \ln n_1 - \ln (N - n_1) = \ln \left(\frac{n_1}{N - n_1} \right) = \ln \left(\frac{n_1}{n_2} \right). \quad (11.96)$$

Because the entropy change of the universe is zero, the entropy increase by the system is equal to the entropy loss by the reservoir. From the first law of thermodynamics we see that the change in entropy of the system is

$$\Delta S = \frac{\Delta U}{T} = -\frac{(E_2 - E_1)}{T}. \quad (11.97)$$

Equating equations 11.96 and 11.97 gives

$$-\frac{(E_2 - E_1)}{T} = k' \ln \left(\frac{n_1}{n_2} \right). \quad (11.98)$$

By inspection, we see that k' must be Boltzmann's constant and equation 11.98 reduces to

$$\frac{n_1}{n_2} = \exp \left[\frac{E_1 - E_2}{k_B T} \right] = \frac{\exp \left[E_1 / k_B T \right]}{\exp \left[E_2 / k_B T \right]}. \quad (11.99)$$

11.4 Quantum Gas

11.4.1 Partition Function

We define the partition function

$$Z(T) = \sum_s \exp^{-E_s/kT} = \sum_s \exp^{-\beta E_s}, \qquad (11.100)$$

where the energy of each state is E_s and the summation is over all states. It is seen that the probability that a system is in the state s is given by

$$P(s) = \frac{\exp^{-\beta E_s}}{Z}, \qquad (11.101)$$

and that the sum over all probabilities is unity. Equivalently, we could express the partition function in terms of a sum over all energy levels as

$$Z(T) = \sum_{E_s} g(E_s) \exp^{-E_s/kT}, \qquad (11.102)$$

where $g(E_s)$, known as the degeneracy of the state E_s, is a measure of the number of states that have the same energy E_s. It follows from equation 11.101 that the probability a system has energy E_s is

$$P(E_s) = g(E_s) \frac{\exp^{-\beta E_s}}{Z}. \qquad (11.103)$$

The mean energy of a system is then

$$\langle E \rangle = \frac{\sum_s E_s \exp^{-\beta E_s}}{Z} = -\frac{\partial \ln Z}{\partial \beta}, \qquad (11.104)$$

and the mean occupation number of a state s is

$$\langle n_s \rangle = -\frac{1}{\beta} \frac{\partial \ln Z}{\partial E_s}. \qquad (11.105)$$

The total energy of a state is given by

$$E = \sum_s n_s E_s, \tag{11.106}$$

and a gas of N particles satisfies

$$N = \sum_s n_s. \tag{11.107}$$

In terms of the occupation numbers, the partition function is

$$Z = \sum_{\text{ASN}} \exp\left(-\beta \sum_s n_s E_s\right), \tag{11.108}$$

where the first summation is over all sets of numbers (ASN) that satisfy equation 11.107.

11.4.2 Bose-Einstein Gas (Bosons)

Consider the situation where the mean occupation number of a given state is independent of the mean occupation numbers of the other states. That is, each occupation number may range from zero to infinity. These statistics apply to photons, pions, and K-mesons. Equation 11.108 reduces to

$$Z = \sum_{n_1=0}^{\infty} \sum_{n_2=0}^{\infty} \cdots \exp\left(-\beta \sum_s n_s E_s\right)$$

$$= \sum_{n_1=0}^{\infty} \sum_{n_2=0}^{\infty} \cdots \exp\left(-\beta(n_1 E_1 + n_2 E_2 + \cdots)\right)$$

$$= \left\{\sum_{n_1=0}^{\infty} \exp(-\beta n_1 E_1)\right\}\left\{\sum_{n_2=0}^{\infty} \exp(-\beta n_2 E_2)\right\} \cdots$$

$$= \prod_{s=0}^{\infty}\left\{\sum_{n_s=0}^{\infty} \exp(-\beta n_s E_s)\right\}$$

$$= \prod_{s=0}^{\infty}\left\{\frac{1}{1-\exp(-\beta E_s)}\right\}, \tag{11.109}$$

or

$$\ln Z = -\sum_{s=0}^{\infty} \ln \left[1 - \exp\left(-\beta E_s\right)\right]. \qquad (11.110)$$

It follows from equation 11.104 that the mean energy is

$$\langle E \rangle = -\frac{\partial \ln Z}{\partial \beta}$$

$$= \frac{\partial}{\partial \beta} \left\{ \sum_{s=0}^{\infty} \ln \left[1 - \exp\left(-\beta E_s\right)\right] \right\}$$

$$= \sum_{s=0}^{\infty} \left\{ \frac{\partial}{\partial \beta} \ln \left[1 - \exp\left(-\beta E_s\right)\right] \right\}$$

$$= \sum_{s=0}^{\infty} \left\{ \frac{E_s}{1 - \exp\left(-\beta E_s\right)} \right\}. \qquad (11.111)$$

Similarly, it follows from equation 11.105 that the mean occupation number of a given state is

$$\langle n_s \rangle = -\frac{1}{\beta} \frac{\partial \ln Z}{\partial E_s}$$

$$= \frac{1}{\beta} \frac{\partial}{\partial E_s} \left\{ \sum_{s'=0}^{\infty} \ln \left[1 - \exp\left(-\beta E_{s'}\right)\right] \right\}$$

$$= \frac{1}{\beta} \frac{\partial}{\partial E_s} \left\{ \ln \left[1 - \exp\left(-\beta E_s\right)\right] \right\}$$

$$= \frac{\exp\left(-\beta E_s\right)}{1 - \exp\left(-\beta E_s\right)}$$

$$= \frac{1}{\exp\left(\beta E_s\right) - 1}. \qquad (11.112)$$

11.4.2.1 Planck's Law

Because photons obey Bose-Einstein statistics, if we make the substitution $E = \hbar\omega$ and note that the group velocity is the speed of light, combining equations 11.81 and 11.112 gives the number density of photons in the frequency range ω to $\omega + d\omega$ as

$$f_{N,BE}(\omega)d\omega = \frac{dN_\omega}{dV} = \frac{1}{\pi^2 c^3} \frac{\omega^2}{\exp(\beta\hbar\omega)-1} d\omega. \qquad (11.113)$$

Similarly, the energy density is obtained by multiplying this expression by the energy of a single photon to obtain

$$f_{E,BE}(\omega)d\omega = \frac{\hbar}{\pi^2 c^3} \frac{\omega^3}{\exp(\beta\hbar\omega)-1} d\omega. \qquad (11.114)$$

Equation 11.114 is known as *Planck's radiation law*.

11.4.3 Fermi-Dirac Gas (Fermions)

Consider the situation where the mean occupation number of a given state is dependent on the mean occupation numbers of the other states. In particular, we consider the situation where the states obey the Pauli exclusion principle. That is, a particular possible state will be either occupied or unoccupied, so that $n_s = 0$ or 1. Electrons, positrons, protons, and neutrons obey Fermi-Dirac statistics. Equation 11.108 reduces to

$$Z = \sum_{n_1=0}^{1} \sum_{n_2=0}^{1} \cdots \exp\left(-\beta \sum_s n_s E_s\right)$$

$$= \sum_{n_1=0}^{1} \sum_{n_2=0}^{1} \cdots \exp(-\beta(n_1 E_1 + n_2 E_2 + \dots))$$

$$= \left\{\sum_{n_1=0}^{1} \exp(-\beta n_1 E_1)\right\}\left\{\sum_{n_2=0}^{1} \exp(-\beta n_2 E_2)\right\} \cdots$$

$$= \prod_{s=0}^{\infty} \{1 + \exp(-\beta E_s)\}, \qquad (11.115)$$

or

$$\ln Z = \sum_{s=0}^{\infty} \ln\left[1 + \exp(-\beta E_s)\right]. \qquad (11.116)$$

It follows from equation 11.104 that the mean energy is

$$\langle E \rangle = -\frac{\partial \ln Z}{\partial \beta}$$

$$= -\frac{\partial}{\partial \beta} \left\{ \sum_{s=0}^{\infty} \ln\left[1 + \exp(-\beta E_s)\right] \right\}$$

$$= -\sum_{s=0}^{\infty} \left\{ \frac{\partial}{\partial \beta} \ln\left[1 + \exp(-\beta E_s)\right] \right\}$$

$$= \sum_{s=0}^{\infty} \left\{ \frac{E_s}{1 + \exp(-\beta E_s)} \right\}. \tag{11.117}$$

Similarly, it follows from equation 11.105 that the mean occupation number of a given state is

$$\langle n_s \rangle = -\frac{1}{\beta} \frac{\partial \ln Z}{\partial E_s}$$

$$= -\frac{1}{\beta} \frac{\partial}{\partial E_s} \left\{ \sum_{s'=0}^{\infty} \ln\left[1 + \exp(-\beta E_{s'})\right] \right\}$$

$$= -\frac{1}{\beta} \frac{\partial}{\partial E_s} \left\{ \ln\left[1 + \exp(-\beta E_s)\right] \right\}$$

$$= \frac{\exp(-\beta E_s)}{1 + \exp(-\beta E_s)}$$

$$= \frac{1}{\exp(\beta E_s) + 1}. \tag{11.118}$$

11.5 Counting Statistics

The number of possible microstates is defined by W_{DE}, where

$$\Omega_{\Delta E} = \sum_{ASN} W(\{n_i\}), \tag{11.119}$$

The sum is over all sets $\{n_i\}$ such that $\sum_i n_i \varepsilon_i = E$, $\sum_i n_i = N$. We need to calculate $W(\{n_i\}) = \prod_j W_j$, where $W(\{n_i\})$ is the number of states corresponding to a given set of occupation numbers of the cells, W_j is the number of ways n_j particles can be assigned to the g_j levels in the jth cell.

If the fluctuations in $\{n_i\}$ about the set $\{\bar{n}_i\}$ which maximizes $W(\{n_i\})$, subject to the constraints, are negligible, then

$$\ln \Omega_{\Delta E}(E,V,N) \approx \ln W(\{\bar{n}_i\}). \qquad (11.120)$$

We can determine $\{\bar{n}_i\}$ by treating $\ln W(\{\bar{n}_i\})$ as a function subject to the constraints. We define

$$f(n_i) = \ln W(\{n_i\}) + \alpha \sum_i n_i - \beta \sum_i \varepsilon_i n_i, \qquad (11.121)$$

by the method of Lagrange multipliers.

11.5.1 Maxwell-Boltzmann Gas (Distinguishable Particles)

$\prod_i g_i^{n_i}$ = number of ways n_i distinguishable objects can occupy g_i levels with no restrictions

$\prod_i n_i!$ = number of ways to place N objects into cells such that there are n_i objects in the ith cell.

$N!$ = number of equivalent permutations

The solution is then

$$W(\{n_i\}) = \frac{N! \prod_i g_i^{n_i}}{\prod_i n_i!}, \qquad (11.122)$$

so that

$$\Omega_{\Delta E} = \sum_{ASN} \frac{N! \prod_i g_i^{n_i}}{\prod_i n_i!}.$$ (11.123)

From equation 11.121,

$$f(n_i) = \ln N! \frac{\prod_i g_i^{n_i}}{\prod_i n_i!} + \alpha \sum_i n_i - \beta \sum_i \varepsilon_i n_i,$$ (11.124)

which reduces to

$$f(n_i) = \ln N! + \sum_i n_i \ln g_i - \sum_i \ln n_i! + \alpha \sum_i n_i - \beta \sum_i \varepsilon_i n_i,$$ (11.125)

and further to

$$f(n_i) \approx (N \ln N - N) + \sum_i \left[n_i \ln g_i - (n_i \ln n_i - n_i) + \alpha n_i - \beta \varepsilon_i n_i \right].$$ (11.126)

Because $\dfrac{\partial f}{\partial n_i} = 0$, we have

$$\ln g_i - \ln n_i - 1 + 1 + \alpha - \beta \varepsilon_i = 0$$ (11.127)

or

$$\frac{g_i}{\bar{n}_i} = \exp^{-\alpha} \exp^{+\beta \varepsilon_i},$$ (11.128)

which has solution

$$\bar{n}_i = g_i \exp^{-\beta(\varepsilon_i - \mu)},$$ (11.129)

where $bm = a$.

11.5.2 Bose-Einstein Gas (Bosons)

There is no restrictions on the occupation numbers for a possible energy state. Each set of occupation numbers $\{n_i\}$ corresponds to a different state.

$$\frac{(n_i + g_i - 1)!}{n_i!(g_i - 1)!} = \text{number of ways to choose } g_i \text{ objects such that the sum}$$

is n_i objects.

The solution is

$$W(\{n_i\}) = \prod_i \frac{(n_i + g_i - 1)!}{n_i!(g_i - 1)!}, \tag{11.130}$$

or

$$\Omega_{\Delta E} = \sum_{ASN} \prod_i \frac{(n_i + g_i - 1)!}{n_i!(g_i - 1)!}. \tag{11.131}$$

From which it follows that

$$f(n_i) = \ln \prod_i \frac{(n_i + g_i - 1)!}{n_i!(g_i - 1)!} + \alpha \sum_i n_i - \beta \sum_i \varepsilon_i n_i, \tag{11.132}$$

or

$$f(n_i) = \sum_i \left[\ln(n_i + g_i - 1)! - \ln n_i! - \ln(g_i - 1)! + \alpha n_i - \beta \varepsilon_i n_i \right]. \tag{11.133}$$

After further reduction,

$$f(n_i) \approx \sum_i \left[(n_i + g_i - 1)\ln(n_i + g_i - 1) - (n_i + g_i - 1) - n_i \ln n_i \right]$$

$$+ \sum_i \left[n_i - (g_i - 1)\ln(g_i - 1) + (g_i - 1) + \alpha n_i - \beta \varepsilon_i n_i \right]. \tag{11.134}$$

Because $\dfrac{\partial f}{\partial n_i} = 0$, we have

$$\ln(n_i + g_i) + 1 - 1 - \ln n_i - 1 + 1 + \alpha - \beta \varepsilon_i = 0, \qquad (11.135)$$

or

$$\frac{\bar{n}_i + g_i}{\bar{n}_i} = \exp^{+\beta(\varepsilon_i - \mu)}, \qquad (11.136)$$

which has solution

$$\bar{n}_i = \frac{g_i}{\exp^{\beta(\varepsilon_i - \mu)} - 1}. \qquad (11.137)$$

11.5.3 Fermi-Dirac Gas (Fermions)

Each energy state has occupation numbers 0 or 1.

$$\frac{g_i!}{n_i!(g_i - n_i)!} = \text{number of ways to choose } n_i \text{ objects out of } g_i \text{ objects.}$$

We have

$$W(\{n_i\}) = \prod_i \frac{g_i!}{n_i!(g_i - n_i)!}, \qquad (11.138)$$

or

$$\Omega_{\Delta E} = \sum_{ASN} \prod_i \frac{g_i!}{n_i!(g_i - n_i)!}. \qquad (11.139)$$

From equation 11.147 we have

$$f(n_i) = \ln \prod_i \frac{g_i!}{n_i!(g_i - n_i)!} + \alpha \sum_i n_i - \beta \sum_i \varepsilon_i n_i, \qquad (11.140)$$

which reduces to

$$f(n_i) = \sum_i \left[(g_i \ln g_i - g_i) - (n_i \ln n_i - n_i) - (g_i - n_i) \ln(g_i - n_i) \right]$$

$$- \sum_i \left[(g_i - n_i) + \alpha n_i - \beta \varepsilon_i n_i \right]. \qquad (11.141)$$

Because $\dfrac{\partial f}{\partial n_i} = 0$, we have

$$-\ln n_i + \ln(g_i - n_i) + \alpha - \beta \varepsilon_i = 0, \qquad (11.142)$$

or

$$\frac{g_i - \bar{n}_i}{\bar{n}_i} = \exp^{+\beta(\varepsilon_i - \mu)}, \qquad (11.143)$$

which has solution

$$\bar{n}_i = \frac{g_i}{\exp^{\beta(\varepsilon_i - \mu)} + 1}. \qquad (11.144)$$

11.6 Equilibrium Ensembles

11.6.1 The Microcanonical Ensemble

Consider a closed isolated system with a configuration space volume V and a fixed number of particles N which is constrained to move within the energy shell $E \to E + \Delta E$. The state of a system is completely specified by the $6N$ independent variables $\left(\vec{p}^{\,N}, \vec{q}^{\,N}\right)$. The state vector $\vec{X}^N\left(\vec{p}^{\,N}, \vec{q}^{\,N}\right)$ specifies a point in phase space. We define the probability density $\rho\!\left(\vec{X}^N\right)$ such that the product $\rho\!\left(\vec{X}^N\right) d\vec{X}^N$ is the probability that the state point $\vec{X}^N$ lies in the volume element $\vec{X}^N \to \vec{X}^N + d\vec{X}^N$ at time t. The best choice for the equilibrium probability density is

$$\rho\!\left(\vec{X}^N\right) = \frac{1}{\Sigma(E)}, \qquad (11.145)$$

for $H\!\left(\vec{X}^N\right) = E$, where $\Sigma(E)$ is the area of the energy surface and $H\!\left(\vec{X}^N\right)$ is the Hamiltonian. We have the constraint

$$\int_{\Gamma} \rho(\bar{X}^N) d\bar{X}^N = 1, \tag{11.146}$$

where the integration is over all phase space. The form of the entropy chosen by Gibbs is

$$S = -k_B \int_{\Gamma} d\bar{X}^N \rho(\bar{X}^N) \ln\left[C^N \rho(\bar{X}^N)\right], \tag{11.147}$$

where C^N is a constant to make the units consistent. Using the method of Lagrange multipliers, we have

$$\delta S = \delta\left[\int_{\Gamma} d\bar{X}^N \left\{-k_B \rho(\bar{X}^N) \ln\left[C^N \rho(\bar{X}^N)\right] + \alpha \rho(\bar{X}^N)\right\}\right], \tag{11.148}$$

or

$$\delta S = \delta \int_{\Gamma} d\bar{X}^N \left\{-k_B \ln\left[C^N \rho(\bar{X}^N)\right] - k_B + \alpha\right\} \delta\rho(\bar{X}^N). \tag{11.149}$$

We require

$$-k_B \ln\left[C^N \rho(\bar{X}^N)\right] - k_B + \alpha = 0. \tag{11.150}$$

The solution is

$$\rho(\bar{X}^N) = \frac{1}{C^N} \exp\left(\frac{\alpha}{k_B} - 1\right) = K \tag{11.151}$$

if the integration is in the range E to $E + \Delta E$, zero otherwise. From the normalization condition, within the shell

$$\rho(\bar{X}^N) = \frac{1}{\Omega_{\Delta E}(E,V,N)}, \tag{11.152}$$

where $\Omega_{\Delta E}(E,V,N) = \Delta E \Sigma(E,V,N)$. Equation 11.121 reduces to

$$S(E,V,N) = k_B \ln\left(\frac{\Omega_{\Delta E}(E,V,N)}{C^N}\right). \tag{11.153}$$

11.6.2 The Canonical Ensemble

A closed system can exchange heat with its surroundings, and as a consequence the total energy may fluctuate. We wish to minimize the entropy subject to the constraints

$$\int_{\Gamma} \rho(\bar{X}^N)\, d\bar{X}^N = 1, \tag{11.154}$$

and

$$\langle E \rangle = \int_{\Gamma} d\bar{X}^N\, H(\bar{X}^N)\rho(\bar{X}^N). \tag{11.155}$$

The variation of S is

$$\delta\left[\int_{\Gamma} d\bar{X}^N \left\{ \alpha_o \rho(\bar{X}^N) + \alpha_E H(\bar{X}^N)\rho(\bar{X}^N) - k_B \rho(\bar{X}^N)\ln\left[C^N \rho(\bar{X}^N)\right]\right\}\right], \tag{11.156}$$

which of course is zero. We find, as in the microcanonical ensemble,

$$\alpha_o + \alpha_E H(\bar{X}^N) - k_B \ln\left[C^N \rho(\bar{X}^N)\right] - k_B = 0, \tag{11.157}$$

which gives

$$\rho(\bar{X}^N) = \frac{1}{C^N} \exp\left[\frac{\alpha_o}{k_B} - 1 + \frac{\alpha_E}{k_B} H(\bar{X}^N)\right]. \tag{11.158}$$

From the normalization condition, we have

$$\exp\left[1 - \frac{\alpha_o}{k_B}\right] = \frac{1}{C^N} \int d\bar{X}^N \exp\left[\frac{\alpha_E}{k_B} H(\bar{X}^N)\right]. \tag{11.159}$$

This is simply the partition function, $Z_N(V, \alpha_E)$, equation 11.134. To determine α_E, we multiply equation 11.131 by $\rho(\bar{X}^N)$ and integrate. We have

$$(\alpha_o - k_B)\int d\vec{X}^N \rho(\vec{X}^N) + \alpha_E \int d\vec{X}^N H(\vec{X}^N)\rho(\vec{X}^N)$$

$$-k_B \int d\vec{X}^N \rho(\vec{X}^N)\ln\left[C^N \rho(\vec{X}^N)\right], \tag{11.160}$$

which is

$$-k_B \ln Z_N(V,\alpha_E) + \alpha_E\langle E\rangle + S = 0. \tag{11.161}$$

This looks similar to the equation for the Helmholtz free energy,

$$A - U + ST = 0. \tag{11.162}$$

We make the identifications

$$\alpha_E = -\frac{1}{T}, \tag{11.163}$$

and

$$A = -k_B T \ln Z_N(T,V). \tag{11.164}$$

The partition function is now given by

$$Z_N(T,V) = \frac{1}{C^N}\int d\vec{X}^N \exp^{-\beta H(\vec{X}^N)}, \tag{11.165}$$

and

$$\rho(\vec{X}^N) = \frac{1}{C^N Z_N(T,V)}\exp^{-\beta H(\vec{X}^N)}. \tag{11.166}$$

The canonical ensemble has

$$A(T,V,N) = E - TS, \tag{11.167}$$

and

$$dA = -S\,dT - P\,dV + \mu\,dN, \tag{11.168}$$

with

$$S = -\left(\frac{\partial A}{\partial T}\right)_{V,N},$$ (11.169)

$$P = -\left(\frac{\partial A}{\partial V}\right)_{T,N},$$ (11.170)

and

$$\mu = -\left(\frac{\partial A}{\partial N}\right)_{T,V}.$$ (11.171)

Example 11.1

A simple model of a parametric solid consists of a system of N noninteracting magnetic ions of spin 1/2 and magnetic moment m_B in an external applied magnetic field B. Calculate: (1) the partition function, (2) the entropy, and (3) the average energy.

(1) For a parametric solid

$$H = -g_s \mu_B \sum_{i=1}^{N} \vec{S}_i \cdot \vec{B} + \frac{p^2}{2m},$$ (11.172)

so that the partition function (eq. 11.165) is given by

$$Z_N(T,V) = \frac{1}{C^N} \int d\vec{X}^N \exp\left[\frac{g_s \mu_B \sum_{i=1}^{N} \vec{S}_i \cdot \vec{B} - \frac{p^2}{2m}}{k_B T}\right].$$ (11.173)

Separating the previous equation into two terms, we find that the first term is given by

$$\prod_i \left[\int \exp\left[\frac{g_s \mu_B}{k_B T} SB \cos\theta\right] r^2 \sin\theta \, dr \, d\theta \, d\varphi\right],$$ (11.174)

or

$$\prod_i \left(\frac{2\pi r^3}{3} \right) \left[\int \exp^{[x \cos\theta]} \sin\theta d\theta \right], \qquad (11.175)$$

where

$$x = \frac{g_s \mu_B}{k_B T} SB. \qquad (11.176)$$

It can be shown that the integral is equivalent to sinh x/x, so that the first term in equation 11.173 is

$$\left(\frac{2\pi r^3}{3} \frac{\sinh x}{x} \right)^N = \left(\frac{V}{2} \frac{\sinh x}{x} \right)^N. \qquad (11.177)$$

The second term from equation 11.173 is given by

$$\int d^3 p_1 \ldots d^3 p_N \exp\left[-\frac{p^2}{2mk_B T} \right], \qquad (11.178)$$

which reduces to

$$[2\pi mk_B T]^{3N/2}. \qquad (11.179)$$

Combining equations 11.177 and 11.179, we see that the partition function is given by

$$Z(T,B,N) = \frac{1}{C^N} \left[\frac{V}{2} \frac{\sinh\left(\frac{g_s \mu_B}{k_B T} SB \right)}{\left(\frac{g_s \mu_B}{k_B T} SB \right)} (2\pi mk_B T)^{3/2} \right]^N. \qquad (11.180)$$

(2) The entropy is given, from equations 11.164 and 11.169, by

$$S = \frac{\partial}{\partial T} \left\{ k_B TN \ln\left[\frac{V}{2} \frac{\sinh\left(\frac{g_s \mu_B}{k_B T} SB \right)}{\left(\frac{g_s \mu_B}{k_B T} SB \right)} (2\pi mk_B T)^{3/2} \right] \right\}, \qquad (11.181)$$

which reduces to

$$S = -\frac{A}{T} + \frac{5}{2}k_B N - k_B N \left(\frac{g_s \mu_B}{k_B T} SB\right) \coth\left(\frac{g_s \mu_B}{k_B T} SB\right). \quad (11.182)$$

(3) The average energy is given by

$$E = A + TS = k_B NT \left[\frac{5}{2} - \left(\frac{g_s \mu_B}{k_B T} SB\right) \coth\left(\frac{g_s \mu_B}{k_B T} SB\right)\right]. \quad (11.183)$$

11.6.3 The Grand Canonical Ensemble

An open system will exchange both heat and matter with its surroundings and, therefore, both heat and matter can fluctuate; that is, energy and the particle number will fluctuate. We wish to minimize the entropy subject to

$$\sum_{n=0}^{\infty} \int d\vec{X}^N \rho\left(\vec{X}^N\right) = 1, \quad (11.184)$$

$$\langle E \rangle = \sum_{n=0}^{\infty} \int d\vec{X}^N H\left(\vec{X}^N\right) \rho\left(\vec{X}^N\right), \quad (11.185)$$

and

$$\langle N \rangle = \sum_{n=0}^{\infty} \int d\vec{X}^N N \rho\left(\vec{X}^N\right). \quad (11.186)$$

The extrema of the entropy satisfies

$$\delta S = \delta \sum_{n=0}^{\infty} \int d\vec{X}^N \left(\alpha_o \rho\left(\vec{X}^N\right) + \alpha_E H\left(\vec{X}^N\right) \rho\left(\vec{X}^N\right)\right.$$

$$\left. + \alpha_N N \rho\left(\vec{X}^N\right) - k_B \rho\left(\vec{X}^N\right) \ln\left[C^N \rho\left(\vec{X}^N\right)\right]\right). \quad (11.187)$$

This leads us to the condition

$$\alpha_o + \alpha_E H\left(\vec{X}^N\right) + \alpha_N N - k_B \ln\left[C^N \rho\left(\vec{X}^N\right)\right] = 0. \quad (11.188)$$

The normalization condition allows us to define the grand partition function,

$$\mathcal{G}_N\left(\alpha_E, V, \alpha_N\right) = \exp^{\left(1 - \frac{\alpha_o}{k_B}\right)}$$

$$= \sum_{n=0}^{\infty} \frac{1}{C^N} \int d\vec{X}^N \exp^{\left(\frac{\alpha_E H\left(\vec{X}^N\right) + \alpha_N N}{k_B}\right)}. \quad (11.189)$$

As before, we multiply equation 11.188 by $\rho\left(\vec{X}^N\right)$ and integrate to obtain

$$-k_B \ln \mathcal{G}\left(\alpha_E, V, \alpha_N\right) + \alpha_E \langle E \rangle + S + \alpha_N \langle N \rangle = 0. \quad (11.190)$$

This is similar to the equation for the grand potential

$$\Omega = U - TS - \mu N, \quad (11.191)$$

with

$$\alpha_E = -\frac{1}{T}, \quad (11.192)$$

$$\alpha_N = \frac{\mu}{T}, \quad (11.193)$$

and

$$\Omega(T, V, \mu) = -k_B T \ln \mathcal{G}(T, V, \mu). \quad (11.194)$$

We then have

$$\mathcal{G}(T, V, \mu) = \sum_{n=0}^{\infty} \frac{1}{C^N} \int d\vec{X}^N \exp^{\left(\frac{-H\left(\vec{X}^N\right) + \mu N}{k_B T}\right)}, \quad (11.195)$$

and

$$\rho\left(\bar{X}^N\right) = \frac{1}{C^N} \exp^{\beta\left[\Omega(T,V,\mu) - H\left(\bar{X}^N\right) + \mu N\right]}.$$
(11.196)

The Grand canonical ensemble has

$$\Omega(T,V,\mu) = E - TS - \mu N,$$
(11.197)

and

$$d\Omega = -S\,dT - P\,dV - N\,d\mu,$$
(11.198)

with

$$S = -\left(\frac{\partial\Omega}{\partial T}\right)_{V,\mu},$$
(11.199)

$$P = -\left(\frac{\partial\Omega}{\partial V}\right)_{T,\mu},$$
(11.200)

and

$$N = -\left(\frac{\partial\Omega}{\partial\mu}\right)_{T,V}.$$
(11.201)

11.7 Phase Transitions

11.7.1 Bose-Einstein Condensation

For an ideal Bose-Einstein gas the grand partition function is of the form

$$\mathcal{Z}_{BE}(T,V,n) = \prod_{l=0}^{\infty}\left(\frac{1}{1 - \exp^{-\beta(\varepsilon_l - \mu)}}\right).$$
(11.202)

As we have previously seen, this gives

$$\langle N \rangle = \sum_{l=0}^{\infty}\left(\frac{1}{\exp^{\beta(\varepsilon_l - \mu)} - 1}\right) = \sum_{l=0}^{\infty}\langle n_{pl} \rangle,$$
(11.203)

where $\langle n_{pl} \rangle$ is the average number of particles in the momentum state $\vec{p}_l$. We have

$$\langle n_{pl} \rangle = \frac{1}{\exp^{\beta \varepsilon_l} Z^{-1} - 1}, \qquad (11.204)$$

and

$$\langle n_0 \rangle = \frac{Z}{1 - Z} \xrightarrow{Z \to 1} \infty . \qquad (11.205)$$

That is, as $Z \to 1$, the zero momentum state can become macroscopically occupied. Isolating all terms belonging to the zero momentum state gives

$$\Omega_{BE}(T,V,\mu) = -k_B T \ln Z_{BE}(T,V,\mu), \qquad (11.206)$$

which reduces to

$$\Omega_{BE}(T,V,\mu) = -k_B T \sum_{l=0}^{\infty} \ln\left(1 - \exp^{-\beta(\varepsilon_l - \mu)}\right), \qquad (11.207)$$

or

$$\frac{\Omega}{V} = \frac{k_B T}{V} \ln(1 - Z) + \frac{k_B T}{V} \sum_{l=0}^{\infty} \ln\left(1 - \exp^{-\beta(\varepsilon_l - \mu)}\right). \qquad (11.208)$$

We may make the supposition that if the volume is large, then the particle energies will be evenly spaced. This justifies the change

$$\sum_l \to \frac{V}{(2\pi)^3} \int d\vec{k}_l = \frac{V}{h^3} \int d\vec{p}_l . \qquad (11.209)$$

Thus,

$$\frac{\Omega}{V} = \frac{k_B T}{V} \ln(1 - Z) + \frac{k_B T}{V} \frac{V}{h^3} \int d\vec{p}_l \ln\left(1 - \exp^{-\beta(\varepsilon_l - \mu)}\right). \qquad (11.210)$$

We make the substitution $x^2 = \beta \dfrac{p^2}{2m}$, which implies that $\varepsilon = \dfrac{p^2}{2m}$ and $z = \exp^{\beta \mu}$. Using this we find that

$$\frac{\Omega}{V} = \frac{k_B T}{V} \ln(1 - Z) + \frac{k_B T}{V} \frac{V}{h^3} \int d^3 \vec{p}_l \ln\left(1 - z \exp^{-\left(\beta p^2 / 2m\right)}\right), \quad (11.211)$$

which reduces to

$$\frac{\Omega}{V} = \frac{k_B T}{V} \ln(1 - Z) + \frac{k_B T}{V} \frac{V}{h^3} \int d^3 \vec{x} \left(\frac{2m}{\beta}\right)^{1/2} \ln\left(1 - z \exp^{-x^2}\right), \quad (11.212)$$

or

$$\frac{\Omega}{V} = \frac{k_B T}{V} \ln(1 - Z) + \frac{k_B T}{h^3} \left(\frac{2m}{\beta}\right)^{1/2} 4\pi \int x^2 \ln\left(1 - z \exp^{-x^2}\right) dx. \quad (11.213)$$

We let $\lambda_T = \left(\dfrac{2\pi \hbar^2}{m k_B T}\right)$, and the previous expression reduces to

$$\frac{\Omega}{V} = \frac{k_B T}{V} \ln(1 - z) + \frac{k_B T}{\lambda_T^3} g_{5/2}(z), \quad (11.214)$$

where

$$g_{5/2}(z) = -\frac{4}{\sqrt{\pi}} \int_0^\infty dx\, x^2 \ln\left(1 - z \exp^{-x^2}\right) = \sum_{\alpha=1}^\infty \frac{z^\alpha}{\alpha^{5/2}}. \quad (11.215)$$

Similarly,

$$\frac{\langle N \rangle}{V} = \frac{\langle n_o \rangle}{V} + \sum_{l=0}^\infty \left(\frac{1}{z \exp^{\beta \varepsilon_l} - 1}\right), \quad (11.216)$$

or

$$\frac{\langle N \rangle}{V} = \frac{\langle n_o \rangle}{V} + \frac{1}{\lambda_T^3} g_{3/2}(z), \qquad (11.217)$$

where

$$g_{3/2}(z) = z \frac{\partial}{\partial z} g_{5/2}(z) = \sum_{\alpha=1}^{\infty} \frac{z^\alpha}{\alpha^{3/2}}. \qquad (11.218)$$

For high values of T and/or low values of N, we can neglect the $p = 0$ terms to find that, when $z = 0$,

$$\left(\frac{\langle N \rangle}{V} \lambda_T^3 \right)_{z=0} = g_{3/2}(0) = 0, \qquad (11.219)$$

and when $z = 1$,

$$\left(\frac{\langle N \rangle}{V} \lambda_T^3 \right)_{z=1} = g_{3/2}(1) = \xi\left(\frac{3}{2}\right) = 2.612. \qquad (11.220)$$

For some critical combination of density and temperature, the *fugacity*, z, will reach its maximum value. If we decrease the temperature or increase the temperature past the critical values, the function $g_{3/2}(z)$ can no longer change since z cannot become larger than 1. Thus, the entire change in $\left(\frac{\langle N \rangle}{V} \lambda_T^3 \right)$ must come from the term $\left(\frac{\langle n_o \rangle}{V} \lambda_T^3 \right)$, and the zero momentum state begins to take on nonzero values. That is,

$$\frac{\langle N \rangle}{V} \lambda_T^3 = g_{3/2}(z), \qquad (11.221)$$

for $\frac{\langle N \rangle}{V} \lambda_T^3 < 2.612$, and

$$\frac{\langle N \rangle}{V} \lambda_T^3 = \frac{\langle n_o \rangle}{V} \lambda_T^3 + g_{3/2}(1), \qquad (11.222)$$

for $\dfrac{\langle N \rangle}{V} \lambda_T^3 \geq 2.612$. Macroscopic occupation of the zero momentum state is called Bose-Einstein condensation. It occurs for values of temperature and density such that

$$\lambda_T^3 \geq 2.612 \frac{V}{\langle N \rangle}. \tag{11.223}$$

The critical values are

$$T_c = \left(\frac{2\pi\hbar^2}{mk_B} \right) \left(\frac{\langle N \rangle}{2.612V} \right)^{2/3}, \tag{11.224}$$

and

$$\left(\frac{\langle N \rangle}{V} \right)_c = 2.612 \left(\frac{mk_B T}{2\pi\hbar^2} \right)^{3/2}. \tag{11.225}$$

11.7.2 The Ising Model

Some systems exhibit a transition from ordered to disordered states. One of the simplest of such systems is that of a lattice composed of two different types of objects, A and B. We assume that the objects interact with only their nearest neighbors. If we raise the temperature of the system at some point, the system will melt and become completely disordered. A mathematical model of such a system was developed by Ising to describe ferromagnetism. If we let the interaction energy be denoted by J, self-energy by H, the total energy is

$$E(\sigma_1, \sigma_2, \ldots, \sigma_n) = -J \sum_{j=1}^{N} \sigma_j \sigma_{j+1} - H \sum_{j=1}^{N} \sigma_j, \tag{11.226}$$

where $\sigma_j = \pm 1$, $\sigma_{j+1} = 1$. The partition function is

$$Z_N = \sum_\sigma \exp^{-\beta E(\sigma)} = \sum_\sigma \exp^{-\beta\left[-J \sum_{j=1}^{N} \sigma_j \sigma_{j+1} - H \sum_{j=1}^{N} \sigma_j \right]}, \tag{11.227}$$

or

$$Z_N = \sum_{\sigma_1=\pm1} \cdots \sum_{\sigma_N=\pm1} \left[\exp^{h\sigma_1/2} \exp^{k\sigma_1\sigma_2} \exp^{h\sigma_2/2} \cdots \exp^{h\sigma_N/2} \right], \quad (11.228)$$

where $k = \dfrac{J}{k_B T}$, and $h = \dfrac{H}{k_B T}$. The previous expression can be written in the form

$$Z_N = \sum_{\sigma_1=\pm1} \cdots \sum_{\sigma_N=\pm1} V(\sigma_1,\sigma_2) V(\sigma_2,\sigma_3) \cdots V(\sigma_N,\sigma_1), \quad (11.229)$$

where

$$V(\sigma,\sigma') = \exp^{k\sigma\sigma' + \frac{h}{2}(\sigma+\sigma')}. \quad (11.230)$$

Because $\sigma = \pm1$, this can be written in matrix notation as

$$V(\sigma,\sigma') = \begin{pmatrix} \exp^{k+h} & \exp^{-k} \\ \exp^{-k} & \exp^{k-h} \end{pmatrix}. \quad (11.231)$$

This expression for Z_N becomes $Z_N = \mathrm{Tr}\, V^N$. The eigenvalues of V are

$$\det \begin{vmatrix} \exp^{k+h} - \lambda & \exp^{-k} \\ \exp^{-k} & \exp^{k-h} - \lambda \end{vmatrix} = \left(\exp^{k+h} - \lambda \right)\left(\exp^{k-h} - \lambda \right) - \exp^{-2k} = 0. \quad (11.232)$$

Solving this gives

$$\lambda^2 - \exp^k \left(\exp^h + \exp^{-h} \right)\lambda + \left(\exp^{2k} - \exp^{-2k} \right) = 0, \quad (11.233)$$

or

$$\lambda = \exp^k \frac{\left(\exp^h + \exp^{-h} \right)}{2}$$

$$\pm \frac{1}{2}\left[\exp^{2k}\left(\exp^h + \exp^{-h} \right)^2 - 4\left(\exp^{2k} - \exp^{-2k} \right) \right]^{1/2}, \quad (11.234)$$

which reduces to

$$\lambda = \exp^k \cosh h \pm \left[\exp^{2k} \cosh^2 h - \left(\exp^{2k} - \exp^{-2k}\right)\right]^{1/2}. \quad (11.235)$$

The two solutions are

$$\lambda_1 = \exp^k \cosh h + \left[\exp^{2k} \sinh^2 h + \exp^{-2k}\right]^{1/2}, \quad (11.236)$$

$$\lambda_2 = \exp^k \cosh h - \left[\exp^{2k} \sinh^2 h + \exp^{-2k}\right]^{1/2}. \quad (11.237)$$

We may write

$$V = P \begin{pmatrix} \lambda_1 & 0 \\ 0 & \lambda_2 \end{pmatrix} P^{-1}, \quad (11.238)$$

so that

$$Z_N = \mathrm{Tr}\left[P\begin{pmatrix} \lambda_1 & 0 \\ 0 & \lambda_2 \end{pmatrix}P^{-1} P\begin{pmatrix} \lambda_1 & 0 \\ 0 & \lambda_2 \end{pmatrix}P^{-1} \cdots P\begin{pmatrix} \lambda_1 & 0 \\ 0 & \lambda_2 \end{pmatrix}P^{-1}\right], (11.239)$$

or

$$Z_N = \mathrm{Tr}\begin{pmatrix} \lambda_1 & 0 \\ 0 & \lambda_2 \end{pmatrix}^N = \lambda_1^N + \lambda_2^N. \quad (11.240)$$

The free energy is

$$A = -k_B T \ln Z = -k_B T \ln\left[\lambda_1^N\left[1 + \left(\frac{\lambda_2}{\lambda_1}\right)^N\right]\right], \quad (11.241)$$

or

$$A = -Nk_B T \ln \lambda_1 - k_B T \ln\left[1 + \left(\frac{\lambda_2}{\lambda_1}\right)^N\right]. \quad (11.242)$$

From this, we can find the entropy and other state functions, noting that

$$\lim_{N \to \infty} \left(\frac{\lambda_2}{\lambda_1}\right)^N \to 0. \tag{11.243}$$

Example 11.2
A rubber band at absolute temperature T is fastened at one end to a peg and supports a weight W of mass M at the other end. Assume that we can model the rubber band as a linked polymer chain consisting of N segments of length L joined end to end. Each segment can be oriented either parallel or antiparallel to the vertical direction. Find an expression for the mean length of the rubber band as a function of T and M. (Neglect the mass of each segment and the interaction between segments.)

The individual segments either add, if they are parallel, or cancel, if they are antiparallel. We have

$$L_{tot} = \left(N_{up} + N_{dn}\right)L. \tag{11.244}$$

The probability of having n_1 segments facing up and $n_2 = N - n_1$ segments facing down is

$$P_N(n_1) = \frac{N!}{n_1!(N - n_1)!} p^{n_1} (1-p)^{(N-n_1)}. \tag{11.245}$$

The multiplicity of systems with length L_{tot} is

$$g(n_1, n_2) = \frac{N!}{n_1! n_2!} = \frac{N!}{n_1!(N - n_1)!}. \tag{11.246}$$

Thus, the entropy is given by

$$S = k_B \ln g = k_B \left\{\ln N! - \ln n_1! - \ln(N - n_1)!\right\}, \tag{11.247}$$

or

$$S = k_B \left\{N \ln N - n_1 \ln n_1 - (N - n_1)\ln(N - n_1)\right\}. \tag{11.248}$$

Entropy is related to energy by

$$dE = T\,dS - dW. \tag{11.249}$$

When the mass is hanging in equilibrium (i.e., when it is at rest), $dE = 0$ and

$$T\,dS = dW = F\,dl = Mg\,dl. \tag{11.250}$$

We rewrite equation 11.244 as

$$L_{tot} = \left(n_1 - \left(N - n_1\right)\right)L = \left(2n_1 - N\right)L, \tag{11.251}$$

so that

$$dl = dL_{tot} = 2\,L\,dn_1. \tag{11.252}$$

From equation 11.250,

$$\frac{dS}{dl} = \frac{Mg}{T}, \tag{11.253}$$

and from equation 11.252,

$$\frac{dS}{dn_1} = \frac{2\,MgL}{T}. \tag{11.254}$$

However, differentiating equation 11.248 gives

$$\frac{dS}{dn_1} = k_B\left\{-\ln n_1 + \ln\left(N - n_1\right)\right\} = k_B \ln\left(\frac{N - n_1}{n_1}\right), \tag{11.255}$$

so that

$$\frac{2\,MgL}{T} = k_B \ln\left(\frac{N - n_1}{n_1}\right), \tag{11.256}$$

or

$$\left(\frac{N - n_1}{n_1}\right) = \exp\left[\frac{2\,MgL}{k_B T}\right]. \tag{11.257}$$

From equation 11.251, the length is given by

$$L_{tot} = n_1 \left(1 - \exp\left[\frac{2MgL}{k_BT} \right] \right) L .$$ (11.258)

However, because equation 11.257 may be rewritten as

$$n_1 = N \left\{ 1 + \exp\left[\frac{2MgL}{k_BT} \right] \right\}^{-1} ,$$ (11.259)

equation 11.258 reduces to

$$L_{tot} = NL \frac{1 - \exp\left[\frac{2MgL}{k_BT} \right]}{1 + \exp\left[\frac{2MgL}{k_BT} \right]} = -NL \tanh\left[\frac{MgL}{k_BT} \right] .$$ (11.260)

11.8 Quantum Statistical Mechanics

The density of states in phase space satisfies

$$\frac{\partial \rho}{\partial t} + \vec{\nabla} \cdot (\rho \vec{v}) = 0,$$ (11.261)

and

$$\frac{d\rho}{dt} = \frac{\partial \rho}{\partial t} + [\rho, H] = 0,$$ (11.262)

which are the equation of continuity and Liouville's equation, respectively. In quantum mechanics, $\langle f \rangle$ is given by

$$\langle f \rangle = \frac{\int f(p,q)\rho(p,q)d^{3N}p\, d^{3N}q}{\int \rho(p,q)d^{3N}p\, d^{3N}q} .$$ (11.263)

For a quantum mechanical system,

$$\hat{H}\Psi^k(x,t) = E\Psi^k(x,t), \tag{11.264}$$

where $\Psi^k(x,t) = \sum_n a_n^k(t)\varphi_k(x)$ is the probability amplitude for the kth system of the ensemble to be in the state $\varphi_k(x)$. The normalization condition is

$$\sum_n \left|a_n^k(t)\right|^2 = 1. \tag{11.265}$$

We define the density operator as

$$\rho_{mn} = \frac{1}{N}\sum_{k=1}^{N} a_m^k(t)a_n^{k*}(t). \tag{11.266}$$

We have

$$\langle f \rangle = \frac{1}{N}\sum_{k=1}^{N}\int \Psi^{k*} f \Psi^k \, d\tau,$$

$$= \frac{1}{N}\sum_{k=1}^{N}\left[\sum_{m,n} a_n^{k*} f_{nm} a_m^k\right],$$

$$= \sum_{m,n}\rho_{mn}f_{nm} = Tr\left(\hat{\rho}\hat{f}\right). \tag{11.267}$$

In the microcanonical ensemble, any one of the microstates is as likely to occur as any other; therefore,

$$\rho_{mn} = \rho_n \delta_{mn}, \tag{11.268}$$

with $r_n = 1/G$ (where G = total number of microstates), if $E \le E_k \le E + \Delta E$, or $r_n = 0$.

In the canonical ensemble, the probability that a system, chosen at random, possesses an energy, E_r, is determined by the Boltzmann factor.

$$\hat{\rho}_n = C\exp^{\left(-\beta\hat{E}_n\right)}, \tag{11.269}$$

$$C = \frac{1}{\sum_n \exp^{\left(-\beta\hat{E}_n\right)}} = \frac{1}{Z_N}, \tag{11.270}$$

and

$$\langle f\rangle = Tr\left(\hat{\rho}\hat{f}\right) = \frac{Tr\left(\hat{f}\exp^{-\beta\hat{H}}\right)}{Tr\left(\exp^{-\beta\hat{H}}\right)}. \tag{11.271}$$

In the grand canonical ensemble, we have

$$\hat{\rho} = \frac{1}{\mathscr{q}(T,V,N)}\exp^{-\beta\left(\hat{H}-\mu\hat{N}\right)}, \tag{11.272}$$

$$\mathscr{q}(T,V,N) = Tr\left(\exp^{-\beta\left(\hat{H}-\mu\hat{N}\right)}\right), \tag{11.273}$$

and

$$\langle f\rangle = \frac{Tr\left(\hat{f}\exp^{-\beta\left(\hat{H}-\mu\hat{N}\right)}\right)}{Tr\left(\exp^{-\beta\left(\hat{H}-\mu\hat{N}\right)}\right)}. \tag{11.274}$$

11.9 Equipartition Theorem

By definition,

$$\left\langle x_i\frac{\partial H}{\partial x_j}\right\rangle = \frac{1}{\Sigma(E)}\int_\Gamma d\mathbf{X}^N\delta\left(H(\mathbf{X}^N)-E\right)x_i\frac{\partial H}{\partial x_j}, \tag{11.275}$$

which, because the surface term vanishes, is equivalent to

$$\left\langle x_i \frac{\partial H}{\partial x_j} \right\rangle = -\frac{1}{\Sigma(E)} \int_\Gamma dx H(x) \frac{\partial}{\partial x_j} \left[x_i \delta(H(x) - E) \right]. \quad (11.276)$$

The integrand is

$$\frac{\partial}{\partial x_j} \left[x_i \delta(H(x) - E) \right] = \delta_{ij} \delta(H(x) - E) - x_i \frac{\partial}{\partial E} \delta(H(x) - E) \frac{\partial H}{\partial x_j}, \quad (11.277)$$

so the previous expression becomes

$$\left\langle x_i \frac{\partial H}{\partial x_j} \right\rangle = -\frac{1}{\Sigma(E)} \int_\Gamma dx \left\{ \delta_{ij} \delta(H(x) - E) + x_i \frac{\partial}{\partial E} \right\} H(x), \quad (11.278)$$

which reduces as follows:

$$\left\langle x_i \frac{\partial H}{\partial x_j} \right\rangle = -\delta_{ij} E + \frac{1}{\Sigma(E)} \frac{\partial}{\partial E} \int_\Gamma dx H(x) x_i \frac{\partial H(x)}{\partial x_j} \delta(H(x) - E),$$

$$= -\delta_{ij} E + \frac{1}{\Sigma(E)} \frac{\partial}{\partial E} \left\{ E \int_\Gamma dx x_i \frac{\partial H(x)}{\partial x_j} \delta(H(x) - E) \frac{\Sigma(E)}{\Sigma(E)} \right\},$$

$$= -\delta_{ij} E + \frac{1}{\Sigma(E)} \frac{\partial}{\partial E} \left\{ E \left\langle x_i \frac{\partial H(x)}{\partial x_j} \right\rangle \Sigma(E) \right\},$$

$$= -\delta_{ij} E + \left\langle x_i \frac{\partial H(x)}{\partial x_j} \right\rangle + \frac{E}{\Sigma(E)} \frac{\partial \Sigma(E)}{\partial E} \left\langle x_i \frac{\partial H(x)}{\partial x_j} \right\rangle$$

$$+ E \frac{\partial}{\partial E} \left\langle x_i \frac{\partial H(x)}{\partial x_j} \right\rangle. \quad (11.279)$$

Therefore,

$$\delta_{ij} E = \frac{E}{\Sigma(E)} \frac{\partial \Sigma(E)}{\partial E} \left\langle x_i \frac{\partial H(x)}{\partial x_j} \right\rangle + E \frac{\partial}{\partial E} \left\langle x_i \frac{\partial H(x)}{\partial x_j} \right\rangle. \quad (11.280)$$

Recall that

$$S = k_B \ln\left[\Omega_{\Delta E}(E)C^{-1}\right] = k_B \ln\left[\Delta E \Sigma(E)C^{-1}\right]. \quad (11.281)$$

From this it follows that

$$\frac{\partial S}{\partial E} = k_B \Delta E C^{-1} \frac{\partial \Sigma}{\partial E} = \frac{1}{T}, \quad (11.282)$$

and that

$$\frac{\partial \Sigma}{\partial E} = \frac{\Sigma(E)}{k_B T}. \quad (11.283)$$

Using this relationship, the previous expression for $d_{ij}E$ gives

$$\delta_{ij}E = \frac{E}{k_B T}\left\langle x_i \frac{\partial H(x)}{\partial x_j}\right\rangle + O\left(\frac{1}{N}\right)_{\to 0}, \quad (11.284)$$

so that

$$\left\langle x_i \frac{\partial H(x)}{\partial x_j}\right\rangle = \delta_{ij}k_B T. \quad (11.285)$$

11.10 Theory of Solids

11.10.1 Einstein Theory of Solids

Einstein assumed that a solid was composed of $3N$ independent distinguishable quantum oscillators. The Hamiltonian for a single quantum harmonic oscillator is

$$\hat{H} = \hbar\omega\left(\hat{N} + \frac{1}{2}\right). \quad (11.286)$$

We have

$$Z_N(T,V) = \text{Tr} \exp\left(-\beta\hbar\omega \sum_{i=1}^{3N}\left(\hat{N}+\frac{1}{2}\right)\right), \tag{11.287}$$

or

$$Z_N(T,V) = \sum_{n_1=0}^{\infty} \ldots \sum_{3N=0}^{\infty} \exp\left(-\beta\hbar\omega \sum_{i=1}^{3N}\left(n_i+\frac{1}{2}\right)\right), \tag{11.288}$$

which is equivalent to

$$Z_N(T,V) = \left[\frac{\exp^{-\beta(\hbar\omega/2)}}{1-\exp^{-\beta\hbar\omega}}\right]^{3N}. \tag{11.289}$$

The Helmholtz free energy is

$$A(T,V,N) = -k_B T \ln Z_N,$$

$$= -3Nk_B T \ln\left[\frac{\exp^{-\beta(\hbar\omega/2)}}{1-\exp^{-\beta\hbar\omega}}\right],$$

$$= \frac{3N\hbar\omega}{2} + 3Nk_B T \ln\left[1-\exp^{-\beta\hbar\omega}\right]. \tag{11.290}$$

From this expression, we can find S, C_v, and other state functions.

11.10.2 Debye Theory of Solids

Debye assumed that the oscillators which composed the solid were coupled. We have

$$H(\vec{p}^N,\vec{q}^N) = \sum_{i=1}^{3N} \frac{p_i^2}{2m} + \sum_{i,j=0}^{3N} A_{ij}q_i q_j. \tag{11.291}$$

The diagonalized form is

$$H\left(\vec{P}^N, \vec{Q}^N\right) = \sum_{i=1}^{3N} \frac{P_i^2}{2m} + \sum_{i=1}^{3N} \frac{m\omega_i^2}{2} Q_i^2 . \qquad (11.292)$$

Where the $w_i's$ are all different, the Hamiltonian may be rewritten as

$$\hat{H} = \sum_{i=1}^{3N} \hbar\omega_i \left(\hat{N} + \frac{1}{2}\right). \qquad (11.293)$$

Thus,

$$Z_N(T,V) = \sum_{n_1=0}^{\infty} \cdots \sum_{n_{3N}=0}^{\infty} \exp\left\{-\beta\hbar\sum_{i=1}^{3N}\left(n_i + \frac{1}{2}\right)\omega_i\right\}, \qquad (11.294)$$

and the Helmholtz free energy is

$$A(T,V,N) = \frac{\hbar}{2}\sum_{i=1}^{3N}\omega_i + k_B T \sum_{i=1}^{3N} \ln\left(1 - \exp^{-\beta\hbar\omega_i}\right). \qquad (11.295)$$

11.11 Bibliography

Kittel, C., and Kromer, H. *Thermal Physics*. 2d ed. San Francisco: W. H. Freeman, 1980.

Mandl, F. *Statistical Physics*. New York: John Wiley & Sons, 1971.

Reichl, L. E. *A Modern Course in Statistical Physics*. Austin: University of Texas Press, 1980.

12 Solid State Physics

12.1 Conductors

12.1.1 Drude Theory of Metals

12.1.1.1 Assumptions

- Positive charge is attached to immobile ions.

- Valence electrons wander freely through the metal.

- Between collisions, the interaction of a given electron is negligible.

- Collisions are instantaneous events that abruptly alter the velocity of an electron.

- The probability of an electron undergoing a collision in a time interval dt is just dt/τ, where τ is the average time between collisions.

- Electrons achieve thermal equilibrium only through collisions.

12.1.1.2 DC Conductivity

Consider a typical electron at time $t = 0$. If t' is defined to be the time elapsed since its last collision, we have

$$\vec{v}(t') = \vec{v}_o - \frac{e\vec{E}t}{m}.$$

(12.1)

Time averaging equation 12.1 gives

$$\langle v \rangle = \vec{v}_{avg} = \left\langle \vec{v}_o - \frac{e\vec{E}t}{m} \right\rangle = -\frac{e\vec{E}\tau}{m},$$

(12.2)

where τ is the relaxation time, and

$$\vec{p} = m\vec{v}_{avg} = -e\vec{E}\tau. \tag{12.3}$$

The current density, j, is defined by $\vec{j} = -ne\vec{v}$, and the resistivity, ρ, is defined by $\vec{E} = \rho\vec{j}$. From these definitions it follows that

$$\vec{j} = \left(\frac{ne^2\tau}{m}\right)\vec{E}, \tag{12.4}$$

and

$$\sigma = \frac{1}{\rho} = \left(\frac{ne^2\tau}{m}\right). \tag{12.5}$$

12.1.1.3 AC Conductivity

The equation of motion of an electron is

$$\frac{d\vec{p}}{dt} = -\frac{\vec{p}}{\tau} - e\vec{E}, \tag{12.6}$$

which, given plane wave solutions, reduces to

$$-i\omega\vec{p} = -\frac{\vec{p}}{\tau} - e\vec{E}. \tag{12.7}$$

The solution for p is now

$$\vec{p} = -\frac{e}{\left(\dfrac{1}{\tau} - i\omega\right)}\vec{E}. \tag{12.8}$$

Using this expression, together with equations 12.3 and 12.4, gives

$$\vec{j} = -\left(\frac{ne}{m}\right)\vec{p} = \left(\frac{ne^2}{m}\right)\frac{\vec{E}}{\left(\dfrac{1}{\tau} - i\omega\right)}. \tag{12.9}$$

We may also define the current density as

$$\vec{j}(\vec{r},\omega) = \sigma(\omega)\vec{E}(\vec{r},\omega), \tag{12.10}$$

so that the AC conductivity is given by

$$\sigma(\omega) = \frac{\sigma_o}{1 - i\omega\tau}, \tag{12.11}$$

with $\sigma_o = \dfrac{ne^2\tau}{m}$.

From Maxwell's equations (chapter 2) it can be shown that

$$\vec{\nabla}\times\left(\vec{\nabla}\times\vec{E}\right) = -\nabla^2\vec{E} = \frac{i\omega}{c}\vec{\nabla}\times\vec{H} = \frac{\omega^2}{c^2}\left(1 + \frac{4\pi i\sigma_o}{\omega}\right)\vec{E}. \tag{12.12}$$

This has the form of the usual wave equation

$$-\nabla^2\vec{E} = \frac{\omega^2}{c^2}\varepsilon(\omega)\vec{E}, \tag{12.13}$$

which has a complex dielectric constant given by

$$\varepsilon(\omega) = 1 + \frac{4\pi i\sigma_o}{\omega}. \tag{12.14}$$

From the definition of σ_o, when $\omega\tau \gg 1$, equation 12.14 reduces to

$$\varepsilon(\omega) = 1 - \frac{\omega_p^2}{\omega^2}, \tag{12.15}$$

with $\omega_p^2 = \dfrac{4\pi ne^2}{m}$. If ε is real and negative, no radiation propagates. If ε is positive, the solutions become oscillatory and radiation propagates.

12.1.1.4 Magnetoresistance: The Hall Effect

In 1879, E. H. Hall wrote, "If the current of electricity in a fixed conductor is itself attracted by a magnet, the current should be drawn to one side of the wire, and therefore the resistance experienced should be increased." The experimental geometry is shown in figure 12.1.

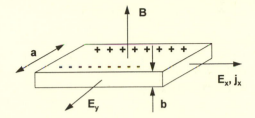

Figure 12.1 The Hall effect geometry.

If an electric field E_x is applied to the wire as shown, and a magnetic field B points in the z-direction, the Lorentz force will deflect electrons in the negative y-direction. As the electrons move, they produce an E-field that balances the Lorentz force. We have

$$f_m = \frac{qvB}{c},$$ (12.16)

and

$$f_e = qE.$$ (12.17)

The current is by definition the charge transported per unit time across a transverse surface of the conductor

$$I = nqabv.$$ (12.18)

Setting $f_e = -f_m$ gives

$$E = -\frac{IB}{nabq} = -\frac{IBR}{ab},$$ (12.19)

where $R = 1/nq$ is the Hall coefficient. Knowing R gives B.

12.2 Ground State Properties of the Electron Gas

In the last chapter it was seen that solving the Schrödinger equation placed a constraint on the value of k for solutions in a periodic potential. The constraint was (eq. 11.79)

$$\vec{k} = \frac{\pi}{L}\left(n_x, n_y, n_z\right).$$ (12.20)

If we alter the boundary conditions slightly so that instead of requiring $\psi = 0$ on all faces of the cube we simply require $\psi(x, y, z) = \psi(x + L, y, z)$, and cyclically, we obtain the constraint

$$k_i = \frac{2\pi}{L}n_i.$$ (12.21)

If the region of k-space is large compared to $2\pi/L$, the number of allowed values of k is simply the volume of k-space contained in the region divided by the volume of k-space per point. That is, the number of allowed values of k is

$$\frac{\Omega}{(2\pi/L)^3} = \frac{\Omega V}{8\pi^3}.$$ (12.22)

Stated differently, the number of allowed values of k per unit volume of k-space is $\frac{V}{8\pi^3}$. When the number of electrons in the ground state is enormous, the occupied region will be essentially a sphere. Consequently, the number of allowed values of k within the sphere is defined by

$$\left(\frac{4\pi k_F^3}{3}\right)\left(\frac{V}{8\pi^3}\right) = \frac{k_F^3}{6\pi^2}V,$$ (12.23)

where k_F is the Fermi wave vector. Because each value of k corresponds to 2 one-electron levels (spin up and spin down), the number of electrons is

$$N = 2\frac{k_F^3}{6\pi^2}V = \frac{k_F^3 V}{3\pi^2},$$ (12.24)

and the electron density is

$$n = \frac{N}{V} = \frac{k_F^3}{3\pi^2}.$$ (12.25)

12.2.1 Definitions

Fermi sphere: Sphere of radius k_F containing the occupied one-electron levels

Fermi surface: Surface of the Fermi sphere, separates occupied from unoccupied levels.

Fermi momentum: $p_F = \hbar k_F$

Fermi energy: $E_F = \dfrac{\hbar^2 k_F^2}{2m}$

Fermi velocity: $v_F = \dfrac{p_F}{m}$

Fermi temperature: $T_F = \dfrac{E_F}{k}$

12.3 Crystalline Structure

12.3.1 Bravais Lattice

The Bravais lattice specifies the periodic array in which the repeated units of the crystal are arranged. That is, the Bravais lattice is an infinite array of discrete points with an arrangement and orientation that appears exactly the same, from whichever of the points of the array it is viewed. A Bravais lattice consists of all points with position vectors R of the form

$$\vec{R} = n_1 \vec{a}_1 + n_2 \vec{a}_2 + n_3 \vec{a}_3, \qquad (12.26)$$

where $\vec{a}_1$, $\vec{a}_2$, $\vec{a}_3$ are any three vectors, called *primitive vectors*, not all in the same plane. A volume of space that, when translated through all the vectors of a Bravais lattice, just fills all of space is called a *primitive cell*. A primitive cell must contain one lattice point, but choices of the primitive cell are not unique. A common choice for the primitive cell is the *Wigner-Seitz cell.* The Wigner-Seitz cell about a lattice point is the region of space that is closer to that point than to any other lattice point.

12.3.2 Reciprocal Lattice

Consider a set of points $\vec{R}$ constituting a Bravais lattice and a plane wave $e^{i\vec{k}\cdot\vec{r}}$. The set of wave vectors $\vec{k}$ that satisfy $e^{i\vec{k}\cdot\vec{r}} = e^{i\vec{k}\cdot(\vec{r}+\vec{R})}$ is the reciprocal lattice. That is, if

$$\vec{k} = k_1\vec{b}_1 + k_2\vec{b}_2 + k_3\vec{b}_3, \tag{12.27}$$

then

$$\vec{k} \cdot \vec{R} = 2\pi(k_1 n_1 + k_2 n_2 + k_3 n_3). \tag{12.28}$$

The reciprocal lattice may be generated by the three primitive vectors

$$\vec{b}_1 = 2\pi \frac{\vec{a}_2 \times \vec{a}_3}{\vec{a}_1 \cdot (\vec{a}_2 \times \vec{a}_3)}, \tag{12.29}$$

and cyclically. The Wigner-Seitz primitive cell of the reciprocal lattice is known as the *first Brillouin zone*.

12.3.3 Measurement of Atomic Structure

12.3.3.1 Bragg Diffraction

Bragg treated the crystal as made out of parallel planes of ions spaced a distance d apart. The path difference between two rays reflecting off different planes is $2d\sin\theta$, as shown in figure 12.2. They interfere constructively if the path length is an integral number of wavelengths. That is, if

$$2d\sin\theta = n\lambda. \tag{12.30}$$

Consequently, experimental measurements of X-ray diffraction patterns can be used to infer atomic spacing distances, which were seen to be on the order of an Angstrom (10^{-10} m). Also, locating the various Bragg planes gives information about the arrangement of atoms within the crystalline lattice.

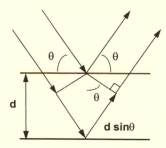

Figure 12.2 X-ray diffraction geometry.

12.3.3.2 Von Laue Formulation

Consider the picture seen in figure 12.3. The condition for constructive interference is

$$d\cos\theta + d\cos\theta' = m\lambda,$$ (12.31)

which gives

$$\vec{d}\cdot\left(\vec{k}-\vec{k}'\right) = \frac{2\pi}{\lambda}\vec{d}\cdot\left(\hat{n}-\hat{n}'\right) = 2\pi m.$$ (12.32)

The lattice sights are displaced by the Bravais lattice vectors R, so that

$$\vec{R}\cdot\left(\vec{k}-\vec{k}'\right) = 2\pi m,$$ (12.33)

which implies

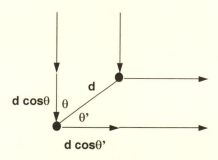

Figure 12.3 Scattering geometry.

$$e^{i(\vec{k}-\vec{k}')\cdot\vec{R}} = 1. \tag{12.34}$$

Constructive interference will occur provided that the change in the wave vector, $\Delta\vec{k} = (\vec{k} - \vec{k}')$, is a vector of the reciprocal lattice.

12.3.4 Brillouin Zones

The $(n+1)$th Brillouin zone is the set of points, in k-space, in the $(n+1)$th zone that can be reached from the nth zone by crossing only one Bragg plane. The second Brillouin zone is the set of points that can be reached from the first zone by crossing only one Bragg plane. The first Brillouin zone is the set of points that can be reached from the origin without crossing any Bragg plane.

12.3.5 Fermi Surfaces

The Fermi surface is a constant energy surface in k-space. When $T = 0$, all electrons are in the ground state and all values of k corresponding to energies less than the Fermi energy are filled. To determine the Fermi surface:

1. Draw the free-electron Fermi sphere.

2. Deform it slightly in the immediate vicinity of the Bragg planes.

3. Take that portion of the surface of the free electron sphere lying within the nth Brillouin zone and translate it through all reciprocal lattice vectors.

Example 12.2
Assume that solid lithium has, on average, one free electron per atom. The density of lithium is 0.542 g/cm³. Lithium metal is a body-centered cubic structure with lattice constant 3.491 Å. The atomic weight of lithium is 6.939. (1) Calculate the size of the first Brillouin zone; (2) find the Fermi energy at 0 K; and (3) determine the approximate shape of the Fermi surface.

(1) The first Brillouin zone is the region of k-space that can be reached from the origin without crossing any Bragg planes. The closest Bragg plane is defined by the smallest reciprocal lattice vector. For a body-centered cubic, there are eight equivalent reciprocal lattice vectors. Consequently the first Brillouin zone will be an eight-sided figure. The lattice constant is 3.491 Å.

The Brillouin zone is an eight-sided figure with the perpendicular from the origin to each side being 1.7455 Å.

(2) The Fermi energy is defined as the energy below which all levels are filled and above which all levels are empty when $T = 0$ K. In k-space this is given by a sphere. The number of orbitals is seen to be given by (at 0 K),

$$N = \left(\frac{4}{3} \pi k_F^3 \right) 2 \left(\frac{L}{2\pi} \right)^3 = \frac{V}{3\pi^2} k_F^3, \qquad (12.35)$$

where the factor of 2 represents the two possible spin states and the factor $L/2\pi$ is used to convert from k-space to Cartesian space. The free particle Schrödinger equation

$$-\frac{\hbar^2}{2m} \nabla^2 \psi = E\psi, \qquad (12.36)$$

has solution

$$E_F = \frac{\hbar^2}{2m} k_F^2 = \left[\frac{N}{V} 3\pi^2 \left(\frac{\hbar^2}{2m} \right)^{3/2} \right]^{2/3}. \qquad (12.37)$$

From the density of lithium we can calculate N/V and, knowing the value of the other constants, we find that $E_F \sim 2.35 \times 10^{-18}$ J.

(3) The Fermi surface is the surface of constant energy in k-space. This is found by calculating the intersection of the Fermi sphere, radius k_F, with the different Brillouin zones. From equation 12.37, $k_F \sim 1.962$ Å, so that by definition $\lambda = 3.2$ Å or $\lambda/2 \sim 1.6$ Å. Thus, the Fermi surface lies entirely within the first Brillouin zone and is a sphere of radius 1.6 Å.

12.4 Bloch's Theorem

Because the ions in a perfect crystal are arranged in a periodic array, the electronic potential should satisfy

$$U(\vec{r}) = U(\vec{r} + \vec{R}). \qquad (12.38)$$

THEOREM: The eigenstates ψ of the one-electron Hamiltonian,

$$H = -\frac{\hbar^2}{2m}\nabla^2 + U(\vec{r}), \tag{12.39}$$

where $U(\vec{r}) = U(\vec{r} + \vec{R})$, can be chosen to have the form of a plane wave times a function with the periodicity of a Bravais lattice. That is,

$$\psi_{nk}(\vec{r}) = e^{i\vec{k}\cdot\vec{r}} u_{nk}(\vec{r}), \tag{12.40}$$

where

$$u_{nk}(\vec{r}) = u_{nk}(\vec{r} + \vec{R}). \tag{12.41}$$

PROOF: Define

$$T_R f(r) = f(r + R). \tag{12.42}$$

By definition,

$$T_R H\psi = H(r + R)\psi(r + R) = H(r)\psi(r + R) = HT_R\psi, \tag{12.43}$$

so that

$$T_R H = HT_R, \tag{12.44}$$

or

$$[T_R, H] = 0. \tag{12.45}$$

Similarly,

$$T_R T_{R'} = T_{R'} T_R = T_{R+R'}. \tag{12.46}$$

We have eigenvalues of T_R such that

$$T_R T_{R'}\psi = c(R)T_{R'}\psi = c(R)c(R')\psi = c(R + R')\psi. \tag{12.47}$$

If a_i is a lattice vector, then $c(a_i) \propto e^{2\pi i x_i}$. This gives

$$c(\vec{R}) = c(\vec{a}_1)^{n_1} c(\vec{a}_2)^{n_2} c(\vec{a}_3)^{n_3}. \tag{12.48}$$

This is equivalent to $c(\vec{R}) = e^{i\vec{k}\cdot\vec{R}}$, where

$$\vec{k} = k_1\vec{b}_1 + k_2\vec{b}_2 + k_3\vec{b}_3. \tag{12.49}$$

Thus, we have shown that the eigenstates ψ of H have the periodicity of the Bravais lattice vector R.

Example 12.1
Consider a linear period potential consisting of delta functions

$$V(x) = \sum_{j=-\infty}^{+\infty} A\delta(x - a_j), \tag{12.50}$$

where $a_j = ja$ (a is the lattice constant) and A is a constant. Using Bloch's theorem, solve the Schrödinger equation for an electron in this potential.

The Schrödinger equation is

$$\left\{ -\frac{\hbar^2}{2m}\frac{\partial^2}{\partial x^2} + \sum_{j=-\infty}^{+\infty} A\delta(x - a_j) \right\}\psi(x) = E\psi(x). \tag{12.51}$$

Using equation 12.40, this expression reduces to

$$\left\{ -\frac{\hbar^2}{2m}\frac{\partial^2}{\partial x^2} + \sum_{j=-\infty}^{+\infty} A\delta(x - a_j) \right\}e^{ikx}u_{nk}(x) = Ee^{ikx}u_{nk}(x). \tag{12.52}$$

Performing the differentiation gives

$$\left\{ -\frac{\hbar^2 k^2}{2m}u_{nk}(x) - \frac{\hbar^2}{2m}u''_{nk}(x) - Eu_{nk}(x) \right.$$

$$\left. + \sum_{j=-\infty}^{+\infty} A\delta(x - a_j)u_{nk}(x) \right\}e^{ikx} = 0, \tag{12.53}$$

or simply

$$-\frac{\hbar^2 k^2}{2m} u_{nk}(x) - \frac{\hbar^2}{2m} u_{nk}''(x) - E u_{nk}(x)$$

$$+ \sum_{j=-\infty}^{+\infty} A \delta(x - a_j) u_{nk}(x) = 0. \tag{12.54}$$

Because u_{nk} is periodic we consider a region of space $-\frac{a}{2} \le x \le +\frac{a}{2}$. In this region, equation 12.54 reduces to

$$u_{nk}''(x) + \left\{ k^2 + \frac{2mE}{\hbar^2} - \frac{2mA}{\hbar^2} \delta(x - 0) \right\} u_{nk}(x). \tag{12.55}$$

For $x \ne 0$ we have

$$u_{nk}''(x) + \left\{ k^2 + \frac{2mE}{\hbar^2} \right\} u_{nk}(x) = 0, \tag{12.56}$$

which has solution

$$u_{nk}(x) = A e^{i\alpha x} + B e^{-i\alpha x}, \tag{12.57}$$

with

$$\alpha = \left(k^2 + \frac{2mE}{\hbar^2} \right)^{1/2}. \tag{12.58}$$

We denote the region where $-\frac{a}{2} \le x \le 0$ with the symbol "−", and the region where $0 \le x \le \frac{a}{2}$ with the symbol "+". With this notation, equation 12.57 is more explicitly written as

$$u_{nk,-}(x) = A_- e^{i\alpha x} + B_- e^{-i\alpha x}, \tag{12.59}$$

and

$$u_{nk,+}(x) = A_+ e^{i\alpha x} + B_+ e^{-i\alpha x}, \qquad (12.60)$$

subject to the constraint

$$u_{nk,-}(0) = u_{nk,+}(0), \qquad (12.61)$$

or

$$A_- + B_- = A_+ + B_+. \qquad (12.62)$$

A second constraint is found by integrating equation 12.55 near $x = 0$. This gives

$$\lim_{\varepsilon \to 0} \int_{-\varepsilon}^{+\varepsilon} \left\{ u''_{nk}(x) + \left\{ k^2 + \frac{2mE}{\hbar^2} - \frac{2mA}{\hbar^2}\delta(x-0) \right\} u_{nk}(x) \right\} dx = 0. \quad (12.63)$$

The respective terms in equation 12.63 reduce to

$$\lim_{\varepsilon \to 0} \int_{-\varepsilon}^{+\varepsilon} \left\{ u''_{nk}(x) \right\} dx = \lim_{\varepsilon \to 0} \left[\frac{du_{nk}(+\varepsilon)}{dx} - \frac{du_{nk}(-\varepsilon)}{dx} \right] = \left. \frac{du}{dx} \right|_+ - \left. \frac{du}{dx} \right|_-, \quad (12.64)$$

$$\lim_{\varepsilon \to 0} \int_{-\varepsilon}^{+\varepsilon} \left\{ \left\{ k^2 + \frac{2mE}{\hbar^2} \right\} u_{nk}(x) \right\} dx = \left\{ k^2 + \frac{2mE}{\hbar^2} \right\} \lim_{\varepsilon \to 0} \int_{-\varepsilon}^{+\varepsilon} u_{nk}(x) dx = 0, \quad (12.65)$$

and

$$\lim_{\varepsilon \to 0} \int_{-\varepsilon}^{+\varepsilon} \left\{ \left\{ -\frac{2mA}{\hbar^2}\delta(x-0) \right\} u_{nk}(x) \right\} dx = -\frac{2mA}{\hbar^2} u(0). \quad (12.66)$$

As a result, a second constraint is

$$\left. \frac{du}{dx} \right|_+ - \left. \frac{du}{dx} \right|_- = \frac{2mA}{\hbar^2} u(0). \qquad (12.67)$$

From equations 12.59, 12.60, and 12.62, this equation reduces to

$$[i\alpha A_+ - i\alpha B_+] - [i\alpha A_- - i\alpha B_-] = \frac{2mA}{\hbar^2}[A_- + B_-]. \quad (12.68)$$

If we envision an electron incident on the barrier from the left, after interaction with the potential there may be a reflection moving to the left in the region $x < 0$ and a transmission moving to the right in the region $x > 0$. However, there can be no motion to the left in the region $x > 0$. This places the further constraint that $B_+ = 0$. Equations 12.62 and 12.68 reduce to

$$A_- + B_- = A_+, \quad (12.69)$$

and

$$[i\alpha A_+] - [i\alpha A_- - i\alpha B_-] = \frac{2mA}{\hbar^2}[A_- + B_-]. \quad (12.70)$$

After reduction, it can be shown that

$$\frac{A_+}{A_-} = \frac{\alpha\hbar^2}{\alpha\hbar^2 - imA}, \quad (12.71)$$

and

$$\frac{B_-}{A_-} = \frac{-mA}{i\alpha\hbar^2 + mA}, \quad (12.72)$$

which satisfies the relation

$$\left|\frac{A_+}{A_-}\right|^2 + \left|\frac{B_-}{A_-}\right|^2 = 1, \quad (12.73)$$

as expected.

12.5 Electrons in a Weak Periodic Potential

When the periodic potential is zero, the solutions to the Schrödinger equation are plane waves. Therefore, we can write the wave function of a Bloch level with crystal momentum $\vec{k}$ in the form

$$\psi_k(r) = \sum_k c_{\kappa-k} e^{i(\kappa-k)r}, \tag{12.74}$$

where the $c_{\kappa-k}$ are determined by

$$\left[-\frac{\hbar^2}{2m}(\kappa-k)^2 - E\right]c_{\kappa-k} + \sum_k U_{\kappa-k} c_{\kappa-k} = 0. \tag{12.75}$$

For the free-electron case, $U_{k'-k} = 0$, so that

$$\left[E^0_{\kappa-k} - E\right]c_{\kappa-k} = 0. \tag{12.76}$$

If there is a group of reciprocal lattice vectors $\vec{k}_1, \ldots, \vec{k}_m$ that satisfy

$$E^0_{\kappa-k_1} = E^0_{\kappa-k_2} = \ldots = E^0_{\kappa-k_m}, \tag{12.77}$$

then when E is equal to the common value of these free-electron energies there are m independent degenerate plane wave solutions. Since any linear combination of degenerate solutions is also a solution, one has complete freedom in choosing the coefficients $c_{\kappa-k}$ for $\vec{k} = \vec{k}_1, \ldots, \vec{k}_m$.

Case I:

Fix k and consider a particular lattice vector k_1 such that the free electron energy, $E^0_{\kappa-k_1}$, is far from the values of $E^0_{\kappa-k}$ for all other k. When compared with U,

$$\left|E^o_{\kappa-k_1} - E^o_{\kappa-k}\right| \gg U. \tag{12.78}$$

Thus,

$$\left[E - E^0_{\kappa-k_1}\right]c_{\kappa-k_1} = \sum_k U_{\kappa-k} c_{\kappa-k}, \tag{12.79}$$

where $U_k = 0$ when $k = 0$. We wish to examine the solution for which $c_{\kappa-k}$ vanishes when $k \neq k_1$ in the limit of vanishing U. The equation is

$$c_{\kappa-k} = \frac{U_{k_1-k}c_{\kappa-k_1}}{E - E^0_{\kappa-k}} + \sum_{k'\neq k_1} \frac{U_{k'-k}c_{\kappa-k'}}{E - E^0_{\kappa-k}}, \qquad (12.80)$$

because $c_{\kappa-k} \to 0$ as $k \neq k_1$, the second term is of second order in U. That is,

$$c_{\kappa-k} = \frac{U_{k_1-k}c_{\kappa-k_1}}{E - E^0_{\kappa-k}} + O(U^2), \qquad (12.81)$$

so that

$$\left(E - E^0_{\kappa-k}\right)c_{\kappa-k} = \sum_k \frac{U_{k-k_1}c_{\kappa-k_1}}{E - E^0_{\kappa-k}} + O(U^2). \qquad (12.82)$$

The shift in energy from the free-electron case is second order in U.

Case II:

Suppose the value of k is such that there are reciprocal lattice vectors $\vec{k}_1, \ldots, \vec{k}_m$ with $E - E^0_{\kappa-k_1}, \ldots, E - E^0_{\kappa-k_m}$, all within order U of each other, but far apart from other $E - E^0_{\kappa-k}$ on the scale of U.

$$\left|E^0_{\kappa-k} - E^0_{\kappa-k_j}\right| \gg U, \qquad (12.83)$$

for $j = 1, \ldots, m$ and $k \neq k_1, \ldots, k_m$. As a result,

$$\left(E - E^0_{\kappa-k_i}\right)c_{\kappa-k_i} = \sum_{j=1}^{m} U_{k_j-k_i}c_{\kappa-k_j} + \sum_{k\neq k_1,\ldots,k_m} U_{k-k_i}c_{\kappa-k}, \qquad (12.84)$$

for $i = 1, ..., m$. Similar to Case I, we find

$$\left(E - E^0_{\kappa-k_i}\right)c_{\kappa-k_i} = \sum_{j=1}^{m} U_{k_j-k_i}c_{\kappa-k_j}$$

$$+ \sum_{j=1}^{m}\left(\sum_{k\neq k_1,\ldots,k_m} \frac{U_{k-k_i}c_{\kappa-k}}{E - E^0_{\kappa-k}}\right)c_{\kappa-k_j} + O(U^2). \qquad (12.85)$$

When two free-electron levels are within order U of each other, but far compared with U from all other levels, the preceding equation reduces to

$$\left(E - E^0_{\kappa-k_1}\right)c_{\kappa-k_1} = U_{k_2-k_1}c_{\kappa-k_2}, \tag{12.86}$$

$$\left(E - E^0_{\kappa-k_2}\right)c_{\kappa-k_2} = U_{k_1-k_2}c_{\kappa-k_1}, \tag{12.87}$$

or if $\vec{q} = \vec{\kappa} - \vec{k}_1$ and $\vec{k} = \vec{k}_2 - \vec{k}_1$,

$$\left(E - E^0_q\right)c_q = U_k c_{q-k}, \tag{12.88}$$

and

$$\left(E - E^0_{q-k}\right)c_{q-k} = U_{-k}c_q = U^*_k c_q. \tag{12.89}$$

We have $E^0_q \propto E^0_{q-k}$, $\left|E^0_q - E^0_{q-k}\right| \gg U$, for $k' \neq k, 0$. This is true if $|\vec{q}| = |\vec{q} - \vec{k}|$, therefore q must lie on the Bragg plane bisecting the line joining the origin of k space to the reciprocal lattice point k,

$$\begin{vmatrix} E - E^0_q & U_k \\ U^*_k & E - E^0_{q-k} \end{vmatrix} = 0, \tag{12.90}$$

so that

$$E = \frac{1}{2}\left(E^0_q - E^0_{q-k}\right) \pm \left[\left(\frac{E^0_q - E^0_{q-k}}{2}\right)^2 + |U_k|^2\right]^{1/2}. \tag{12.91}$$

12.6 Semiconductor Devices

12.6.1 The pn Junction

12.6.1.1 Qualitative Description

If the energy gap between the highest-filled *valence band* (VB) and the lowest empty band is not too great, then a few electrons in the VB may be thermally excited into the lowest empty band, which is called the *conduction band* (CB). If the magnitude of the energy gap between the VB and the CB is about 1 eV then the material is called a *semiconductor* (SC). There are two types of semiconducting materials:

n-type: Excess of free electrons, Fermi level close to VB,

p-type: Excess of free holes, Fermi level close to CB.

If we place an n-type SC next to a p-type SC, as shown in figure 12.4, free electrons will diffuse from the n-type SC into the p-type SC. Likewise, free holes will diffuse from the p-type SC into the n-type SC. Some free electrons will recombine with free holes and be annihilated.

If the annihilation occurs in the n-type SC, the region where this occurs will become a region of fixed positive charge. Similarly, if the annihilation occurs in the p-type SC, it will become a region of fixed negative charge. These regions form what is known as the "space charge region".

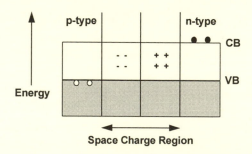

Figure 12.4 The space charge region in a pn semiconductor.

The separation of charges in the space charge region creates an electric field which opposes diffusion and brings about equilibrium. If we assume that the electric field is confined to the space charge region, then the electrostatic potential will be constant outside of the region. This indicates that there is an energy barrier of height eV_o to the electron flow from the n-type SC into the p-type SC. Forward biasing the pn junction with a battery will decrease the height of the energy barrier and more current will flow as shown. Reverse biasing the junction will increase the energy barrier and decrease current flow.

12.6.1.2 Quantitative Treatment

To obtain a quantitative idea of the pn junction we will examine the junction after equilibrium has been obtained. We define

$\vec{J}_{n^1}$ = current density of electrons from n to p due to diffusion,

$\vec{J}_{n^2}$ = current density of electrons from p to n due to the electric field.

We define similar expressions $\vec{J}_{p^1}$, $\vec{J}_{p^2}$ for the holes. The equilibrium conditions are

$$\vec{J}_{n^1} + \vec{J}_{n^2} = 0, \tag{12.92}$$

and

$$\vec{J}_{p^1} + \vec{J}_{p^2} = 0. \tag{12.93}$$

The current density to diffusion is, in general, given by

$$\vec{J} = q\vec{F} = -qD\nabla C, \tag{12.94}$$

where q is the charge on the electron/hole, F is the flux of electrons/holes across the junction, D is the diffusion coefficient, and C is the concentration of electrons/holes. Recalling that the current density of charges q of concentration n and drift velocity v is equal to qnv, we obtain

$$\vec{J}_{n^1} = eD_n \nabla n, \tag{12.95}$$

and

$$\vec{J}_{n^2} = en\vec{v}_n = en\mu_n \vec{E}, \tag{12.96}$$

where μ_n is the electron mobility. Using similar expressions for $\vec{J}_{p^1}$, $\vec{J}_{p^2}$, the one-dimensional equilibrium conditions are

$$eD_n \frac{dn(x)}{dx} + en(x)\mu_n E(x) = 0, \tag{12.97}$$

and

$$-eD_p \frac{dp(x)}{dx} + ep(x)\mu_p E(x) = 0. \tag{12.98}$$

If we make use of the Einstein relationship, $eD = \mu kT$, and the fact that $E(x) = \dfrac{dV(x)}{dx}$, the equations reduce to

$$\frac{1}{n}\frac{dn(x)}{dx} - \frac{e}{kT}\frac{dV(x)}{dx} = 0, \tag{12.99}$$

and

$$\frac{1}{p}\frac{dp(x)}{dx} + \frac{e}{kT}\frac{dV(x)}{dx} = 0. \tag{12.100}$$

Using the coordinate system shown below, the boundary conditions are

$$n = n_p, \tag{12.101}$$

$$p = p_p, \tag{12.102}$$

and

$$V = V_p, \tag{12.103}$$

for $x \le -x_p$, and

$$n = n_n, \tag{12.104}$$

$$p = p_n, \tag{12.105}$$

and

$$V = V_n, \tag{12.106}$$

for $x \ge x_n$. The equilibrium conditions specify that

$$\int_{p_p}^{p_n} \frac{dp}{p} = -\frac{e}{kT} \int_{V_p}^{V_n} dV, \tag{12.107}$$

or

$$\frac{p_n}{p_p} = \exp\left(-\frac{eV_o}{kT}\right) = \frac{n_p}{n_n}. \tag{12.108}$$

Electrical neutrality requires that $N_d x_n = N_a x_p$. The space charge density $\rho(x)$ is given by

$$\rho(x) = -en(x) + ep(x) - eN_a + eN_d. \tag{12.109}$$

Poisson's equation is, in general,

$$\frac{d^2V(x)}{dx^2} = -\frac{\rho(x)}{\varepsilon}. \tag{12.110}$$

We can approximate

$$\rho(x) \approx -eN_a, \tag{12.111}$$

for $-x_p \le x \le 0$, and

$$\rho(x) \approx eN_d, \tag{12.112}$$

for $0 \leq x \leq x_n$. Consequently, in the region $0 \leq x \leq x_n$, Poisson's equation is

$$\frac{d^2V(x)}{dx^2} = -\frac{dE(x)}{dx} = -\frac{eN_d}{\varepsilon}. \qquad (12.113)$$

If we make use of the boundary conditions, we obtain

$$E(x) = \frac{e}{\varepsilon}N_d(x - x_n), \qquad (12.114)$$

for $0 \leq x \leq x_n$ and

$$E(x) = -\frac{e}{\varepsilon}N_a(x + x_p), \qquad (12.115)$$

for $-x_p \leq x \leq 0$. If we integrate the expressions for $E(x)$ and apply the boundary conditions again, we obtain

$$V(x) = -\frac{e}{2\varepsilon}N_d(x - x_n)^2 + V_n, \qquad (12.116)$$

for $0 \leq x \leq x_n$, and

$$V(x) = \frac{e}{2\varepsilon}N_a(x + x_p)^2 + V_p, \qquad (12.117)$$

for $-x_p \leq x \leq 0$. Requiring continuity at $x = 0$ gives

$$V_o = V_n - V_p = \frac{e}{2\varepsilon}\left(N_d x_n^2 + N_a x_p^2\right), \qquad (12.118)$$

or

$$V_o = \frac{e}{2\varepsilon}N_d x_n\left(x_n + x_p\right) = \frac{e}{2\varepsilon}N_d x_n w, \qquad (12.119)$$

where w is the width of the space charge region. Rewriting V_o in terms of p_p and p_n gives

$$V_o = \left\{ \frac{ekT}{8\pi\varepsilon^2} \frac{N_a + N_d}{N_a N_d} \ln\left(\frac{N_a N_d}{n_i^2} \right) \right\}^{1/2}, \tag{12.120}$$

where $n_i^2 = pn = $ constant. Similarly, we can show that

$$x_p = \frac{w}{1 + \dfrac{N_a}{N_d}}, \quad x_n = \frac{w}{1 + \dfrac{N_d}{N_a}}. \tag{12.121}$$

12.6.2 The pnp Junction

12.6.2.1 Qualitative Description

A pnp junction may be constructed from two pn junctions, as shown in figure 12.5. We assume that one of the p-type semiconductors is more heavily doped than the other and label it p_+. The circuit shown forward biases the p_+n junction and reverse biases the pn junction.

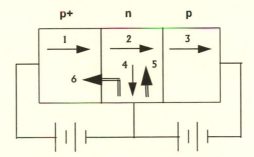

Figure 12.5 Current flow in a pnp semiconductor.

By convention, the p_+-type material is called the *emitter* (E), the n-type material is called the *base* (B), and the p-type material is called the *collector* (C). Because of the bias voltages, the energy barrier between p_+ and n for holes has decreased, while the energy barrier to the diffusion of holes from p to n has increased.

The particle flow is composed of

1) Holes injected into n from p_+

2) Holes from p_+ that diffuse across n

3) Holes swept into p by the induced electric field

4) Holes from p that recombine with electrons in n

5) Electrons that are supplied by the external circuit to replace those lost due to recombination in 4)

6) Electrons that diffuse from n into p_+.

The external currents are $I_E = I_{En} + I_{Ep}$, I_B, and $I_C = I_{Cp}$. The collector current, I_C, is a flow of holes across the base. The base current, I_B, is a flow of electrons into the base. If we increase I_B, the additional electrons will cause additional holes to leave the emitter in order to maintain space charge neutrality. Some of these holes will diffuse into the collector and I_C will increase. If we decrease I_B, the presence of fewer electrons in the base implies that fewer holes will leave the emitter and diffuse into the collector, therefore I_C will decrease. This tells us that the value of I_B can control the value of I_C. If the time required for one hole to diffuse from E to C is less than the lifetime of an electron in the base, then one electron can cause several holes to diffuse into C and produce current gain. We define

$$I_C = B \, I_{Ep}, \tag{12.122}$$

and

$$\gamma = I_{Ep}/I_E. \tag{12.123}$$

From the figure describing particle flow, we see that

$$I_B = I_{En} + (1 - B)I_{Ep}. \tag{12.124}$$

The current gain, $\beta = I_C/I_B$, is given by

$$\beta = B\gamma/(1 - B\gamma). \tag{12.125}$$

12.7.2.2 Quantitative Treatment

Under the assumption that the pnp junction is to be used as an amplifier, (which is not always the case), it is possible to obtain a theoretical estimate for the current gain. The assumptions needed are

1) There is no electric field in the base, holes diffuse from the emitter to the collector

2) The emitter current is composed only of holes

3) The reverse saturation current in the collector junction is small enough to be neglected

4) The system is one dimensional

5) The number of carriers in the space charge region of the emitter and collector junctions is constant

6) All currents and voltages are steady-state values.

The geometry is defined such that the p + n intersection defines $x = 0$. Previously, it was seen that

$$\frac{p_p}{p_n} = \frac{n_n}{n_p} = \exp\left\{\frac{eV}{kT}\right\}, \tag{12.126}$$

where $V = V_o - V_a$, and V_a is the applied potential. If we use

$$p_p = p_p(-x_p) \approx p_{po}, \tag{12.127}$$

then the expression for $p_n(x_n)$ is

$$p_n(x_n) = p_{no} \exp\left\{\frac{eV_a}{kT}\right\}. \tag{12.128}$$

Making use of this expression, we can define the excess hole density at the edge of the emitter junction at $x = 0$ by Δp_E and the excess hole density at the edge of the collector junction at $x = w_b$ by Δp_C, where

$$\Delta p_E = p_{no}\left[\exp\left\{\frac{eV_E}{kT}\right\} - 1\right], \tag{12.120}$$

and

$$\Delta p_C = p_{no}\left[\exp\left\{\frac{eV_C}{kT}\right\} - 1\right]. \tag{12.130}$$

Note that V_E will be positive and V_C will be negative.

The equation of continuity for electric charge is given by

$$\frac{\partial \rho}{\partial t} + \vec{\nabla} \cdot \vec{J} = G - R, \tag{12.131}$$

where ρ is the electric charge density, J is the current density, and G/R are the rates of generation and recombination per unit volume, respectively. By definition, $\rho = ep$ and, using the previously developed expression for J_p and bearing in mind assumptions 1) and 6), we find

$$-eD_p \frac{d^2 p}{dx^2} = G_p - R_p. \tag{12.132}$$

It may be assumed that $G_p = 0$ and that

$$R_p = e \frac{p - p_0}{\tau_p}, \tag{12.133}$$

where τ_p is the hole lifetime and $p - p_0$ is the excess hole density in the nonequilibrium situation. The equation of continuity gives

$$\frac{d^2 p}{dx^2} - \frac{1}{D_p \tau_p}(p - p_0) = 0. \tag{12.134}$$

If we define $L_p^2 = D_p \tau_p$, the equation reduces to

$$\frac{d^2}{dx^2}(p - p_0) = \frac{1}{L_p^2}(p - p_0). \tag{12.135}$$

This equation may be solved by making use of the boundary conditions,

$$p(0) = p_0 + \Delta p_E, \tag{12.136}$$

and

$$p(w_b) = p_0 + \Delta p_C. \tag{12.137}$$

After we obtain an expression for $p(x)$, we may solve for $J(x)$ to show that

$$J_E = J_p(0) = e\frac{D_p}{L_p}\left[\Delta p_E \coth\left(\frac{w_b}{L_p}\right) - \Delta p_C \operatorname{csch}\left(\frac{w_b}{L_p}\right)\right] \quad (12.138)$$

$$J_C = J_p(w_b) = e\frac{D_p}{L_p}\left[-\Delta p_C \coth\left(\frac{w_b}{L_p}\right) + \Delta p_E \operatorname{csch}\left(\frac{w_b}{L_p}\right)\right], \quad (12.139)$$

and

$$J_B = J_E - J_C = e\frac{D_p}{L_p}\left[(\Delta p_E + \Delta p_C)\tanh\left(\frac{w_b}{2L_p}\right)\right]. \quad (12.140)$$

Normally, pnp transistors are biased such that $\Delta p_E \gg \Delta p_C$. Consequently, the current gain may be approximated by

$$\alpha = \frac{J_C}{J_E} = \frac{1}{\cosh\left(\dfrac{w_b}{L_p}\right)}, \quad (12.141)$$

and

$$\beta = \frac{J_C}{J_B} = \frac{\operatorname{csch}\left(\dfrac{w_b}{L_p}\right)}{\tanh\left(\dfrac{w_b}{2L_p}\right)}. \quad (12.142)$$

For large values of L_p/w_b the expression for β reduces to

$$\beta = \frac{2L_p^2}{w_b^2} = \frac{2D_p\tau_p}{w_b^2}. \quad (12.143)$$

12.8 Bibliography

Ashcroft, N. W., and Mermin, N. D. *Solid State Physics*. Philadelphia: Saunders College, 1976.

■ Appendix A

Vector Identities

$$\vec{\nabla} = \frac{\partial}{\partial x}\hat{i} + \frac{\partial}{\partial y}\hat{j} + \frac{\partial}{\partial z}\hat{k}$$

$$\vec{\nabla}\varphi = \frac{\partial \varphi}{\partial x}\hat{i} + \frac{\partial \varphi}{\partial y}\hat{j} + \frac{\partial \varphi}{\partial z}\hat{k}$$

$$\vec{\nabla}\cdot\vec{v} = \frac{\partial v_x}{\partial x} + \frac{\partial v_y}{\partial y} + \frac{\partial v_z}{\partial z}$$

$$\vec{\nabla}\times\vec{v} = \begin{vmatrix} \hat{i} & \hat{j} & \hat{k} \\ \frac{\partial}{\partial x} & \frac{\partial}{\partial y} & \frac{\partial}{\partial z} \\ v_x & v_y & v_z \end{vmatrix} = \left(\frac{\partial v_z}{\partial y} - \frac{\partial v_y}{\partial z}\right)\hat{i} + \left(\frac{\partial v_x}{\partial z} - \frac{\partial v_z}{\partial x}\right)\hat{j} + \left(\frac{\partial v_x}{\partial y} - \frac{\partial v_y}{\partial x}\right)\hat{k}$$

$$\vec{\nabla}\cdot\vec{\nabla}\varphi = \frac{\partial^2\varphi}{\partial x_i^2} = \nabla^2\varphi$$

$$\vec{\nabla}\times\vec{\nabla}\varphi = 0$$

$$\vec{\nabla}\vec{\nabla}\cdot\vec{v} = \vec{\nabla}\cdot\vec{\nabla}v_x\hat{i} + \vec{\nabla}\cdot\vec{\nabla}v_y\hat{j} + \vec{\nabla}\cdot\vec{\nabla}v_z\hat{k}$$

$$\vec{\nabla}\cdot\left(\vec{\nabla}\times\vec{v}\right) = 0$$

$$\vec{\nabla}\times\left(\vec{\nabla}\times\vec{v}\right) = \vec{\nabla}\vec{\nabla}\cdot\vec{v} - \vec{\nabla}\cdot\vec{\nabla}\vec{v}$$

$$\vec{\nabla}\times\left(\varphi\vec{v}\right) = \vec{v}\cdot\vec{\nabla}\varphi + \varphi\vec{\nabla}\cdot\vec{v}$$

$$\vec{\nabla}\left(\vec{a}\cdot\vec{b}\right) = \left(\vec{a}\cdot\vec{\nabla}\right)\vec{b} + \left(\vec{b}\cdot\vec{\nabla}\right)\vec{a} + \vec{a}\times\left(\vec{\nabla}\times\vec{b}\right) + \vec{b}\times\left(\vec{\nabla}\times\vec{a}\right)$$

$$\vec{\nabla}\cdot\left(\vec{a}\times\vec{b}\right) = \vec{b}\cdot\left(\vec{\nabla}\times\vec{a}\right) - \vec{a}\cdot\left(\vec{\nabla}\times\vec{b}\right)$$

$$\vec{\nabla}\times\left(\vec{a}\times\vec{b}\right) = \vec{a}\left(\vec{\nabla}\cdot\vec{b}\right) - \vec{b}\left(\vec{\nabla}\cdot\vec{a}\right) + \left(\vec{b}\cdot\vec{\nabla}\right)\vec{a} - \left(\vec{a}\cdot\vec{\nabla}\right)\vec{b}$$

$$\vec{a}\cdot\left(\vec{b}\times\vec{c}\right) = \vec{b}\cdot\left(\vec{c}\times\vec{a}\right) = \vec{c}\cdot\left(\vec{a}\times\vec{b}\right)$$

$$\vec{a}\times\left(\vec{b}\times\vec{c}\right) = (\vec{a}\cdot\vec{c})\vec{b} - (\vec{a}\cdot\vec{b})\vec{c}$$

$$\left(\vec{a}\times\vec{b}\right)\cdot\left(\vec{c}\times\vec{d}\right) = (\vec{a}\cdot\vec{c})(\vec{b}\cdot\vec{d}) - (\vec{a}\cdot\vec{d})(\vec{b}\cdot\vec{c})$$

■ Appendix B

Vector Derivatives

Cartesian Coordinates

Unit vector: $d\vec{l} = dx\hat{i} + dy\hat{j} + dz\hat{k}$

Volume element: $dV = dxdydz$

Gradient: $\vec{\nabla}a = \dfrac{\partial a}{\partial x}\hat{i} + \dfrac{\partial a}{\partial y}\hat{j} + \dfrac{\partial a}{\partial z}\hat{k}$

Divergence: $\vec{\nabla}\cdot\vec{v} = \dfrac{\partial v_x}{\partial x} + \dfrac{\partial v_y}{\partial y} + \dfrac{\partial v_z}{\partial z}$

Curl: $\vec{\nabla}\times\vec{v} = \left(\dfrac{\partial v_z}{\partial y} - \dfrac{\partial v_y}{\partial z}\right)\hat{i} + \left(\dfrac{\partial v_x}{\partial z} - \dfrac{\partial v_z}{\partial x}\right)\hat{j} + \left(\dfrac{\partial v_x}{\partial y} - \dfrac{\partial v_y}{\partial x}\right)\hat{k}$

Laplacian: $\nabla^2 a = \dfrac{\partial^2 a}{\partial x^2} + \dfrac{\partial^2 a}{\partial y^2} + \dfrac{\partial^2 a}{\partial z^2}$

Spherical Coordinates

Unit vector: $d\vec{l} = dr\hat{r} + rd\theta\,\hat{\theta} + r\sin\theta\,d\phi\,\hat{\phi}$

Volume element: $dV = r^2\sin\theta\,drd\theta d\phi$

Gradient: $\vec{\nabla}a = \dfrac{\partial a}{\partial r}\hat{r} + \dfrac{1}{r}\dfrac{\partial a}{\partial\theta}\hat{\theta} + \dfrac{1}{r\sin\theta}\dfrac{\partial a}{\partial\phi}\hat{\phi}$

Divergence: $\vec{\nabla}\cdot\vec{v} = \dfrac{1}{r^2}\dfrac{\partial}{\partial r}\left(r^2 v_r\right) + \dfrac{1}{r\sin\theta}\dfrac{\partial}{\partial\theta}\left(\sin\theta v_\theta\right) + \dfrac{1}{r\sin\theta}\dfrac{\partial v_\phi}{\partial\phi}$

Curl: $\vec{\nabla} \times \vec{v} = \dfrac{1}{r\sin\theta}\left(\dfrac{\partial}{\partial\theta}\left(\sin\theta v_\phi\right) - \dfrac{\partial v_\theta}{\partial\phi}\right)\hat{r}$

$$+ \dfrac{1}{r}\left(\dfrac{1}{\sin\theta}\dfrac{\partial v_r}{\partial\phi} - \dfrac{\partial}{\partial r}\left(rv_\phi\right)\right)\hat{\theta} + \dfrac{1}{r}\left(\dfrac{\partial}{\partial r}\left(rv_\theta\right) - \dfrac{\partial v_r}{\partial\theta}\right)\hat{\phi}$$

Laplacian: $\nabla^2 a = \dfrac{1}{r^2}\dfrac{\partial}{\partial r}\left(r^2 \dfrac{\partial a}{\partial r}\right) + \dfrac{1}{r^2 \sin\theta}\dfrac{\partial}{\partial\theta}\left(\sin\theta\dfrac{\partial a}{\partial\theta}\right)$

$$+ \dfrac{1}{r^2 \sin^2\theta}\dfrac{\partial^2 a}{\partial\phi^2}$$

Cylindrical Coordinates

Unit vector: $d\vec{l} = dr\,\hat{r} + rd\phi\,\hat{\phi} + dz\,\hat{z}$

Volume element: $dV = r\,dr\,d\phi\,dz$

Gradient: $\vec{\nabla}a = \dfrac{\partial a}{\partial r}\hat{r} + \dfrac{1}{r}\dfrac{\partial a}{\partial\theta}\hat{\theta} + \dfrac{\partial a}{\partial z}\hat{z}$

Divergence: $\vec{\nabla}\cdot\vec{v} = \dfrac{1}{r}\dfrac{\partial}{\partial r}\left(rv_r\right) + \dfrac{1}{r}\dfrac{\partial v_\phi}{\partial\phi} + \dfrac{\partial v_z}{\partial z}$

Curl: $\vec{\nabla}\times\vec{v} = \dfrac{1}{r}\left(\dfrac{\partial v_z}{\partial\phi} - \dfrac{\partial v_\phi}{\partial z}\right)\hat{r} + \left(\dfrac{\partial v_r}{\partial z} - \dfrac{\partial v_z}{\partial r}\right)\hat{\phi}$

$$+ \dfrac{1}{r}\left(\dfrac{\partial}{\partial r}\left(rv_\phi\right) - \dfrac{\partial v_r}{\partial\phi}\right)\hat{z}$$

Laplacian: $\nabla^2 a = \dfrac{1}{r}\dfrac{\partial}{\partial r}\left(r\dfrac{\partial a}{\partial r}\right) + \dfrac{1}{r^2}\dfrac{\partial^2 a}{\partial\phi^2} + \dfrac{\partial^2 a}{\partial z^2}$

Appendix C

Physical Constants

Symbol	Constant	Value	Units
a_o	Bohr Radius	5.292×10^{-11}	m
c	Speed of Light	2.998×10^8	$m\ s^{-1}$
e	Elementary Charge	1.602×10^{-19}	C
F	Faraday Constant	9.648×10^4	$C\ mol^{-1}$
G	Gravitational Constant	6.672×10^{-11}	$m^3\ s^{-2}\ kg^{-1}$
h	Planck Constant	6.626×10^{-34}	J s
k	Boltzmann Constant	1.381×10^{-23}	$J\ K^{-1}$
m_e	Electron Rest Mass	9.110×10^{-31}	kg
m_n	Neutron Rest Mass	1.675×10^{-27}	kg
m_p	Proton Rest Mass	1.673×10^{-27}	kg
N_A	Avogadro's Number	6.022×10^{23}	mol^{-1}
R	Gas Constant	8.314	$J\ mol^{-1}\ K^{-1}$
R	Rydberg Constant	1.097×10^7	m^{-1}
ε_o	Permitivity Constant	8.854×10^{12}	$F\ m^{-1}$
μ_o	Permeability Constant	$4\pi \times 10^{-7}$	$H\ m^{-1}$
μ_B	Bohr Magneton	9.274×10^{-24}	$J\ T^{-1}$
μ_e	Electron Magnetic Moment	9.285×10^{-24}	$J\ T^{-1}$
μ_p	Proton Magnetic Moment	1.411×10^{-26}	$J\ T^{-1}$
μ_N	Nuclear Magneton	5.051×10^{-27}	$J\ T^{-1}$
σ	Stefan-Boltzmann Constant	5.670×10^{-8}	$W\ m^{-2}\ K^{-4}$

Periodic Table of the Elements

1 H 1.008																	2 He 4.003
3 Li 6.939	4 Be 9.012											5 B 10.81	6 C 12.01	7 N 14.01	8 O 16.00	9 F 19.00	10 Ne 20.18
11 Na 22.99	12 Mg 24.31											13 Al 26.98	14 Si 28.09	15 P 30.97	16 S 32.06	17 Cl 35.45	18 Ar 35.45
19 K 39.10	20 Ca 40.08	21 Sc 44.96	22 Ti 47.90	23 V 50.94	24 Cr 52.00	25 Mn 54.94	26 Fe 55.85	27 Co 58.93	28 Ni 58.71	29 Cu 63.54	30 Zn 65.37	31 Ga 69.72	32 Ge 72.59	33 As 74.92	34 Se 78.96	35 Br 79.91	36 Kr 83.80
37 Rb 85.47	38 Sr 87.62	39 Y 88.91	40 Zr 91.22	41 Nb 92.91	42 Mo 95.94	43 Tc 98.91	44 Ru 101.1	45 Rh 102.9	46 Pd 106.4	47 Ag 107.9	48 Cd 112.4	49 In 114.8	50 Sn 118.7	51 Sb 121.8	52 Te 127.6	53 I 126.9	54 Xe 131.3
55 Cs 131.9	56 Ba 137.3	57† La 138.9	72 Hf 178.5	73 Ta 181.0	74 W 183.9	75 Re 186.2	76 Os 190.2	77 Ir 192.2	78 Pt 195.1	79 Au 197.0	80 Hg 200.6	81 Tl 204.4	82 Pb 207.2	83 Bi 209.0	84 Po (210)	85 At (210)	86 Rn (222)
87 Fr (223)	88 Ra (226)	89‡ Ac (227)	104	105													

Lanthanides†

58 Ce 140.1	59 Pr 140.9	60 Nd 144.2	61 Pm (145)	63 Sm 150.4	63 Eu 152.0	64 Gd 157.3	65 Tb 158.9	66 Dy 162.5	67 Ho 164.9	68 Er 167.3	69 Tm 168.9	70 Yb 173.0	71 Lu 175.0

Actinides‡

90 Th 232.0	91 Pa (231)	92 U 238.0	93 Np (237)	94 Pu (242)	95 Am (243)	96 Cm (247)	97 Bk (249)	98 Cf (251)	99 Es (254)	100 Fm (253)	101 Md (256)	102 No (253)	104 Lr (257)

■ Index